Springer Actuarial

This is a series on actuarial topics in a broad and interdisciplinary sense, aimed at students, academics and practitioners in the fields of insurance and finance.

Springer Actuarial informs timely on theoretical and practical aspects of topics like risk management, internal models, solvency, asset-liability management, market-consistent valuation, the actuarial control cycle, insurance and financial mathematics, and other related interdisciplinary areas.

The series aims to serve as a primary scientific reference for education, research, development and model validation.

The type of material considered for publication includes lecture notes, monographs and textbooks. All submissions will be peer-reviewed.

More information about this series at http://www.springer.com/series/15681

Etienne Dupourqué · Frédéric Planchet ·
Néfissa Sator

Editors

Actuarial Aspects of Long Term Care

 Springer

Editors
Etienne Dupourqué
FSA, MAAA
Bellows Falls, VT, USA

Néfissa Sator
IA, CERA, MAAA
ISUP
New York, NY, USA

Frédéric Planchet
IA, ASA
ISFA
Université Lyon
Lyon, France

ISSN 2523-3262   ISSN 2523-3270   (electronic)
Springer Actuarial
ISBN 978-3-030-05659-9   ISBN 978-3-030-05660-5   (eBook)
https://doi.org/10.1007/978-3-030-05660-5

Library of Congress Control Number: 2019930158

Mathematics Subject Classification (2010): 62P05, 60G25

This Springer imprint is published by the registered company Springer Nature Switzerland AG.
The registered company address is: Gewerbestrasse 11, 6330 Cham, Switzerland

Foreword

You should absolutely read this book right now! Stop anything else you are doing! Indeed, there are many reasons for you to start turning the pages of this book and I am ready to bet that you will not stop before coming to the end.

Long Term Care (LTC) is one of the most difficult societal issues we are facing. The assistance to the elderly people used to be provided by families. Today, families are smaller whereas the number of elderly people is growing bigger every day. To add to the difficulty, these people are living longer and at a higher cost. Fewer and fewer individuals have now the financial capacity to support their old parents or grandparents. Therefore, increasingly, insurance appears to be the only way to fix the LTC issue.

If you are an insurer, that is a good news... until you start thinking about how to price it.

There are indeed plenty of hurdles to pricing LTC: lack of historical data, uncertainty coming from the long tail development of claims, and incapacity of traditional deterministic methods to capture all the features of the risks. The logical solution then is to go stochastic to get a better understanding of all possible developments. This is far more easily said than done: which model is the best? The book teaches us that the answer to this question is not the same on one side or the other of the Atlantic.

Suppose now that you have chosen the appropriate model for pricing, then you are still not out of the woods, as the model is not everything: you need data assumptions for it to run. You have to define (i) the distribution curves of each of the three states of life (autonomy, LTC, and death) you want to model, (ii) the probabilities of transition from one of these states to another one, and (iii) the future evolution of the costs of assistance. At this point, I am sure that you wonder whether being a fortune-teller would not be more efficient than being a simple actuary!

The book is illuminating the path for you to make educated decisions on pricing. It also addresses the issue of the capital requirement, as once you are in the LTC business you need to set apart enough capital to ensure your solvency. In Europe, the Solvency II Directive has no specific answer; the Standard Formula is silent about LTC and its specificities. Again, you might decide to find your salvation in a model and go for a (partial) Internal Model. My bet is that it will certainly be a hard battle, as you will have to set/document/simulate the future management actions.

How could you do that when new technologies including artificial intelligence and medical progress will revolutionize the way LTC is managed? Is it an actuarial prospective exercise or a Sci-Fi fictional work? We are lucky that this book gives some clues to keep it in the field of the actuarial techniques.

In 2011, when BNP Paribas Cardif decided to support the Excellence Chair *Data Analytics and Models for Insurance*, we were convinced that models had a significant role to play in our increasingly fast-changing world, helping stakeholders to apprehend complex situations and make educated decisions. This could not be more true than in the field of LTC.

We need academics to do their most-needed researches on complex topics and periodically make a holistic status report of where they stand. This book is the first of such holistic report on LTC. It is a very stimulating book.

I would suggest that it is regularly updated as the LTC landscape is expected to change very fast.

Jean-Paul Félix
BNP Paribas Cardif RISK COO

Contents

Part I
Dependence: Definitions and Facts

Bob Yee

Introduction

France and the United States are no different from many other countries in facing the challenge of long-term services and supports for an aging population. Part of the difficulties in financing for long term care such as through insurance is due to its dynamic nature. Historically, elderly care was provided informally by family members and relatives. Organized care such as nursing facilities started after the Second World War. In recent years, home health care has supplanted institutional care as the preferred care setting. The improving health status of the general population has profoundly changed the supply and demand of long term care. New generations are likely to be healthier and medical advance has altered the prevalence of chronic diseases relating to aging. In the future, technology and artificial intelligence will undoubtedly replace labor-intensive care.

Actuaries are uniquely qualified to provide financing solutions against the risks of long term care in light of past instability and future unknowns. In order to manage existing programs or develop new alternatives however, a thorough mastery of the background and drivers of long term care is a prerequisite. Part I of the book sets the stage on the nature of the long term care risks, the current delivery system, and financing method in each country. Chapter 1 presents the nature of disability status and its relationship with mortality. This relationship illustrates the difficulties in estimating future morbidity risks. Chapters 2 and 3 describe the long term care insurance systems in France and in the United States. France has a system of partnering public and private insurance, where the public system is characterized by fairly low benefit level of cash payments and a relatively high benefit eligibility threshold. In the United States, the public program is mean-tested and is intended only for the indigent. Unlike its counterpart in France, private long term care insurance in the United States so far has a relatively low market penetration.

The foundational materials in the Part I are invaluable in identifying the risks being protected through the public programs and private insurance products, the impetus for their designs as well as the perils inherent in such products. These are the topics for the remaining chapters of this book.

Chapter 1
Interactions of LTC Morbidity and Mortality

Eric Stallard

1.1 Introduction

Among the most difficult challenges facing the practicing LTC insurance (LTCI) actuary is validating the assumptions of his/her model, individually and in aggregate, for the purposes of pricing and/or valuation of LTCI products. The difficulties arise from the unique characteristic of LTC insurance whereby the majority of benefit payments are made many years, even decades, after issuance of the respective policies combined with the inherent uncertainties of the cumulative incidence and severity of claims over the life of each policy form. This differs significantly from single-premium whole-life insurance, for example, where the death benefit and the occurrence of death are certain, leaving only the timing uncertain. The challenges with LTCI are detailed in an American Academy of Actuaries publication entitled "Understanding Premium Rate Increases on Private Long Term Care Insurance Policyholders" [1]. Among these are the need to make valid assumptions regarding long-term future morbidity and mortality patterns, and their interactions.

Thus, the aspiring LTCI actuary needs to develop sufficient expertise with respect to morbidity and mortality patterns for insured lives to adequately support his/her modeling assumptions. This can be done by an in-depth study of the models presented in this book and its cited references. Additional important insight can be gained through the study of concurrent changes in morbidity and mortality in the general population through the use of publicly available data. Given that only about 11% of the general population aged 65 and older in the U.S. currently has LTCI coverage [23]—with this percentage varying by education and income levels—it follows that general population results should be used provisionally, or possibly not at all, depending on whether the results are replicated in insured data; the quantitative parameter values may differ materially.

E. Stallard (✉)
Duke University, PO Box 90408, Durham, NC 27708-0408, USA
e-mail: eric.stallard@duke.edu

© Springer Nature Switzerland AG 2019
E. Dupourqué et al. (eds.), *Actuarial Aspects of Long Term Care*,
Springer Actuarial, https://doi.org/10.1007/978-3-030-05660-5_1

3

The purpose of this chapter is to introduce the actuarial reader to basic findings regarding morbidity and mortality in the general elderly population over the period 1982–2015, mostly using results from the U.S., but supplementing those results with other results from European countries to reflect a broader set of outcomes. The 1982–2015 period is sufficiently lengthy that long-term trends in morbidity and mortality could be reliably detected. Long-term trends are not so readily detectable for LTCI insureds due to changes in regulations and product designs over time, most notably, following the 1996 HIPAA regulations governing tax-qualified LTC services and insurance in the U.S. [22].

1.2   Basic Concepts

Human longevity increased steadily during the past century and is expected to continue increasing this century. Chronic illnesses late in life are likely to continue as the primary causes of morbidity, disability, and mortality; notable among these are Alzheimer's disease and related dementias, heart disease, cancer, stroke, and diabetes.

The distinction between *morbidity* and *disability* in the epidemiological/gerontological literature is not made in LTCI applications. Instead, *morbidity* and *disability* are treated as equivalent concepts; they are used to characterize persons eligible for benefits under a given LTCI policy form. The HIPAA rules governing tax-qualified LTC insurance in the U.S define such persons as "chronically ill individuals" and require such individuals to meet specified conditions relating to the severity of activity-of-daily-living disabilities and/or cognitive impairment [22]. Hence, chronic morbidity and chronic disability have the same meaning in this chapter; they are defined to include HIPAA-consistent limitations in activities of daily living (ADLs) and cognitive impairment (CI), the two most important risks with respect to the loss of independence in the elderly population and the only risks covered by tax-qualified LTC insurance in the U.S. [22]. Other definitions of chronic morbidity may include diagnosed diseases without concurrent ADL and/or CI (denoted ADL/CI) disability. However, without ADL/CI disability, such diseases may represent earlier, less severe, stages in the disablement process [43]. They may exhibit more complex patterns of temporal change with different temporal trends than exhibited by HIPAA-consistent ADL/CI disability [9].

1.3   Compression of Mortality

Compression of mortality is a reduction over time in the variance of age-at-death, leading to progressively more "rectangular" survival functions in the life tables constructed for successive calendar years [30, 33]. Although substantial rectangularization occurred during the first half of the 20th century, the effect had largely played out by the latter half of the 20th century [30]. Edwards and Tuljapurkar [11] noted that the effect could be completely eliminated in the latter period if the survival calculations were restricted to age 10 and older (denoted age 10+). Moreover, theoretical limits on the lower bounds of the variances of ages at death imply that there will be limited potential for future rectangularization [35, 42]; hence, future mortality changes will mostly comprise approximately parallel shifts of survival functions at age 65+ without further compression of mortality [35, 42, 44].

The life expectancy rankings in 2010 for the U.S. at age 65 (males 18th, females 26th) [31] indicate a potential for large gains in U.S. life expectancy, without effective biological constraints, as the U.S. catches up with the rest of the developed world. The expectation that the gains will occur without further compression of mortality has implications for the patterns of mortality likely to be observed in future years and, hence, for the range of valid mortality assumptions in LTCI actuarial models.

1.4 Compression of Morbidity

Compression of morbidity [15] is a reduction over time in total lifetime duration of chronic morbidity, or chronic disability, reflecting a balance between morbidity/disability incidence rates and case-continuance rates—generated by case-fatality and case-recovery rates. The definition of morbidity compression in this chapter focuses on reductions in lifetime ADL/CI disability duration using HIPAA-consistent disability criteria; this definition is presently most relevant in the U.S. because of the large declines over the period 1984–2004 at age 65+ in the National Long Term Care Survey (NLTCS) in both types of lifetime disability durations and the continuation of the CI disability declines through 2012 in the Health and Retirement Study (HRS) and through 2015 in the National Health and Aging Trends Study (NHATS).

Our analysis of the 1984–2004 NLTCS [37] demonstrated, using the Sullivan [41] life table method detailed in Chap. 2, that the relative declines in lifetime ADL disability durations were similar and very substantial for males and females (17.5% and 19.0%, respectively); the relative declines in lifetime CI disability durations were even larger (27.7% and 36.1%, respectively). Moreover, the absolute levels of lifetime ADL and CI disability durations were 1.9 and 2.0 times larger, respectively, for females in 2004 (vs. 1.9 and 2.2 times larger in 1984), implying that substantial additional morbidity compression may still be achieved by narrowing the sex differentials. The quantitative estimates are shown in Table 1.1.

The *Change* column represents the difference between the *2004* and *1984* columns. The *Change* column is further decomposed into the difference between the *Survival Increment* and *Morbidity Decrement* columns, where the survival increment is the change that would have occurred if the age-specific ADL/CI, ADL, or CI disability prevalence rates had remained constant over time and the morbidity decrement is the change that would have occurred if the age-specific life-table survival functions had remained constant over time. Each morbidity decrement was larger than the corresponding survival increment, so that the resulting change was negative

Table 1.1 Components of change in life expectancy and HIPAA ADL/CI, ADL, and CI durations (in years at age 65), United States 1984 and 2004, by sex

At age 65	Year		Change	Survival increment	Morbidity decrement
	1984	2004			
Males					
Life expectancy	14.41	16.67	2.26	2.26	–
ADL/CI duration	1.64	1.26	−0.39	0.44	0.83
ADL duration	1.19	0.98	−0.21	0.32	0.53
CI duration	1.09	0.79	−0.30	0.31	0.62
Females					
Life expectancy	18.66	19.50	0.84	0.84	–
ADL/CI duration	3.26	2.29	−0.97	0.24	1.21
ADL duration	2.32	1.88	−0.44	0.17	0.61
CI duration	2.43	1.55	−0.88	0.18	1.06

Source Stallard and Yashin [37; Tables 1.12, 1.13, 2.25, 2.26, 2.28, 2.29]

in all cases. Thus, the absolute numbers of years expected to be lived while meeting the ADL/CI, ADL, or CI disability criteria were smaller in 2004 than 1984. This reduction in lifetime disability duration constitutes the compression of morbidity.

In addition to the large overall compression of morbidity during 1984–2004, Table 1.1 also shows that there were substantial differences between males and females in the baseline levels and in the subsequent dynamics of lifetime ADL versus CI disability durations—differences that constrain the range of valid morbidity assumptions in LTCI actuarial models.

The results imply—contrary to Fries' original formulation of the morbidity compression hypothesis [16]—that mortality compression is not necessary for morbidity compression [34, 35] because the morbidity compression shown in Table 1.1 occurred in a period during which mortality compression was no longer operating (at least, above age 10); see Fries et al. [18] for a discussion of the reformulated morbidity compression hypothesis without the assumption of mortality compression. The hypothesis-specification issue is now fully resolved, but the resolution implies that there are important interactions between morbidity and mortality.

1.5 Interactions of LTC Morbidity and Mortality

Mortality reduction, with static age-specific morbidity prevalence rates, would lead to increased morbidity, generating the counterfactual survival increments to lifetime morbidity shown in Table 1.1. For example, for ADL/CI disability in the U.S. during 1984–2004, the counterfactual survival-increments to lifetime morbidity were 0.44 year for males and 0.24 year for females. In actuality, mortality improvement in the U.S. during 1984–2004 was counterbalanced by an even greater reduction in age-specific ADL/CI morbidity prevalence rates, generating offsetting morbidity decrements of 0.83 year for males and 1.21 year for females—sufficient to generate the observed ADL/CI morbidity compression of 0.39 year for males and 0.97 year for females during 1984–2004 (Table 1.1).

Kreft and Doblhammer [25] conducted a similar analysis—using the Sullivan [41] life table method—of LTC morbidity at the "severe care level" in Germany at age 65+ during 2001–2009; they found survival increments of 0.142 and 0.244 years, respectively, for males and females compared to morbidity decrements of 0.101 and 0.207 years, implying a relatively small morbidity expansion—i.e., decompression, not compression—of 0.040 and 0.037 years, respectively—with a stable trend in relative lifetime LTC duration for males and a relative compression for females. Thus, continued morbidity compression is not guaranteed; the morbidity reduction must compensate for the natural increase in lifetime morbidity that would occur if mortality reductions were operating with static age-specific morbidity prevalence rates.

The specific criteria used to define LTC morbidity triggers may also influence the outcomes. For example, Crimmins and Beltrán-Sánchez [9] reported that the compression of morbidity reversed (i.e., decompressed) at ages 20+ and 65+ during 1998–2008 in a study where morbidity was defined as a loss of mobility functioning in a non-institutionalized U.S. sample from the National Health Interview Survey. Similarly, Kreft and Doblhammer [25] reported a sizeable expansion of LTC morbidity in Germany at age 65+ at the "any care level" of 0.214 and 0.326 years, respectively, for males and females, 5.4 and 8.8 times larger, respectively, than the 0.040 and 0.037 years expansion for the "severe care level" noted above.

The fact that such divergent results can be obtained from seemingly minor changes in study populations and/or disability criteria underscores the need to scrutinize the specific details of each study when using general population data to inform the assumptions of an LTCI actuarial model.

The results were more consistent for the stringent disability criteria—i.e., using the HIPAA-consistent definitions in the U.S. and the "severe care level" in Germany. Both sets of results support an interaction between morbidity and mortality. The U.S. results indicated that the morbidity prevalence declines more than compensated for the increases in disability that would otherwise have been induced by the improved survival, consistent with the morbidity compression hypothesis [18]. The German results indicated that the morbidity prevalence declines were approximately equal to the increases in disability that would otherwise have been induced by the improved survival, consistent with the alternative "dynamic equilibrium" hypothesis [28]. Interestingly, the rates of decline in ADL disability prevalence—using less stringent criteria—were substantially higher in Northern Europe (Denmark, Sweden) than in Central Europe (Germany, Belgium, Netherlands) at age 50+ between 2004–05 and 2013 in a study conducted by Ahrenfeldt et al. [2], suggesting that these countries may have experienced ADL morbidity compression during this study period; unfortunately, the authors did not provide the Sullivan [41] life table calculations needed to confirm this conclusion.

None of these results support an independence hypothesis under which the morbidity and mortality assumptions can be set independently in a valid LTCI actuarial model. This means that the practicing LTCI actuary needs to understand the mechanisms driving the interactions of morbidity and mortality and to use this understanding in formulating his/her respective morbidity and mortality assumptions.

1.6   Mechanisms Underlying Morbidity Compression

The primary mechanisms driving the compression of morbidity are hypothesized to involve combinations of medical advances and healthier lifestyles (e.g., reduced cigarette smoking; exercise; diabetes, hypertension, and cholesterol control; with a debit for increased obesity) which lead to longevity increases associated with correspondingly greater delays in onset of late-life disability [17].

The University of Pennsylvania Alumni Study and the Runners Study were designed to test such mechanisms in the U.S. during the past 30 years; as hypothesized, they demonstrated that healthy-aging lifestyles were effective in postponing late-life disability in longitudinally-followed study participants by amounts that substantially exceeded corresponding increases in life expectancy [18]. Andersen et al. [4] presented evidence on delayed ages of onset of physical and cognitive disability among centenarians and supercentenarians supporting this same hypothesis. The benefits of healthier lifestyles appear to extend to patients with type 2 diabetes [27] while the debits for increased obesity appear to be stabilizing or possibly beginning to shrink [8].

The implications for LTCI applications are straightforward: Risk selection through compatible underwriting protocols can be expected to yield new cohorts of LTCI policy purchasers with substantially lower than average lifetime morbidity risks.

1.7   ADL Versus CI Morbidity

Our analysis of the NLTCS [37] identified substantial and statistically significant differences over the 1984–2004 study period of about 1% per year in the annual rates of decline of ADL and CI disability prevalence rates; the overall rates of decline were 1.7% and 2.7% per year, respectively. Application of the Sullivan [41] life table method to the component age- and sex-specific prevalence rates for 1984 and 2004 yielded the expected lifetime disability duration estimates shown in Table 1.1.

A critical question is: What happened to these measures in the period after 2004? Answering this question turns out to be more difficult than expected. The problem lies in the need for comparable procedures and instrumentation to obtain valid cross-temporal estimates. These requirements were built into the design of successive waves of the NLTCS. With the termination of the NLTCS in 2004, these requirements are no longer met. The closest match is provided by the 2011–2015 NHATS, which might be deemed the "successor" to the NLTCS.

Two reports from the NHATS are relevant. The first concluded that neither a compression nor expansion of morbidity had taken place over the entire period 1982–2011 using data from the NLTCS and ADL disability criteria less stringent than the HIPAA criteria [14]; this finding appears to contradict the results in Table 1.1. Inspection of their Table 1 [14] showed, however, that the age-standardized prevalence of ADL disability for males declined from 10.7% in 1982 to 7.0% in 2004, followed by an increase to 7.3% in 2011; the corresponding values for females were 13.2%, 9.8%, and 10.2%, respectively. The upticks for 2004–2011 imply an expansion of morbidity although the size of the expansion cannot be determined from the report because the Sullivan [41] life table calculations were not presented for 2004.

Another report by these authors examined trends in ADL prevalence in five studies for the period 2000–2008, concluding that the trends were generally flat for age 65–84, with some decline for age 85+ [13]. The authors noted that in addition to the NLTCS the only other study covering both community and institutional persons was the Medicare Current Beneficiary Survey (MCBS). Inspection of their Table 3 [13] showed that the prevalence of ADL disability for age 65+ declined at a relative rate of 2.0% per year between 2004 and 2008. Although their ADL disability criteria were less stringent than the HIPAA criteria, this decline was faster than the 1.7% rate of decline found for the HIPAA-consistent ADL disability measure for age 65+ for 1984–2004 in the NLTCS [37], providing support for a continuation of the NLTCS trend in ADL morbidity compression through 2008. This support should be considered tentative, however, because the results from the MCBS were inconsistent with the results from the NHATS—illustrating why it is difficult to draw firm conclusions for the post-2004 period.

A second report from the NHATS [12] indicated that the prevalence of HIPAA-consistent CI disability (which they termed "probable dementia") at age 70+ declined at a relative rate of 1.7% per year between 2011 and 2015. This value can be compared with corresponding rates of decline for age 65+ of 2.7% per year from the 1984–2004 NLTCS [37] and 2.5% per year from the 2000–2012 HRS [26]; however, our reanalysis of the HRS [3] indicated that the latter rate of decline for CI disability prevalence was more likely to have been in the range 1.5–2.0% per year, considering alternative sets of cutpoints and procedures for imputing missing CI assessment data. While all three surveys indicated that substantial declines in CI disability occurred for at least three decades, the HRS and NHATS suggest that the rate of decline may have slowed by about 1% per year in the post-2004 period. This lower rate of decline is closer to the 1.4% per year decline reported for dementia in the U.K for the period 1991–2011 [29].

The above results imply a slowdown in the rates of morbidity compression in the post-2004 period, albeit, one in which the ADL disability compression may have halted while the CI disability compression continued at a slower rate. Given the existence of an absolute lower bound at zero-years duration for each form of disability, one should not be surprised by such a slowdown. Moreover, the finding that CI disability duration continued to shrink in the post-2004 period is consistent with its approximately 1% greater rate of prevalence decline in the 1984–2004 period. Looking forward, one might expect that the current rate of decline will slow down further as some nonzero lower bound is approached. To a large extent, this scenario will depend on the success of current and new research initiatives combatting Alzheimer's disease.

Given the above uncertainties, a reasonably conservative approach for LTCI applications would assume that the compression of morbidity had run its course in the general population, leading to adoption of the alternative dynamic equilibrium hypothesis [28] as a working assumption. This alternative hypothesis assumes that morbidity and mortality move in tandem, not independently, in generating the observed lifetime ADL and CI disability durations.

A substantially more conservative approach would assume that the age-specific prevalence rates will hold constant over time, leading to adoption of the morbidity expansion hypothesis [5, 20, 24]. This approach is routinely used in projections of the future burden of Alzheimer's disease [6, 21]. Thus, the emerging evidence for large and continuing declines in CI, overall dementia, and Alzheimer's disease is not reflected in existing projections of the future burden of Alzheimer's disease; nor is it reflected in projections of the potential for primary and secondary prevention of Alzheimer's disease [7, 10, 32]. Projected mortality improvement with static Alzheimer's disease morbidity rates will likely lead to overestimates of the future Alzheimer's disease burden and misestimates of the impact of primary and secondary prevention measures. The actuary will need to consider these potential biases when using such studies to inform the morbidity and mortality assumptions of his/her LTCI models.

1.8   ADL Versus CI Mortality

The total lifetime duration of ADL or CI disability reflects a balance between disability incidence rates and case-continuance rates—generated by case-fatality and case-recovery rates. The concepts of incidence and continuance are well-developed in LTCI actuarial models, as evidenced by the various models presented in this book. The reader may have noticed that all of the estimates presented in this chapter have been based on cross-sectional prevalence rates. That is, they summarize sets of age- and sex-specific rates of current disability in a given study population at a given point in time or over a very short interval of time (e.g., from one to several months). The reason for using such prevalence rates is their ease of measurement in one-off surveys and their compatibility with repeated surveys with new participants or with longitudinal follow-up of earlier participants.

Fries [16] specifically predicted that reductions in case survival rates would occur as consequences of the dynamics hypothesized to underlie morbidity compression, i.e., longevity increases associated with correspondingly greater delays in onset of late-life disability. The 1984–2004 NLTCS employed a longitudinal design which affords us the opportunity to assess survival/mortality changes for disabled persons over two decades. This is important because morbidity compression may result from reductions in disability incidence rates and/or reductions in case survival rates; likewise, morbidity expansion may result from increases in disability incidence rates and/or increases in case survival rates. Thus, we can use the NLTCS survival/mortality data to gain insight into the dynamics of Fries' paradigm.

To do so, we generated age- and sex-specific 5-year survival probabilities for disabled participants using mortality data for the five years following the 1984 and 2004 NLTCS, respectively. These were generated overall (ADL/CI) and for three combinations of ADL and CI disability status in the respective years—those with: [1] ADL disability with no concurrent CI disability (ADL Only), [2] CI disability with no concurrent ADL disability (CI Only), and [3] concurrent ADL and CI disability (ADL & CI). We age-standardized the results for each year using the overall age-specific disability counts for 1984 to generate the requisite weighted averages of the age-specific disability probabilities, overall and for the three disability groups, by sex. Age-standardization was used to control for differences in the age structure of the disabled population over time and between sexes in order to generate valid comparisons [37]. Standard errors of the estimated age-standardized survival probabilities were also generated as described in Stallard and Yashin [37]. Using these standard errors, we conducted two-tailed t-tests of the differences in survival probabilities, overall and for the three disability groups between 1984–1989 and 2004–2009, by sex. The results are displayed in Table 1.2, with the t-statistics converted to the corresponding p-values.

The overall change (ADL/CI) was downward for males and females but only the female decline was statistically significant ($p < 0.05$; boldface font). The decline for concurrent ADL and CI disability (ADL & CI)—the highest level of severity considered—was statistically significant for both males and females, as predicted by

Table 1.2 Age-standardized 5-year survival probabilities for HIPAA disability groups following the 1984 and 2004 NLTCS, age 65 and over, by sex

Group	5-year period		Change	Standard error	p-Value
	1984–89	2004–09			
Males					
ADL/CI	0.300	0.262	−0.038	0.023	0.106
ADL only	0.268	0.260	−0.008	0.037	0.836
CI only	0.414	0.424	0.010	0.055	0.859
ADL & CI	0.245	0.170	−0.075	0.034	**0.026**
Females					
ADL/CI	0.415	0.344	−0.071	0.018	**<0.001**
ADL only	0.393	0.395	0.002	0.032	0.950
CI only	0.567	0.472	−0.095	0.043	**0.029**
ADL & CI	0.316	0.244	−0.072	0.030	**0.014**

Source Author's calculations based on the 1984–2004 NLTCS

Fries [16]. The decline for CI disability with no concurrent ADL disability (CI Only) was statistically significant for females but not males. Interestingly, there was no evidence of an upward trend in 5-year survival probabilities—i.e., the two positive values shown in the table are effectively zero—implying that there was no support in these data for the expansion of morbidity hypothesis.

The declines in 5-year survival for both CI groups for females and for the ADL & CI group for males are consistent with the theory of "cognitive reserve" [38–40]. Under this theory, higher levels of cognitive reserve reflect greater ability to tolerate neuropathology before test scores and other signs/symptoms of Alzheimer's disease and related dementias are impacted. As a consequence, when such individuals present with dementia signs/symptoms, their pathology is further advanced and their residual survival time is correspondingly shorter. Thus, increases in cognitive reserve over time (through increases in educational attainment and more cognitively challenging occupations) could account, in part, for the CI component of the survival declines in Table 1.2. The same mechanism could account for the relatively greater declines in CI versus ADL disability durations in Table 1.1.

The LTCI actuary will need to carefully consider the implications of these results for the mortality assumptions of his/her actuarial models. The NLTCS data provided cross-sectional estimates of disability prevalence in the respective survey years. Almost all disability episodes were ongoing at the time of the in-person interview. Hence, the results in Table 1.2 may be substantially different from applications that follow disability episodes from the time of onset to the time of recovery or death. Moreover, the results in Table 1.2 run counter to the general trend of earlier diagnoses of specific disabling medical conditions (e.g., heart disease, cancer, stroke, diabetes), with extended survival post diagnosis. One mortality lookback study [19] indicated that 27–80% of decedents from non-dementia causes of death—i.e., cancer (54%), organ failure (27%), frailty (33%), sudden death (80%), and other conditions (58%)—had either no disability or catastrophic disability in the last year of life, neither of which would trigger LTCI benefits under a policy with a 3-month elimination period. This contrasted markedly with advanced-dementia deaths for which 85% of decedents had either progressive disability (17%) or persistently severe disability (68%) in the last year of life. Our own analysis of newly diagnosed Alzheimer's disease patients enrolled in the Predictors 2 Cohort Study during 1997–2011 indicated that 55% would survive sufficiently long to need full-time care with an average duration of 3.7 years [36]. Thus, the most critical issue to be considered by the LTCI actuary is the timing of onset of HIPAA-consistent levels of disability among such patients and the changes in survival post onset.

1.9 Discussion

This chapter built on the morbidity compression findings reported in Stallard and Yashin [37] and further elaborated in Stallard [35]. Those publications evaluated the declines in HIPAA ADL and CI disability prevalence rates and lifetime durations at age 65+ between 1984 and 2004 using data from the NLTCS. Large declines were observed in lifetime disability durations at age 65+ over the measurement period 1984–2004 (Table 1.1). Compared to males, ADL/CI disability duration was nearly twice as large for females in 1984 and 1.8 times larger in 2004. The relative decline in ADL/CI disability duration during 1984–2004 was larger for females (30% vs. 24% for males). The declines in CI disability durations were much larger than the ADL disability duration declines for both sexes; the differences were highly statistically significant and substantively meaningful. The absolute decline in CI disability duration during 1984–2004 was 43% larger for males and 100% larger for females than for ADL disability duration.

Why were the declines in CI disability durations so much larger than for ADL disability durations, and especially so for females? The results presented in Table 1.2 demonstrated that the more rapid declines in CI versus ADL disability prevalence rates, and hence disability durations, were due at least in part to significant and substantial declines in the 5-year survival rates for CI disability, with substantial and consistent declines for both sexes for CI disability in combination with ADL disability.

The overall declines in 5-year survival rates for disabled persons following the 1984 and 2004 NLTCS supported the hypothesized compression of morbidity. The different trends in survival for the groups having ADL disability with versus without concurrent CI indicated that morbidity compression was heterogeneous, with:

• Reduced survival for both sexes for concurrent ADL and CI disability (ADL & CI), but no reduction for ADL disability without concurrent CI disability (ADL Only);
• Reduced survival for females, but no reduction for males, for CI disability without concurrent ADL disability (CI Only).

Such reductions could occur if CI-associated clinical signs/symptoms were manifested at later points in the underlying neuropathological processes in 2004 than in 1984—consistent with the hypothesis of improved cognitive reserve for successive cohorts of the elderly [40].

1.10 Conclusion

Morbidity and mortality exhibited complex patterns of change in the general population over the last three decades. Having an informed perspective on the nature of these patterns of change and their likely future directions will be important in developing and validating corresponding morbidity and mortality assumptions for LTCI pricing and valuation models. The results and citations discussed in this chapter constitute only an introduction to this rapidly expanding area of research.

Acknowledgements Support for the research presented in this chapter was provided by the National Institute on Aging, through grant numbers P01-AG043352, R01-AG007370, R01-AG046860, and R56-AG047402-01A1. We gratefully acknowledge use of services and facilities of the Center for Population Health and Aging at Duke University, funded by NIA Center Grant P30-AG034424. David L. Straley provided programming support.

References

1. AAA: Understanding Premium Rate Increases on Private Long-Term Care Insurance Policyholders. Issue Brief. American Academy of Actuaries, Washington, DC (2016)
2. Ahrenfeldt, L.J., Lindahl-Jacobsen, R., Rizzi, S., Thinggaard, M., Christensen, K., Vaupel, J.W.: Comparison of cognitive and physical functioning of Europeans in 2004–05 and 2013. Int. J. Epidemiol. **47**(5), 1518–1528 (2018)
3. Akushevich, I., Yashkin, A.P., Kravchenko, J., Ukraintseva, S., Stallard, E., Yashin, A.I.: Time trends in the prevalence of neurocognitive disorders and cognitive impairment in the United States: the effects of disease severity and improved ascertainment. J. Alzheimer's Dis.137–148 (2018)
4. Andersen, S.L., Sebastiani, P., Dworkis, D.A., Feldman, L., Perls, T.T.: Health span approximates life span among many supercentenarians: compression of morbidity at the approximate limit of life span. J. Gerontol. Ser. A Biol. Sci. Med. Sci. **67**, 395–405 (2012)
5. Boyd, J.H.: Are Americans getting sicker or healthier? J. Relig. Health **45**, 559–585 (2006)
6. Brookmeyer, R., Johnson, E., Ziegler-Graham, K., Arrighi, H.M.: Forecasting the global burden of Alzheimer's disease. Alzheimer's Dement. **3**, 186–191 (2007)
7. Brookmeyer, R., Kawas, C.H., Abdallah, N., Paganini-Hill, A., Kim, R.C., Corrada, M.M.: Impact of interventions to reduce Alzheimer's disease pathology on the prevalence of dementia in the oldest-old. Alzheimer's Dement. **12**, 225–232 (2016)
8. Chang, V.W., Alley, D.E., Dowd, J.B.: Trends in the relationship between obesity and disability, 1988–2012. Am. J. Epidemiol. **186**, 688–695 (2017)
9. Crimmins, E.M., Beltrán-Sánchez, H.: Mortality and morbidity trends: is there compression of morbidity? J. Gerontol. Ser. B **66B**, 75–86 (2011)
10. Crous-Bou, M., Minguillón, C., Gramunt, N., Molinuevo, J.L.: Alzheimer's disease prevention: from risk factors to early intervention. Alzheimers Res. Ther. **9**, 1–9 (2017)
11. Edwards, R.D., Tuljapurkar, S.: Inequality in life spans and a new perspective on mortality convergence across industrialized countries. Popul. Dev. Rev. **31**, 645–674 (2005)
12. Freedman, V.A., Kasper, J.D., Spillman, B.C., Plassman, B.L.: Short-term changes in the prevalence of probable dementia: an analysis of the 2011–2015 National Health and Aging Trends Study. J. Gerontol. Ser. B **73**, S48–S56 (2018)
13. Freedman, V.A., Spillman, B.C., Andreski, P.M., Cornman, J.C., Crimmins, E.M., Kramarow, E., Lubitz, J., Martin, L.G., Merkin, S.S., Schoeni, R.F.: Trends in late-life activity limitations in the United States: an update from five national surveys. Demography **50**, 661–671 (2013)
14. Freedman, V.A., Wolf, D.A., Spillman, B.C.: Disability-free life expectancy over 30 years: a growing female disadvantage in the US population. Am. J. Public Health **106**, 1079–1085 (2016)
15. Fries, J.F.: Aging, natural death, and the compression of morbidity. N. Engl. J. Med. **303**, 130–135 (1980)
16. Fries, J.F.: The compression of morbidity: miscellaneous comments about a theme. Gerontol. **24**, 354–359 (1984)
17. Fries, J.F.: Compression of Morbidity: In Retrospect and in Prospect. Issue Brief. Alliance for Health & the Future, Paris (2005)
18. Fries, J.F., Bruce, B., Chakravarty, E.: Compression of morbidity 1980–2011: a focused review of paradigms and progress. J. Aging Res. **2011**, 1–10 (2011)

19. Gill, T.M., Gahbauer, E.A., Han, L., Allore, H.G.: Trajectories of disability in the last year of life. N. Engl. J. Med. **362**, 1173–1180 (2010)
20. Gruenberg, E.M.: The failures of success. Milbank Mem. Fund Q. Health Soc. 3–24 (1977)
21. Hebert, L.E., Weuve, J., Scherr, P.A., Evans, D.A.: Alzheimer disease in the United States (2010–2050) estimated using the 2010 census. Neurology **80**, 1778–1783 (2013)
22. IRS: Long-Term Care Services and Insurance: Notice 97–31. Internal Revenue Bulletin. Internal Revenue Service, U.S. Government Printing Office, Washington, DC (1997)
23. Johnson, R.W.: Who is Covered by Private Long-Term Care Insurance? Issue Brief. Urban Institute, Washington, DC (2016)
24. Kramer, M.: The rising pandemic of mental disorders and associated chronic diseases and disabilities. Acta Psychiatr. Scand. **62**, 382–397 (1980)
25. Kreft, D., Doblhammer, G.: Expansion or compression of long-term care in Germany between 2001 and 2009? A small-area decomposition study based on administrative health data. Popul. Health Metr. **14**, 1–15 (2016)
26. Langa, K.M., Larson, E.B., Crimmins, E.M., et al.: A comparison of the prevalence of dementia in the United States in 2000 and 2012. JAMA Intern. Med. **177**, 51–58 (2017)
27. Liu, G., Li, Y., Hu, Y., Zong, G., Li, S., Rimm, E.B., Hu, F.B., Manson, J.E., Rexrode, K.M., Shin, H.J.: Influence of lifestyle on incident cardiovascular disease and mortality in patients with diabetes mellitus. J. Am. Coll. Cardiol. **71**, 2867–2876 (2018)
28. Manton, K.G.: Changing concepts of morbidity and mortality in the elderly population. Milbank Mem. Fund Q. Health Soc. 183–244 (1982)
29. Matthews, F.E., Arthur, A., Barnes, L.E., Bond, J., Jagger, C., Robinson, L., Brayne, C.: A two-decade comparison of prevalence of dementia in individuals aged 65 years and older from three geographical areas of England: results of the Cognitive Function and Ageing Study I and II. The Lancet **382**, 1405–1412 (2013)
30. Myers, G.C., Manton, K.G.: Compression of mortality: myth or reality. Gerontologist **24**, 346–353 (1984)
31. NCHS: Health, United States, 2013: With Special Feature on Prescription Drugs. National Center for Health Statistics, U.S. Government Printing Office, Washington, DC (2014)
32. Norton, S., Matthews, F.E., Barnes, D.E., Yaffe, K., Brayne, C.: Potential for primary prevention of Alzheimer's disease: an analysis of population-based data. Lancet Neurol. **13**, 788–794 (2014)
33. Olshansky, S.J., Carnes, B.A., Desesquelles, A.: Prospects for human longevity. Science **291**, 1491–1492 (2001)
34. Stallard, E.: New perspectives on the compression of morbidity and mortality. Contingencies. American Academy of Actuaries, Washington, DC (2014)
35. Stallard, E.: Compression of morbidity and mortality: new perspectives. N. Am. Actuar. J. **20**, 341–354 (2016)
36. Stallard, E., Kinosian, B., Stern, Y.: Personalized predictive modeling for patients with Alzheimer's disease using an extension of Sullivan's life table model. Alzheimers Res. Ther. **9**, 1–15 (2017)
37. Stallard, P.J.E., Yashin, A.I.: LTC Morbidity Improvement Study: Estimates for the Non-Insured U.S. Elderly Population Based on the National Long Term Care Survey 1984–2004. Society of Actuaries, Schaumburg, IL (2016)
38. Steffener, J., Habeck, C., O'Shea, D., Razlighi, Q., Bherer, L., Stern, Y.: Differences between chronological and brain age are related to education and self-reported physical activity. Neurobiol. Aging **40**, 138–144 (2016)
39. Stern, Y.: Cognitive reserve. Neuropsychologia **47**, 2015–2028 (2009)
40. Stern, Y.: Cognitive reserve in ageing and Alzheimer's disease. Lancet Neurol. **11**, 1006–1012 (2012)
41. Sullivan, D.F.: A single index of mortality and morbidity. HSMHA Health Rep. **86**, 347–354 (1971)
42. Tuljapurkar, S., Edwards, R.D.: Variance in death and its implications for modeling and forecasting mortality. Demogr. Res. **24**, 497–525 (2011)

43. Verbrugge, L.M., Jette, A.M.: The disablement process. Soc. Sci. Med. **38**, 1–14 (1994)
44. Yashin, A.I., Begun, A.S., Boiko, S.I., Ukraintseva, S.V., Oeppen, J.: New age patterns of survival improvement in Sweden: do they characterize changes in individual aging? Mech. Ageing Dev. **123**, 637–647 (2002)

Chapter 2
Long Term Care in the United States

Etienne Dupourqué

2.1 Introduction

- In 2016 Genworth, a Virginia-based insurance company with over \$100 billion in assets and over one million Long Term Care insurance (LTCI) policyholders, announced it was in the process of being acquired by China Oceanwide Holding Group Co., for \$2.7 billion. Genworth, which became a publicly traded company in 2004, started marketing LTCI policies in 1974.
- In 2017, the Pennsylvania Commonwealth Court approved the liquidation (bankruptcy) of Penn Treaty Network America Insurance Company and its subsidiary, American Network Insurance Company, with 67,000 LTCI policies. The rehabilitation and liquidation process had started in 2009.[1]
- In 2018, General Electric, a global company with over \$100 billion in market value, announced that it will add \$15 billion over seven years, mostly to the LTCI reserves of 300,000 policies reinsured by its reinsurance unit, Employer Reassurance Corporation.

From a peak of 750,000 Long Term Care insurance policies issued in 2002, to 100,000 in 2016, individual stand-alone lifetime Long Term Care coverage is being replaced by policies providing limited duration, and products combined with life insurance or annuities which mitigate the risk borne by the insurer.

However, the Long Term Care risk impacts everyone, whether insured or not, and must be properly understood and measured to become a viable insurance product as well as to allow public policy-makers to meet the increasing challenges to protect the population against the risk.

This book concentrates on the measurement of biometric risks that bear on incidence and continuance of the Long Term Care risk. Some of the major LTCI risks are not inherent to the risks insured but are due to the different regulatory structure

E. Dupourqué (✉)
106 Atkinson Street, #2, Bellows Falls, VT 05101, USA
e-mail: etienne@dupourque.com

© Springer Nature Switzerland AG 2019
E. Dupourqué et al. (eds.), *Actuarial Aspects of Long Term Care*,
Springer Actuarial, https://doi.org/10.1007/978-3-030-05660-5_2

of jurisdictions, to regional and global economic and financial developments, and to medical and technical advances, among other factors. As of 2018, in the United States, most if not all policy values, such as premium and reserves, are calculated on a nationwide and level premium basis. This is to be contrasted with auto or medical insurance, which are based on regional experience and annually renewable premium. While LTCI rates can differ by gender and risk levels, assumptions are usually based on national data: 30-year treasury bonds, the 2000A mortality table, Society of Actuaries (SOA) Intercompany claim experience studies, and Life Insurance Marketing Research Association (LIMRA) persistency rates reports. However, insurance is not directly regulated nationwide but by laws and regulations in each constitutional jurisdiction (50 states, 1 district, 5 territories). Most policies were marketed to cover a lifetime risk, both for benefit and premium payments. As experience developed, it became clear that aggregate reserves for policies issued ten to twenty years earlier would not be sufficient to fund projected claims. Policy contract reserves (active life and claim) did not reflect current and developing experience. Early policies priced on asset-share projection models assumed voluntary termination (lapse) rates comparable to individual life and annuities policies and were priced during the 1980s' double digit interest rate environment. Premium from lower than expected lapses are not sufficient to fund expected claims for remaining policies, due to smaller reserve release and higher claim exposure. Lower investment income from declining interest rates aggravated the deficiency trend. Statutory reserves[2] are calculated based on a regulatory fixed interest rate. Regulations prescribing changes in interest rates are rather inflexible and modifying these regulations requires legislative and regulatory actions that can take several years.

Individual policies are contractually guaranteed renewable, allowing a company to increase premiums on a class-wide basis. Insurance contracts are not only subject to applicable laws, but to regulations which give regulators the right of approval to rate increase requests. Many policies offer a Cost of Living Adjustment (COLA) rider, as required to be Tax Qualified[3] for Federal income tax purposes. The most popular (and expensive) COLA being a 5% compound annual increase. For early policies, daily benefits (DB) were mostly for Nursing Home coverage, reimbursing about $75 per day, with 40% offering a COLA rider, for an annual premium of about $1,000.[4] That premium would be based on an annual claim cost, the present value of the expected duration and severity of a claim. Much like mortality rates for life insurance, an incidence rate determines the likelihood of a claim. The present value of the expected benefit payments would be applied to this incidence rate to derive the expected claim cost at the time of incidence. High annual mortality rates and lapse rates would contribute to lower future claim exposure, lowering the ultimate loss ratio[5] (LR). That discounted loss ratio would also be reduced by high interest rates since claims increase by duration while the premium is level. Active life mortality rates and lapse rates remained substantially lower than expected, and many dementia related claims were substantially longer and costlier. To give a rough estimate, by halving expected mortality, lapse, and interest rates, and increasing the average claim by 10% while keeping incidence rates as expected, in fifteen years a company would find its investment income half as expected while its claims doubled. Its remaining exposure to future claims would be twice as expected, with many policies having their Daily Benefit increased through the automatic COLA (a $100 DB would have risen to $189).

Faced with increasingly deficient portfolios that threatened their own solvency or the soundness of their other lines of insurance, companies filed for rate increases. Reactions to these requests from state regulators have been varied: from 'deemed' approvals, without review, to outright disapproval. The number of rate increases as well as their magnitude generated widespread adverse media coverage.

The different state reactions to rate increase requests have led to an increasingly balkanized regional environment for premium rates which have little actuarial relationship to the risk insured.

This is compounded by the long-term nature of the risk, both during the active/premium paying period and the claim continuance. Long-Term Services and Supports (LTSS) costs vary widely across the US, impacting both premiums and claims since most policies reimburse the policyholder for LTC daily costs, up to a stated amount in the policy. In 2017 the average daily Nursing Home cost in Louisiana was $174[6] and 7,000 km north, the average cost in Alaska was $800.[6] While policyholders will purchase coverage commensurate with the LTSS costs in his or her own state, inter-state mobility rates should be taken into account. While the average US inter-state mobility has declined to about 1.5% per year,[7] the longer a policy is inforce, the higher the variance around the average claim. This variance behavior is rendered more complex as states have different rate reviewing regulations and policies, which means that two policyholders from different states buying an identical policy will likely pay substantially different premiums over the duration of same premium period. Of course, the relationship of regional inflation to Cost of Living riders[9] should not be ignored. This regional disparity is probably more relevant to policyholders than to companies as their Long Term Care policy may not reflect costs in their new state of residence. Even if the Actual to Expected ratio of original assumptions for mortality, morbidity, persistency, or interest rate remained stable, premiums and reserves would not reflect the current environment. Demographic, LTSS cost, and regulatory heterogeneity add to external factors which strongly impact the LTCI risk.

This chapter will attempt to describe the environment which a US actuary must consider for the LTC risk.

The risk is not new, but the insurance market is relatively new compared to other insured risks, such as life, disability, or even cars, which have been mass produced for about one hundred years. Most personal insurance is based on well defined, tabulated and formulated events which trigger claims. LTCI aims to reimburse an individual for costs arising from loss of autonomy. Early policies were extensions of Medicare Supplemental (Med Sup) policies which reimbursed medical costs not covered by the federal old age social insurance program. Medicare[10] reimburses nursing home care for only a few days of residence, as it is meant to insure acute care, as opposed to chronic care. Med Sup policies regularly increase their premium based on annual experience. LTCI policies differed from Medicare Supplement policies in that they were less like annually renewable health insurance, and more like level premium Long-Term Disability (LTD) insurance. Over the years LTCI evolved to cover all service providers, through so-called comprehensive policies: Nursing Home Care (NHC), Assisted Living Facilities (ALF), Adult Day Care (ADC), and Home Care (HC), as long as the policyholder met the ADL[11] or cognitive minimum claim threshold. ALF differs from NHC in that they offer only paramedical services.

As Table 2.1 indicates, ALF requires about half the cost of NHC, but incidence, continuance, and utilization rates must be taken into account to determine its proper share of expected aggregate claim from which premium and reserve are derived.

Most policyholders and policymakers prefer the Home Care option, as long as possible. This option presents its own feasibility and risk measurement challenges, such as caregiver costs and availability.

The era in which Long Term Care risk is now studied is global, with Big Data feeding predictive models. Pre-1970, the main tools to compute insurance values were annual mortality and disability rates, commutation functions, mathematical formulas and constant loads based on few variables and static assumptions applied to net present values. The approach was deterministic, based on fixed expected values. Expected loss ratio[5] would be calculated, with a 65% loss ratio[5] being standard for Long-Term Disability insurance. Most insurance was principally governed through state valuation and marketing regulations. The advent of computers allowed asset-share models to project expected cash flows over the maximum duration of an insurance policy or a portfolio. Variables not inherent to the covered risk could be introduced to calculate present values of cash flow scenarios. Stochastic models were developed to account for the variability of averages as well as around averages. Instead of profits being imbedded as premium load, premium was set to meet Internal Rate of Return (IRR) or Return on Equity (ROE) targets. Rapid increase in computer capacity has allowed the application of Markov processes and complex distributions to produce predictive models. The internet allowed access to a volume of data not previously readily available. Globalization of information and data, as well as the growth of multi-national insurers, has brought forward complex analytic and accounting tools and standards such as Principle Based Reserving, Solvency 2, and International Accounting Standards. The Long Term Care risk started to be approached for insurance purposes on the tail end of the commutation function era and its complexity can now be approached with the newer tools.

Table 2.2 shows that private insurance contributes about 10% of Long-Term Services and Supports costs, its annual growth rates have been in the single digits since 1997, although it has been increasing sporadically since 2010. It should be noted

Table 2.1 2017 LTC provider annual cost[6]

2017 LTC provider annual cost	
Adult day care	$18,204
Assisted living facility	$45,000
Home care	$48,558
Nursing home care	$91,614

Table 2.2 Long-term services and supports costs, private insurance and gross domestic poduct[8]

	LTSS as % of		Private insurance as % of LTSS	Annual growth (%)		
	Health expenditures	GDP		LTSS	Private insurance	GDP
1980	7	1	3	16	38	9
1981	7	1	3	15	30	12
1982	7	1	4	13	31	4
1983	7	1	5	13	31	9
1984	7	1	5	11	29	11
1985	7	1	6	10	27	8
1986	7	1	7	10	30	6
1987	7	1	8	6	16	6
1988	7	1	9	14	34	8
1989	8	1	10	14	17	8
1990	8	1	10	17	21	6
1991	8	1	10	12	11	3
1992	8	1	10	11	12	6
1993	9	1	10	10	13	5
1994	9	1	11	9	19	6
1995	9	1	13	13	29	5
1996	10	1	13	9	16	6
1997	10	1	13	6	6	6
1998	9	1	14	2	9	6
1999	9	1	14	0	−1	6
2000	9	1	13	4	−6	6
2001	8	1	12	7	−5	3
2002	8	1	11	5	1	3
2003	8	1	10	7	−4	5
2004	8	1	9	7	−6	7
2005	8	1	9	7	4	7
2006	8	1	8	4	−2	6
2007	8	1	8	9	3	4
2008	8	1	8	6	3	2
2009	8	1	7	5	4	−2
2010	8	1	8	5	7	4
2011	8	1	8	4	4	4
2012	8	1	8	2	10	4
2013	8	1	8	2	6	3
2014	8	1	9	3	8	4

(continued)

Table 2.2 (continued)

	LTSS as % of		Private insurance as % of LTSS	Annual growth (%)		
	Health expenditures	GDP		LTSS	Private insurance	GDP
2015	8	1	9	4	11	4
2016	8	1	10	3	5	3
Average				4	7	3

that in the table LTSS includes only Nursing Homes and Home Care, the total LTSS related expenses is higher, and the ratio to Health Expenditure is closer to 10% than 8%; private insurance includes all insurance sources, such as Medicare Supplement, and the ratio of LTCI to total LTSS expenses is closer to 7% than 10%.

2.1.1 Legal Environment

Long Term Care is highly regulated. Both at the federal and state level, laws are created through an executive (president/governor) and a bicameral legislature (except Nebraska, which is unicameral). Most regulatory jurisdictions have an insurance supervisory office and several supervisors, or commissioners, are elected by the voting population. Many LTSS providers must be licensed by the state in which they operate; LTCI agents must comply with continuing education programs, the American Academy of Actuaries publishes Actuarial Standard Of Practice[12] guidelines; the federal government, through its taxation code,[3] has defined what is known as tax qualified Long Term Care policies; most states have defined what is known as partnership Long Term Care policies,[13] which can be used in conjunction with their Medicaid[14] programs. The 1945 McCarran–Ferguson Act legislated the rights of states to regulate insurance, but federal laws increasingly impact how insurance products are designed, valued, and marketed. The 1933 Glass–Steagall law separated investment and commercial banking and made interstate banking subject to federal supervision. Glass–Steagall prevented banks from fully participating in the insurance market and this has led to the quasi absence of bancassurance. While Glass–Steagall was repealed in 1999, the 2008 global financial crisis has brought further national and international oversight of the financial market. LTSS providers and LTC insurers must comply with the Health Insurance Portability and Accountability Act (HIPAA)[15] a 1996 federal law which sets privacy rules and defines minimum requirements for a LTCI policy to allow its premium to be included in the annual medical expense tax deduction and its benefits excludable from taxation. HIPAA also sets the maximum number of Activities of Daily Living[11] which an insurer can require to pay full benefits: 2 out of 5. Other federal laws impact Long Term Care Insurance. For instance, the Employee Retirement Income Security Act (ERISA, 1974), which regulates pensions, allows some group insurance plans to bypass state mandates.

The legal environment plays an important role. LTSS is a newly identified sector of the economy which affects a segment of the population considered particularly vulnerable, but increasingly politically powerful. It involves complex, not purely financial, transactions between consumers and providers. At the national level, the federal judiciary consists of a Supreme Court, twelve regional circuits and ninety-four judicial districts. In general, states have a similar judicial structure, down to municipal courts. Common law, which relies heavily on past rulings, applies throughout the US, except in Louisiana, which relies on a mixture of common law, French (so-called Napoleonic code), and Spanish law. Insurance companies usually market their product on a nationwide basis, but are domiciled in one state, while the individual policyholder resides in the smallest judicial level, so litigation can be complex. Several states require extra territorial jurisdiction, where group insurance contracts covering one of its residents must comply with the resident insurance laws and regulations. It is also likely for a local lawsuit to be broadcasted by national news or social media and impact the image of the insurer. Since there is little LTCI history, how it will fit into the fabric of the US society remains to be seen.

2.1.2 Community Living Assistance Services and Supports Act

In 2010 the Community Living Assistance Services and Supports Act[16] (CLASS Act), part of the Patient Protection and Affordable Care Act (ACA), authorized the first nationwide Long Term Care social insurance initiative. It would have created a voluntary but Guaranteed Issue (no underwriting) long term care insurance option for active employees. Participation would have been optional but would have required the employee to decline the coverage, instead of electing the coverage. The spouse of an enrolled employee would also have been eligible subject to some underwriting requirement. A five-year waiting period would apply before any benefit would be paid. Uni-gender premiums varied by issue age. Premium increases were allowed but were restricted for participants over 65. The minimum daily cash benefits were to be $50 per day, based on 2 ADLs out 6, with higher benefits payable for more severe impairments, based on nationwide criteria.

A critical provision of the law required the actuarial adequacy of the program over a 75-year period, i.e. premiums received over that period must fund benefits, to be certified annually. The voluntary and Guaranteed Issue aspects, as well as limits on premiums, created concerns about the actuarial viability of the program. That provision was withdrawn in 2013 without being implemented.

2.2   Social Environment

2.2.1   Economic Dependency

With a rapidly aging population, measurements such as healthy life expectancy and dependency ratios attract more attention. Dependency ratios measure the ratio of the non-working population to the labor force. Table 2.3 points to the need to address growing economic dependency among the elderly.

According to the Bureau of Labor statistics, in 2016, for every 65-and-older dependent, about 4 persons were in the labor force, and by 2026, about 3 persons are projected to be in the labor force for each older dependent. Another way to measure the age dependency ratio is to compare the 65-and-older population with the 16-to-64 year old population, but as Table 2.3 shows, that population has its own dependent population. Economic dependency depends on the country's definition of labor force, which in the US includes the unemployed but excludes the military.

Dependency ratios and trends have a significant impact on LTCI viability, which is faced with a decreasing pool of potential policyholders and caregivers, as well as an increasing population in need of the coverage.

Long Term Care is concerned with health-related dependence, a subset of economic-related dependency. While dependency in the Long Term Care context is mainly concerned with incidence and continuance, economic dependency greatly impacts how LTC behaves: for insurance, most policies are offered to non-economic dependent individuals, and for policy-makers LTSS expenses have much different implications than unemployment or education. The whole state of dependency should, at a minimum, not be ignored.

Table 2.3 Economic dependency[17]

Economic dependency				
	Total population	Labor force	Dependent population	Dependency ratio (%)
Total 2026	347,304,498	169,582,274	177,722,224	105
Under 16	72,065,076		65,967,505	39
16–64	208,301,911		59,353,796	35
65 and older	66,937,511		52,400,923	**31**
Total 2016	322,400,000	159,524,988	162,875,012	102
Under 16	69,432,000		64,448,095	40
16–64	204,488,000		58,705,195	37
65 and older	48,480,000		39,721,722	**25**

2.2.2 *Age and Gender*

While female to male ratios remain rather even through age 60, they climb to 1.1 at 70 and 2 at 90, ages of particular importance to Long Term Care due to higher female LTSS utilization (Fig. 2.1).

2.2.3 *Income*

The following three graphs[19] display differences in household income levels between ages, genders, and regions (Figs. 2.2, 2.3 and 2.4).

Income plays an important role in LTCI affordability and need. While LTCI is regarded as asset protection insurance, the higher the income, the more affordable it

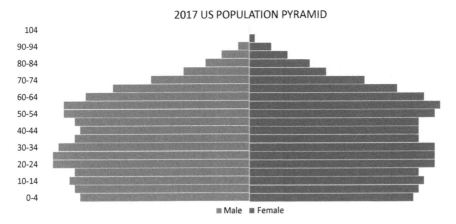

Fig. 2.1 2017 US population pyramid[18]

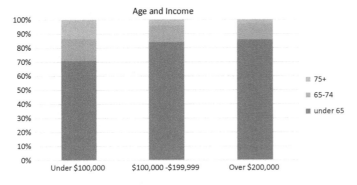

Fig. 2.2 Age and income

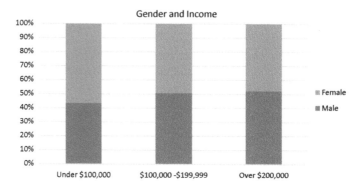

Fig. 2.3 Gender and income

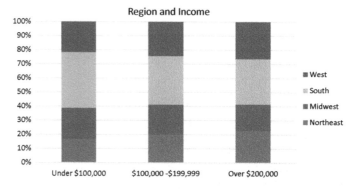

Fig. 2.4 Region and income

becomes but the less the need for such an expensive insurance. The older the person, the more expensive a policy is, but also that person is likely to be in the lower income range. Females comprise a majority of the under $100,000 income range but are in the segment of the population that may need LTCI the most. Paradoxically, the Northeast, with about 18% of households and over 20% of over $200,000 income, accounts for about 28% of Medicaid[14] spending, Medicaid[14] being the largest LTSS payor. Lower income areas tend to have lower LTSS costs and individuals buying LTCI would not need to purchase a policy with as high a Daily Benefit maximum as an individual living in a higher cost region.

2.2.4 Assets

As Table 2.4 shows, assets for people aged 65 and over are skewed at the higher range. While this may be a good argument to buy LTCI for asset protection, careful estate planning must now account for the shorter LTCI benefit periods currently available in the US market.

Actuarial models should not ignore asset and income levels as the need for Long Term Care is closely linked to health, and health is correlated to income and wealth.[21] Income is also correlated to education.

2.2.5 Affordability of Long-Term Services and Supports

The four tables below, derived from a 2016 report,[22] 'Long-Term Services and Supports for older Americans: risks and financing' summarize another challenge LTC presents. Table 2.5 estimates expected costs of lifetime LTSS, per payer, per person. At first glance LTSS seems a manageable personal expense, which is borne out by the fact that over 50% of the costs are paid by individuals. But when the cost per population is shifted to cost per user of services (Table 2.6), the cost doubles, which strains individual financing and indicates the need for insurance. The last two Tables 2.7 and 2.8, show the corresponding average values for females and males, with expenses almost doubling for women.

2.3 Measurement Tools

2.3.1 The Sullivan Index

In 1971 Daniel Sullivan introduced a life expectancy index in an article ("A Single Index of Mortality and Morbidity", 1971)[23] which defined healthy and disabled life expectancy. The following is a brief description of the index, which is based on the 1965 US Life table, a public table, and assuming a stationary (instead of generational) population:

l_x: Total population at age x.

L_x: Average stationary population within observation interval.

T_x: Person-years lived at and over age x, calculated by summing L_x values at and over age x.

w_x: Number of days of disability per person per year in the interval beginning at age x.

I_x: Disability-free prevalence factor, or Sullivan index $= 1 - (w_x/365)$.

e_x: Life expectancy at x.

L_x^h, T_x^h, e_x^h, are defined as above, but for healthy (disability free) lives (Table 2.9).

Table 2.4 2013 asset distribution by household[20]

Head of household age	Zero or negative	$1–$4,999	$5,000–$9,999	$10,000–$24,999	$25,000–$49,999	$50,000–$99,999	$100,000–$249,999	$250,000–$499,999	$500,000 or over
Less than 35 years (%)	32	14	8	11	9	9	10	4	3
35–44 years (%)	21	9	5	8	8	10	18	10	11
45–54 years (%)	16	7	3	6	7	10	18	14	19
55–64 years (%)	12	7	3	5	5	9	18	16	26
65 years and over (%)	6	7	2	5	5	10	21	18	27
65–69 years (%)	7	6	3	6	6	9	19	14	30
70–74 years (%)	6	6	3	5	5	10	19	21	28
75 and over (%)	6	8	2	4	5	10	23	19	24

Table 2.5 Average lifetime LTSS expenditures per adult turning 65 in 2015–2019

Average sum (2015$) of LTSS expenditures from age 65 through death projected for adults turning 65 in 2015–2019

	Payer							
	Medicare	Medicaid	Other public	Total public	Out-of pocket	Private insur-ance	Total private	Total
Nursing facility								
Dollars	$7,200	$33,300	$700	$41,200	$22,500	$1,600	$24,100	$65,300
Percentage	11.03%	51.00%	1.07%	63.09%	34.46%	2.45%	36.91%	100.00%
Community-based								
Dollars	$6,600	$14,100	$400	$21,100	$49,600	$2,100	$51,700	$72,800
Percentage	9.07%	19.37%	0.55%	28.98%	68.13%	2.88%	71.02%	100.00%
Total expenditures								
Dollars	$13,700	$47,400	$1,100	$62,200	$72,200	$3,700	$75,900	$138,100
Percentage	9.92%	34.32%	0.80%	45.04%	52.28%	2.68%	54.96%	100.00%

Table 2.6 Average lifetime LTSS expenditures per adult turning 65 in 2015–2019 and using LTSS

Average sum (2015$) of expenditures from age 65 through death projected for users of paid, formal LTSS who turn 65 in 2015–2019

	Payer							
	Medicare	Medicaid	Other public	Total public	Out of pocket	Private insur-ance	Total private	Total
Nursing facility								
Dollars	$14,000	$64,000	$1,000	$79,000	$44,000	$3,000	$47,000	$126,000
Percentage	11.11%	50.79%	0.79%	62.70%	34.92%	2.38%	37.30%	100.00%
Community-based								
Dollars	$12,000	$27,000	$1,000	$40,000	$96,000	$4,000	$100,000	$140,000
Percentage	8.57%	19.29%	0.71%	28.57%	68.57%	2.86%	71.43%	100.00%
Total expenditures								
Dollars	$26,000	$91,000	$2,000	$119,000	$140,000	$7,000	$147,000	$266,000
Percentage	9.77%	34.21%	0.75%	44.74%	52.63%	2.63%	55.26%	100.00%

2.3.2   1985 National Nursing Home Survey

Early insurers in the LTCI field relied on National Nursing Home Surveys[24] for actuarial estimates of claim rates using incidence and termination tables published in the Transactions of the Society of Actuaries 1988–1990 Reports, Report of the Long Term Care Experience Committee, 1985 National Nursing Home Survey, Utilization data.[25] The survey estimated an insurance population by excluding individuals whose diagnoses would not qualify them for individual insurance, which is almost always underwritten.

Table 2.7 Average lifetime LTSS expenditures per female turning 65 in 2015–2019

Average sum (2015$) of expenditures from age 65 through death projected for female turning age 65 in 2015–2019

	Payer							
	Medicare	Medicaid	Other public	Total public	Out of pocket	Private insurance	Total private	Total
Nursing facility								
Dollars	$8,800	$48,200	$800	$57,800	$27,500	$2,300	$29,800	$87,600
Percentage	10.05%	55.02%	0.91%	65.98%	31.39%	2.63%	34.02%	100.00%
Community-based								
Dollars	$7,900	$18,800	$400	$27,100	$64,800	$2,700	$67,500	$94,600
Percentage	8.35%	19.87%	0.42%	28.65%	68.50%	2.85%	71.35%	100.00%
Total expenditures								
Dollars	$16,700	$67,000	$1,100	$84,800	$92,400	$5,000	$97,400	$182,200
Percentage	9.17%	36.77%	0.60%	46.54%	50.71%	2.74%	53.46%	100.00%

Table 2.8 Average lifetime LTSS expenditures per male turning 65 in 2015–2019

Average sum (2015$) of expenditures from age 65 through death projected for men turning age 65 in 2015–2019

	Payer							
	Medicare	Medicaid	Other public	Total public	Out of pocket	Private insurance	Total private	Total
Nursing facility								
Dollars	$5,500	$17,400	$600	$23,500	$17,100	$900	$18,000	$41,500
Percentage	13.25%	41.93%	1.45%	56.63%	41.20%	2.17%	43.37%	100.00%
Community-based								
Dollars	$5,100	$9,100	$400	$14,600	$33,600	$1,400	$35,000	$49,600
Percentage	10.28%	18.35%	0.81%	29.44%	67.74%	2.82%	70.56%	100.00%
Total expenditures								
Dollars	$10,600	$26,600	$1,000	$38,200	$50,600	$2,300	$52,900	$91,100
Percentage	11.64%	29.20%	1.10%	41.93%	55.54%	2.52%	58.07%	100.00%

Table 2.10 illustrates two annual continuance calculations, from Tables 11 and 14 of the 1985 NNHS report. The first part (Proportion of Admissions at End of Period) shows the annual compound survival rates from admission, which is used to calculate the Average Length Of Stay (ALOS), or the disabled life expectancy. The second part (Proportion of Days Left After Period) displays the likelihood that a stay will exceed a number of years, which can be useful to calculate variances.

Table 2.9 Life expectancy

Computation of the approximate expectation of life free of disability (e_x^h) for white males, civilian resident population, United States, mid-1960s

Age group	Exact initial age	Disability						
		1965 abridged life table value		Days/year	Weighting factor	Life table values, weighted for disability		
	x	l_x	L_x	w_x	l_x	L_x^h	T_x^h	e_x^h
Under 15	0	100,000	1,457,411	12	0.967	1,409,316	6,252,783	62.5
15–44	15	96,767	2,830,657	13	0.964	2,728,753	4,843,467	50.1
45–64	45	90,639	1,623,962	31	0.915	1,485,925	2,114,713	23.3
65–74	65	65,901	532,960	72	0.802	427,434	628,788	9.5
75 and over	75	39,665	318,095	134	0.633	201,354	201,354	5.1

The report includes tables for prevalence rate, the fraction of total dependent population at a point in time. A relationship formula between incidence and prevalence is sometimes used:

$$(\text{incidence rate}) \times (\text{ALOS}) = \text{prevalence rate}.$$

Prevalence rates can be used to estimate outstanding claims in cash flow analysis, but the authors of the paper found that the above relationship does not bear out if the

Table 2.10 Continuance and utilization

Years from admission	Proportion of admissions at end of period			Proportion of days left after period		
	All	Female	Male	All	Female	Male
0	1.00000	1.00000	1.00000	1.0000	1.0000	1.0000
1	0.38621	0.42884	0.31949	0.7974	0.6915	0.5839
2	0.26299	0.31527	0.20227	0.6720	0.5073	0.3891
3	0.19243	0.23023	0.14193	0.5708	0.3779	0.2685
4	0.14841	0.18193	0.10511	0.4850	0.2843	0.1928
5	0.11384	0.14371	0.08717	0.4084	0.2136	0.1358
6	0.08344	0.11043	0.06174	0.2982	0.1604	0.0949
7	0.06070	0.08289	0.04540	0.2955	0.1225	0.0680
8	0.04454	0.06176	0.03459	0.2204	0.0963	0.0490
9	0.03334	0.04719	0.02647	0.2226	0.0780	0.0349
10	0.02549	0.03856	0.02016	0.1679	0.0644	0.0245

(continued)

Table 2.10 (continued)

Years from admission	Proportion of admissions at end of period			Proportion of days left after period		
	All	Female	Male	All	Female	Male
11	0.01969	0.03036	0.01513	0.1711	0.0539	0.0169
12	0.01525	0.02474	0.01113	0.1296	0.0456	0.0114
13	0.01181	0.02049	0.00801	0.1329	0.0389	0.0076
14	0.00921	0.01741	0.00557	0.1013	0.0334	0.0049
15	0.00731	0.01521	0.00376	0.1043	0.0287	0.0031
16	0.00584	0.01360	0.00251	0.0794	0.0246	0.0020
17	0.00474	0.01236	0.00170	0.0815	0.0209	0.0013
18	0.00387	0.01136	0.00114	0.0615	0.0175	0.0008
19	0.00320	0.01053	0.00079	0.0624	0.0144	0.0005
20	0.00264	0.00979	0.00053	0.0458	0.0116	0.0003
21	0.00223	0.00916	0.00037	0.0453	0.0089	0.0002
22	0.00187	0.00859	0.00026	0.0309	0.0065	0.0001
23	0.00157	0.00806	0.00017	0.0273	0.0041	0.0001
24	0.00131	0.00759	0.00013	0.0142	0.0020	0.0000
25	0.00114	0.00714	0.00009	0.0000	0.0000	0.0000
ALOS	20 months	24 months	16 months			

populations and the observation periods for incidence and prevalence do not exactly match.

For premium rate calculations, models closely followed life insurance assumptions and tables, such as mortality rates, lapse rates, and interest rates. Usually a present value of continuance rates, using a fixed interest rate, would be multiplied by the appropriate incidence rate to calculate a claim rate, and then, to fit existing models, would be applied like a life insurance benefit to active lives at the end of a policy year. In some instances, incidence rates being small compared to active life termination rates, its count would not be excluded from the active life exposure of the next year. While this may seem a conservative approach by increasing the exposure for future years, differences in mortality rates between active and dependent lives make this assumption less appropriate, as it distorts active life termination rates in unpredictable ways.

2.3.3 Long Term Care Intercompany Experience Studies

Since 1993, the Society of Actuaries has published five LTCI intercompany studies for claims[26] and since 2002, with the Life Insurance Marketing Research Associa-

Table 2.11 Society of actuaries long term care intercompany experience study—aggregate database 2000–2011 report—incidence rates

Age group	Incidence rates											
	Total				Female				Male			
	Total (%)	Nursing facility (%)	Assisted living facility (%)	Home health care (%)	Total (%)	Nursing facility (%)	Assisted living facility (%)	Home health care (%)	Total (%)	Nursing facility (%)	Assisted living facility (%)	Home health care (%)
0–49	0.05	0.00	0.00	0.03	0.05	0.00	0.00	0.03	0.04	0.00	0.00	0.03
50	0.07	0.01	0.00	0.05	0.08	0.01	0.00	0.05	0.05	0.01	0.00	0.03
55	0.09	0.01	0.01	0.06	0.09	0.01	0.01	0.06	0.07	0.01	0.01	0.05
60	0.13	0.02	0.01	0.08	0.15	0.02	0.01	0.08	0.10	0.02	0.01	0.07
65	0.27	0.06	0.03	0.15	0.32	0.06	0.04	0.15	0.21	0.05	0.03	0.10
70	0.67	0.19	0.11	0.31	0.78	0.21	0.13	0.31	0.51	0.17	0.09	0.22
75	1.62	0.56	0.31	0.64	1.87	0.60	0.38	0.64	1.28	0.51	0.22	0.46
80	3.51	1.36	0.76	1.18	3.89	1.42	0.90	1.18	2.93	1.28	0.54	0.92
85	6.44	2.79	1.39	1.93	6.79	2.82	1.55	1.93	5.82	2.74	1.10	1.67
90+	9.55	4.72	1.85	2.58	9.70	4.76	1.98	2.57	9.20	4.64	1.55	2.60

tion (LIMRA), three policy persistency studies.[27] Reliable uses of the data are held back due to lack of standard classifications for termination rates such as deaths, lapse and recoveries, and clear and uniform reporting among different companies (and sometimes within a company) is difficult to achieve. The volume of the data, however helps to analyze trends. Table 2.11 shows incidence rates derived from the 2000–1011 report. These rates aggregate incurral dates, Waiting Periods,[28] and Elimination Periods.[28]

As the continuance Table 2.12 indicates, data credibility declines with claim duration as claims occur many years after a policy is issued, and may take its full course over many more years. The lack of reliable reporting for diagnostic and termination types, and dates, requires great care in the use of these rates. The 2000–2011 report mentions that, even with the large amount of data collected, only the first four years are reliable. The Average Length of Stay (ALOS) shown here is the average claim duration for the 17-year observation period, not the expected Length of Stay of a policyholder.

Continuance rates are calculated from claim termination rates, much as survival rates are derived from mortality rates, except that while continuance means survival in the 'dependent' status, termination rates may include deaths, recoveries, or transitions to another dependent status or provider. Provider and ADL[11] specific termination rates may be required to obtain an accurate measure of claim continuance rates.

The majority of stand-alone LTCI policies reimburse claimants on actual expenses incurred, subject to a maximum daily benefit. Table 2.13 displays the amount of reimbursement relative to maximum daily benefit, or claim utilization rate ('utilization' has multiple meanings depending on the context), a key morbidity assumption for modeling long term care policies with reimbursement provisions.

Since a model will usually take a maximum daily benefit amount as its assumed benefit, utilization materially impacts the validity of a model's results. Especially when annual cost of living increases are applied to benefits.

Much of LTC cost modeling assumes simplified transitions from one dependent state to another, but a reimbursement model should take into account transitions to and from providers which may reflect transitions from one dependent state (partial) to another (total).

Active life termination rates, when a policyholder ceases to pay premiums, have a critical impact on the ability of an insurer to fund its future liabilities. While active life reserves and claim reserves are calculated on a per policy basis, a termination causes actual reserves to be released as well as reducing the risk exposure and contributing to the overall balance of cash flows for the portfolio. Variations in termination rates outside a narrow margin from expected cause a set of circumstances that greatly impact reserve adequacy and the ability of future premium to fund such reserves, especially when investment income drops below expectations. Higher than expected terminations, especially with lapses, may not symmetrically improve financial results due to anti-selection. Table 2.14 was taken from the 2015 inter-company termination study.[27]

Table 2.12 Society of actuaries long term care intercompany experience study—aggregate database 2000–2011 report—claim termination rates—all causes

Claim duration	Claim termination rates			Recovery			Mortality		
	All terminations								
	Total (%)	Female (%)	Male (%)	Total (%)	Female (%)	Male (%)	Total (%)	Female (%)	Male (%)
1	48.81	45.38	55.27	15.19	17.24	11.50	33.62	28.13	43.77
2	24.49	21.35	31.13	3.43	3.74	2.80	21.05	17.61	28.32
3	26.21	23.99	31.47	4.60	5.07	3.54	21.60	18.92	27.93
4	26.78	24.50	32.61	4.19	4.36	3.74	22.60	20.14	28.87
5	28.29	26.75	32.55	5.84	6.17	4.98	22.44	20.58	27.57
6	26.27	24.96	30.15	3.74	3.76	3.70	22.53	21.20	26.45
7	24.79	23.90	27.57	2.98	3.05	2.79	21.81	20.85	24.79
8	24.25	24.16	24.58	2.08	2.19	1.71	22.17	21.97	22.87
9	23.42	24.53	19.91	2.05	2.20	1.54	21.38	22.33	18.37
10	25.30	26.65	21.35	2.43	2.56	2.05	22.87	24.09	19.30
11	21.44	21.35	21.75	1.40	1.30	1.68	20.03	20.05	20.07
12	23.66	25.38	18.73	0.44	0.32	0.82	23.22	25.06	17.91
13	32.27	34.80	24.48	3.89	3.75	4.26	28.38	31.05	20.22
14	20.54	26.32	5.11	1.25	0.64	2.95	19.30	25.67	2.16
15	35.16	33.38	41.44	3.76	3.84	3.71	31.40	29.54	37.73
16	20.90	25.68	0.00	6.56	7.98	0.00	14.34	17.70	0.00
17	37.92	49.57	0.00	11.46	13.14	0.00	26.46	36.42	0.00
ALOS	24 months	27 months	18 months						

Table 2.13 Society of actuaries long term care intercompany experience study—aggregate database 2000–2011 report—utilization rate

	Claim utilization rate			
	Total (%)	Nursing facility (%)	Assisted living facility (%)	Home health care (%)
Female and male	75	81	93	60
Female	75	82	94	60
Male	73	80	91	59

Active life experience mortality rates are at about 2/3 of current annuity mortality rates, moreover most statutory reserves are based on older mortality tables which would make that ratio lower.

The availability of credible data for incidence, continuance, and dependent mortality is scarce compared to data available to evaluate other personal insurance, such as life and annuities.[29]

Table 2.14 Policy terminations—aggregate database—2000–2011 report

Policy year	Aggregate voluntary lapse (%)	Active life mortality rates					
		Mortality rates			Actual to expected to 2012 IAM		
		Female (%)	Male (%)	Aggregate (%)	Female (%)	Male (%)	Aggregate (%)
1	5.8	0.15	0.24	0.19	22.17	26.59	24.44
2	4.3	0.26	0.38	0.31	33.88	37.93	35.93
3	3.3	0.34	0.52	0.42	40.40	46.68	43.55
4	2.8	0.43	0.65	0.52	44.59	52.05	48.29
5	2.6	0.53	0.80	0.65	49.62	57.73	53.59
6	2.4	0.64	0.97	0.78	53.61	62.63	57.99
7	2.2	0.76	1.16	0.93	57.42	67.69	62.35
8	2.2	0.92	1.37	1.11	62.03	71.46	66.52
9	2.1	1.11	1.64	1.33	66.44	76.42	71.14
10	2.2	1.32	1.93	1.57	68.73	79.06	73.52
11	2.3	1.56	2.28	1.84	72.02	82.45	76.82
12	2.3	1.79	2.73	2.16	73.71	87.56	80.02
13	2.4	2.09	3.08	2.48	77.28	88.14	82.16
14	2.6	2.36	3.54	2.81	79.33	91.49	84.73
15	2.7	2.63	3.92	3.11	81.43	93.45	86.70
16	2.8	2.87	4.31	3.40	82.55	95.59	88.20

(continued)

Table 2.14 (continued)

Policy year	Aggregate voluntary lapse (%)	Active life mortality rates					
		Mortality rates			Actual to expected to 2012 IAM		
		Female (%)	Male (%)	Aggregate (%)	Female (%)	Male (%)	Aggregate (%)
17	3.1	3.13	4.61	3.67	84.71	96.42	89.72
18	3.4	3.44	4.95	3.99	87.45	97.41	91.64
19	4.0	3.78	5.36	4.34	89.72	98.68	93.43
20	4.4	4.01	5.77	4.62	87.86	98.63	92.22
Aggregate	3.0	0.97	1.34	1.12	63.42	71.37	67.15

IAM: Individual Annuity Mortality

2.4 Diagnosis

Diagnosis may be the most critical actuarial aspect of Long Term Care, as current research in healthy versus disabled life expectancy and the relationship between mortality and morbidity indicate.

Pricing and reserving a Long Term Care risk must recognize the impact in severity (utilization) of different providers and continuance (duration) of different ailments, especially cognitive and physical. As Table 2.15 shows, over 50% of reimbursements made of claims for identifiable diagnoses went to cognitive related claims.

Tables 2.16, 2.17 and 2.18 show different results between cognitive and physical ailments that lead to Long Term Care Insurance claims. Termination rates are also divided by gender and recovery and death. The Average Length of Stay over the 17 years is more than double for cognitive versus physical diagnostics. Of 214,967 claims, 94,291 were either unknown or not clearly identifiable. While identifiable cognitive and physical claim numbers are almost equal, 61,617 and 59,059, the cost of cognitive claims is likely to be more than twice as costly. In this report, data after two years becomes increasingly less credible. As mentioned above, the intercompany study is very useful to study emerging trends and identify emerging risks not taken into account while averaging claim rates for the insured population. This data alone should not be used to price or reserve a Long Term Care insurance product.

Overall, reliable tools specific to the Long Term Care risk are scarce for pricing and reserving, leaving actuaries to use company and reinsurer experience for products that provide long-term coverage.

Table 2.15 2000–2011 LTC experience study claim utilization analysis per diagnosis

Claim paid by diagnosis ($million)

Diagnosis	Amount paid	Share (%)	Cumulative	
			Amount paid	Share (%)
Alzheimer's	**1,156**	**25**	**1,156**	**25**
Mental	**635**	**14**	**1,791**	**39**
Stroke	**543**	**12**	**2,334**	**51**
Arthritis	527	11	2,861	62
Nervous system and sense organs	401	9	3,262	71
Injury	394	9	3,656	79
Circulatory	361	8	4,017	87
Respiratory	198	4	4,215	91
Cancer	149	3	4,364	95
Diabetes	75	2	4,439	96
Digestive system	55	1	4,494	97
Genitourinary system	54	1	4,548	99
Hypertension	28	1	4,576	99
Endocrine/Immunity system	21	0	4,597	100
Skin and subcutaneous tissue	12	0	4,609	100
Pregnancy disorders	4	0	4,613	100
Congenital	4	0	4,617	100

2.5 Evolution of the Insurance Market

As we have seen earlier (Table 2.2), private insurance benefits accounts for less than 10% of LTSS expenditures. A 2016 National Association of Insurance Commissioners[4] report, 'The State of Long Term Care Insurance', states that at the beginning of this century over 100 insurers marketed LTCI, fifteen years later, about a dozen were still in the stand-alone market. As a result, claims are rapidly overtaking premium income, as Fig. 2.5 indicates.

Group insurance products for employees offer products with lower benefits and premium (significantly lower average issue age) than individual LTCI. Coverage is predominantly elective and requires minimal underwriting, if any, for an individual actively at work. Sometimes the employer matches employee contributions. Like stand-alone individual LTCI, group LTCI has steadily declined, with few insurers offering coverage. Group insurance sales as a proportion of total LTCI sales dropped by over 50% in ten years. By 2014 Group insurance accounted for 20% of new sales and 30% of outstanding policies.[4]

The US federal government offers group coverage as part of its Federal Employees Health Benefits (FEHB) program. The program currently has 278,000 participants.

Table 2.16 Claim termination rates for identified cognitive and physical ailments

Cognitive and physical

Claim termination rates

Claim duration	All terminations			Recovery			Mortality		
	Total (%)	Female (%)	Male (%)	Total (%)	Female (%)	Male (%)	Total (%)	Female (%)	Male (%)
1	49.45	46.68	54.60	17.94	20.36	13.65	31.51	26.32	40.94
2	25.75	22.56	32.35	4.15	4.49	3.49	21.60	18.07	28.86
3	26.80	24.54	32.03	4.86	5.31	3.85	21.95	19.23	28.18
4	27.79	25.46	33.64	4.63	4.81	4.17	23.16	20.65	29.48
5	28.97	27.21	33.77	5.72	6.00	4.97	23.25	21.21	28.80
6	27.17	25.68	31.51	3.30	3.34	3.18	23.87	22.34	28.32
7	25.51	24.68	28.06	2.36	2.34	2.43	23.14	22.35	25.63
8	25.53	25.15	26.76	2.20	2.18	2.27	23.33	22.97	24.49
9	24.25	25.45	20.34	2.21	2.22	2.21	22.04	23.23	18.13
10	26.98	28.29	22.91	2.22	2.17	2.38	24.76	26.12	20.53
11	23.32	23.59	22.61	1.19	0.96	1.89	22.14	22.63	20.72
12	24.10	26.44	16.95	0.00	0.00	0.00	24.10	26.44	16.95
13	32.50	36.40	20.31	3.15	2.80	4.21	29.35	33.61	16.10
14	23.15	30.61	4.14	1.70	0.88	4.14	21.45	29.74	0.00
15	40.36	34.37	64.86	1.20	0.00	3.25	39.17	34.37	61.61
16	20.50	22.89	0.00	9.71	10.78	0.00	10.79	12.11	0.00
17	49.58	55.17	0.00	13.16	13.45	0.00	36.42	41.72	0.00
ALOS	23 months	26 months	17 months						

Table 2.17 Claim termination rates for identified cognitive ailment diagnoses

Cognitive[a]

Claim duration	Claim termination rates			Recovery			Mortality		
	All terminations								
	Total (%)	Female (%)	Male (%)	Total (%)	Female (%)	Male (%)	Total (%)	Female (%)	Male (%)
1	28.79	24.80	35.24	8.89	9.24	8.30	19.90	15.56	26.95
2	23.01	19.57	29.26	3.06	3.22	2.78	19.95	16.35	26.49
3	26.15	23.49	31.51	4.01	4.40	3.27	22.14	19.09	28.24
4	27.56	25.26	32.62	4.10	4.29	3.69	23.46	20.98	28.93
5	28.70	26.80	33.22	5.21	5.54	4.47	23.49	21.26	28.75
6	26.81	25.03	31.39	2.87	2.92	2.76	23.94	22.11	28.63
7	25.39	24.62	27.50	2.02	2.06	1.92	23.37	22.56	25.57
8	26.28	26.10	26.80	1.90	1.79	2.16	24.38	24.31	24.64
9	24.53	26.09	20.19	1.85	1.70	2.30	22.68	24.39	17.90
10	26.67	29.02	20.52	1.67	1.85	1.19	25.00	27.17	19.32
11	21.99	22.62	20.48	0.44	0.33	0.71	21.55	22.29	19.78
12	21.66	24.60	14.37	0.00	0.00	0.00	21.66	24.60	14.37
13	32.08	36.62	21.24	1.69	1.48	2.35	30.39	35.14	18.90
14	20.57	28.90	4.96	2.43	1.39	4.96	18.14	27.51	0.00
15	39.53	30.35	63.60	1.73	0.00	3.72	37.80	30.35	59.89
16	15.98	19.06	0.00	0.00	0.00	0.00	15.98	19.06	0.00
17	33.82	40.86	0.00	0.00	0.00	0.00	33.82	40.86	0.00
ALOS	33 months	37 months	26 months						

[a] Alzheimer, mental, stroke, nervous system and sense organ

Table 2.18 Claim termination rates for identified physical ailment diagnoses

Physical[a]

Claim duration	Claim termination rates								
	All terminations			Recovery			Mortality		
	Total (%)	Female (%)	Male (%)	Total (%)	Female (%)	Male (%)	Total (%)	Female (%)	Male (%)
1	64.57	61.29	71.36	22.46	25.78	15.97	42.11	35.51	55.39
2	30.00	26.72	38.38	5.81	6.22	4.80	24.19	20.50	33.59
3	27.93	26.18	33.22	6.33	6.74	5.18	21.60	19.44	28.04
4	28.20	25.77	36.12	5.59	5.67	5.31	22.61	20.11	30.81
5	29.47	27.89	35.20	6.65	6.76	6.25	22.82	21.12	28.95
6	27.84	26.80	31.86	4.10	4.05	4.31	23.73	22.76	27.55
7	25.74	24.81	29.53	3.00	2.82	3.70	22.74	21.99	25.83
8	24.10	23.52	26.72	2.80	2.86	2.56	21.30	20.66	24.16
9	23.75	24.38	20.78	2.87	3.06	1.95	20.88	21.32	18.83
10	27.54	27.04	30.05	3.25	2.72	5.70	24.29	24.32	24.34
11	26.06	25.25	30.23	2.61	1.97	5.56	23.45	23.28	24.67
12	29.95	30.30	29.50	0.00	0.00	0.00	29.95	30.30	29.50
13	33.57	35.90	18.12	6.99	5.55	18.12	26.57	30.35	0.00
14	29.95	34.88	0.00	0.00	0.00	0.00	29.95	34.88	0.00
15	48.08	41.88	58.53	0.77	0.00	0.00	47.31	41.88	58.53
16	0.00	0.00	0.00	0.00	0.00	0.00	0.00	0.00	0.00
17	30.00	42.87	0.00	0.00	0.00	0.00	30.00	42.87	0.00
ALOS	15 months	17 months	10 months						

[a]Arthritis, cancer, circulatory, congenital, diabetes, digestive system, endocrine/immunity system, hypertension, injury, pregnancy disorders, respiratory, skin and subcutaneous tissue

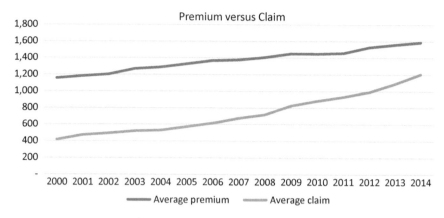

Fig. 2.5 Premium and claim[4]

California offers coverage to its state employees and dependents through its Calpers Long Term Care program, with 128,000 participants. Both plans were subject to significant premium increases. Individual insurance is also marketed to members of associations, such as the American Association of Retired Person (AARP), with full underwriting.

Table 2.19[4] indicates that the LTC risk is increasingly being insured through combination, or hybrid, products. While these products offer many advantages, such as not losing one's premium if the need for LTSS does not arise, benefits are limited and do not usually cover what can be catastrophic expenses; and total policy premiums can even be higher than the stand-alone LTCI policy to provide for the life or annuity coverage. Other products being introduced are underwritten immediate annuities, so-called Immediate Needs Annuities. Other insurance-like products are reverse mortgages and accelerated benefits on a life insurance policy. Many of these products offer risk hedging opportunities to the insurer but are partial solutions to the Long Term Care risk for the insured. Overall the reaction of insurers has been to reduce the coverage period and limit the amount of benefit, while increasing premiums.

2.6   Salient Differences Between the United States and France

Several chapters describe the French Long Term Care environment. Following are further remarks to contrast the US and French systems.

Table 2.19 Stand-alone versus combination

Year	Sales of stand-alone LTCI and combination products						Percent of total sales				
	$million						Stand alone (%)	Combination products			
	All products	Stand alone	Combination products					All (%)	Annuity (%)	Life (%)
			All	Annuity	Life					
2012	3,190	580	2,610	210	2,400		18	82	7	75
2013	3,266	406	2,860	260	2,600		12	88	8	80
2014	3,156	316	2,840	440	2,400		10	90	14	76
2015	3,831	261	3,570	470	3,100		7	93	12	81
2016	4,306	226	4,080	480	3,600		5	95	11	84

2.6.1 Jurisdictions

Insurance products in both countries are priced and marketed nationwide, but the United States companies have operated for a long time in a dual statewide jurisdiction and nationwide market. French insurance companies, until the second half of the twentieth century, have operated in a highly regulated, quasi monopolistic nationwide market. The European Union has opened the insurance market but brought new regulations, such as uni-gender premiums. Long Term Care services standards and regulations are rather uniform in the US and in France, but not in the European Union, which has longstanding cultural and political differences. French insurance companies must now take these differences into account. For social responses to the LTSS risk, there is now a similarity between US states and the French national government. They mostly have the same fiscal tools to fund programs, but the French government is now restricted by EU budget deficit limit (3% of GDP). The monetary tools are also limited as the French government cannot issue Euros and the US states cannot issue Dollars.

2.6.2 Actuarial Memorandum

Insurance companies are required to file in each insurance jurisdiction marketing, contract, and rate information. That filing must be approved, within a period after which a filing is 'deemed' approved, usually 30 days, before a product can be marketed. The same applies to any revision to the policy, including rates, before any change can be introduced. Actuarial memoranda are included in such filing, certified by an actuary (sample document in end-notes).[30] In France, the policy information and documentation for the actuarial basis of the premium rates and reserve calculations are kept on file and are subject to audit. Insurance regulations in the US are rather specific to a product, in France the three insurance regulations (see Section 4.d 'Prospects for Medical and Social Evolutions' note 5) are comparatively vague, leaving much leeway to the insurance company and the actuary to price and reserve a product. The flip side of this freedom is that an audit can radically change the marketing viability of a product and may require its withdrawal from the market. US states have joined an interstate insurance compact[31] to streamline the insurance filing process, but the basic requirements have not changed.

2.6.3 Gender

Starting in 2012, insurance policies sold in the European Union cannot vary premium or benefits by gender, after a 2011 ruling by the European Court of Justice. At that time, LTCI products in France and the United States were mostly offering uni-gender premium rates. But US companies have started offering different rates for males and females, with higher premium rates for females. As mentioned earlier, there are wide differences in biometric and utilization rates between genders, and performing projections over a long horizon combining genders for active life, and continuance distributions can be challenging, if not an added risk in itself.

2.6.4 Social Insurance

The social insurance approach taken in each country is very different but offers some similarities. The social and political environment in the United States, which shapes its insurance markets, has been rather stable since its 19th century Civil War and the early 20th century Great Depression. The brunt of LTSS financing is carried by the government through the federal old age insurance program, Medicare,[6] which has an eligibility age of 65 (with exceptions), and the federal/state health insurance program, Medicaid,[12] which is means-tested (for income and assets). Both programs primarily reimburse medical and nursing care expenses where Long Term Care is ancillary. Medicare is funded through a payroll tax and premiums, Medicaid is Pay-As-You-Go, each state funding its share as its sees fit. Most states have income taxes, a source of revenue not available to French administrative regions responsible for the APA program.

France's current social and political environment derives a great deal from mid-twentieth century events, with the end of World War II and the introduction of the European Common Market. Long Term Care is partly covered through its social security program (*Securité Sociale*), which has four main programs: health, workers' compensation, retirement, and family. For several years, Long Term Care has been proposed to be added as a fifth branch. France has a relatively new program for LTSS, APA (*Allocation Personnalisée d'Autonomie*, or Allowance for Personal Autonomy). It is Pay-As-You-Go partially funded nationally through an additional 0.3% payroll tax generated by a forfeited holiday (*Pentecôte*). Participation is universal and is income based through a graded co-pay, but has no asset requirement. Like Medicaid, funding and payments are shared by national and regional governments, but the administrative regions have fewer fiscal tools to fund the benefits than US states. Like French LTCI products, benefits take the form of cash benefits.

2.6.5 Cost-of-Living Adjustments

Cost-of-Living Adjustments (COLAs) are a common feature of US LTCI products, the more so since this option must be offered for its premium and benefits to be income tax qualified.[3] This option is also a requirement in partnership plans.[13] Many early products offered a 5% annually compounded benefit increase. In the 1980s when the first products were introduced this feature could reasonably be priced as the average

CPI for that decade was 5% and the average long-term interest was still about 10%. Projections usually used fixed assumptions over the expected duration of the policy. One of these assumptions was a higher than expected voluntary policy termination. By 2000 the average CPI was about 3% while the average long-term interest rate was about 5%. A $100 daily benefit in 1985 was worth $265 by 2015, while its level premium earned a much lower rate of return. That divergence between costs and income is mitigated by the reimbursement nature of the benefit: as LTSS inflation was lower than expected (although health inflation is higher than average inflation), claims would not reach the maximum benefit level. But as claim exposures became much higher than expected, daily benefit amounts became higher as well, increasing expected utilization, to be added to a much longer continuance of Alzheimer claims. This concordance of worse than expected experience is greatly exacerbated by the presence of the 5% compound COLA. Companies have responded by offering a lower annual COLA increase, such as 3% or CPI, and simple annual increases. Future Guaranteed Issue purchases are also available and must be purchased at the insured's attained age. Overall, upward, guaranteed, benefit adjustments are complex features which are not only challenging to price and reserve, but also to administer.

2.6.6 Premium Persistency

French pricing and reserving do not take voluntary termination into account, whereas US models take lapse rates into account. Another peculiarity is that French products include a reduced paid up benefit after a policy has been retained for at least eight years, a feature that should increase lapse rates; US products offer such features as an optional rider. Lapse rate assumptions have played a key role in US insurance companies' drive to seek premium rate increases. At the introduction of Long Term Care products, companies used assumptions based on the experience of life insurance, annuity, and disability products, using early lapse rates close to 10%, grading to 5% per year. As Table 2.14 indicates, actual lapse rates turned out to be close to half the assumed rates. This additional premium income should be welcome if all other variable assumptions and their distributions behaved as priced. If, on a per-policy basis, biometric distributions, discount rates, and expense assumptions were appropriate, and a sufficient profit margin was incorporated, why would higher than expected retention worsen the financial viability of a portfolio? Most products were priced on a deterministic basis, with a long-term horizon. When higher retention is combined with lower interest rates, lower mortality rates, longer than expected cognitive type of claims, and higher utilization, it rendered early products deeply deficient. By the time companies were aware of the deficiencies, many years had elapsed, making a premium adjustment much more difficult, notwithstanding the reluctance of state regulators to grant high rate increases. All risk aspects of a Long Term Care insurance policy should be taken into account, but, lacking credible information, a conservative approach, such as ignoring lapse, may be called for; however this is not a permanent solution, as a well-priced product should withstand lower than expected, as well as higher than expected, lapse rates.

2.6.7 Cash Versus Reimbursement

Most private or social insurance Long Term Care benefits in France take the form of cash monthly annuities. The cash approach treats Long Term Care benefits as a lifetime monthly annuity with an incidence age as a starting age, and a continuance rate equal to the mortality rate. The benefit paid does not vary by region, nor is it directly dependent on inflation. The cash approach does not require the occurrence of payments for services after the claim has been approved, that is, no change in benefit status is assumed when expenses are lower or no longer required while the claimant is alive. The introduction of Partial versus Total dependence (see Section 12) has changed that situation somewhat, but monthly payments are still based on an ADL trigger as opposed to expenses incurred. In the US, most private and social Long Term Care insurance benefits take the form of reimbursement. For private insurance this means that the base benefit is defined as a maximum daily amount, ranging from $100 per day to $500. Another popular type of benefit is indemnity, which is the payment of a stated amount if an expense was incurred, whether lower or higher. The difference between an indemnity and cash benefit is that an expense must be incurred to receive a benefit payment for the indemnity type. The indemnity amount can be stated as a daily, weekly, or monthly amount. For the weekly or monthly type, only one occurrence of an expense needs to occur to receive the whole period's amount. It is common to price and reserve a policy on an indemnity basis, using a well defined annual amount at the start of a projection. But these estimates should be subject to regional and utilization variations.

2.6.8 Reserves

Methodologies and most assumptions for reserves are not prescribed in French regulations but are required to be sufficient to cover future liabilities. The interest rate, however, is prescribed (as a function of average French treasury bonds rates). This approach to reserving is very similar to what is known in the United States as Principle Based Reserving,[32] where the actuary is called to use calculations and assumptions appropriate to the risks reserved for.

US contract reserves[2] are defined by state regulations. For instance, the current Active Life Reserve method is One Year Preliminary Term as prescribed by the Commissioner Reserve Valuation Method (CRVM).[33] Mortality rates, lapse rates, and interest are prescribed and require a change in regulation. The major risks reserved for, incidence, continuance, and utilization, are not defined, however. Actuaries apply Provisions for Adverse Deviation (PAD) to claim rates, but these are fixed and are small compared to the magnitude of premium and reserve increases. Modifying these regulations require legislative and regulatory actions that can take several years, by which time the Long Term Care risk environment may have changed significantly enough to make these new regulations obsolete.

An important reserve for Long Term Care in the US is the Incurred But Not Reported (IBNR) reserve. This reserve accounts for claims that are not yet known to the insurer. This is critical for an insurance which covers old age risks, especially Alzheimer-type ailments where the insured may not know if he or she has a Long Term Care Insurance policy.

Besides insurance regulatory reserves, other important reserves are gross premium reserves, which reflect gross premium and expenses and are used to determine the financial soundness of a portfolio; and tax reserves for income tax purposes, as prescribed by the Internal Revenue Services (IRS). Since they are prescribed and are maximum reserves, they tend to limit the amount of reserves a company is willing to set up in a particular year due to its tax status. Publicly owned companies calculate Generally Accepted Accounting Principles (GAAP) reserves as prescribed by the Financial Accounting Standard Board (FASB); this method incorporates an expense reserve and the Deferred Acquisition Costs (DAC) asset, an important feature for US companies as commissions are usually front loaded. Multi-national companies are subject to International Financial Reporting Standards (IFRS) as prescribed by the International Accounting Standards Board (IASB). Most of these reserve standards are not specific to the Long Term Care risk and are defined for general insurance contracts.

One major difference between the current reserving approach used in the US for LTC is that it is deterministic, whereas the proposed PBR technique is stochastic.

In France, reserving and pricing are closely linked through the revalorization process, which contractually gives the right of a company to reassess premium and benefits based on developing experience.

2.6.9   Facultative Versus Mandatory

Long Term Care insurance is predominantly underwritten, even for group insurance, which is mostly facultative. In France about 20% of Long Term Care is through mandatory, guaranteed issue, group insurance.[34]

LTC insurance in France is mainly sold through 'mutuelles', (see 'mutuals and mutuelles' below). In 2010, 2 million members of a 'Mutuelle' connected with the educational system, MGEN (Mutuelle Générale de l'Education Nationale), were insured for Long Term Care through their group contract, which almost overnight brought the number of insureds to 5.5 million. The automatic or mandatory addition of the insurance to an individual's existing coverage has a great impact on the actuarial soundness of the Long Term Care plan as the number of covered individuals is larger and anti-selection is avoided.

Most Long Term Care coverage in group contracts in France are part of a portfolio of savings, retirement, medical, and disability benefits and are usually integrated with these benefits. For instance, in some plans, it is possible to earn 'points' which accumulate in the participant's account and can be used by the participant upon leaving employment, through retirement or otherwise.

The Group/Individual, mandatory/elective, Guaranteed Issue/Underwritten distinction, however, is not clear-cut since *'mutuelles'*, such as MGEN, offer Long Term Care insurance to other employers (non-education) and to individuals with various levels of election options and underwriting requirements.

2.6.10 Mutuals and Mutuelles

In the US there has been a trend to demutualize insurance companies, but several mutual companies market Long Term Care insurance. The major difference between non-mutual products and mutual products are dividends. This difference is significant in the premium increase environment LTC insurance has found itself in. While premium increases have sometimes been in the 100% range and dividends are in the 10% range, the fact that mutual insurers can reflect developing experience on an annual basis, as opposed to the extended process of a rate increase state approval, means that mutual companies are much better equipped to manage the LTC risk.

In France the public function of the *'mutuelles'* has its own legal and regulatory framework. Mutual companies have kept the trade aspect of mutual societies whose main purpose is to help its members, through services and mutualization of risks. Many mutual companies offer Long Term Care services and provide such services instead of cash payments when a member becomes dependent. Several mutual companies have research centers to develop new products, services, and technologies.

2.6.11 Loss Ratios

Loss Ratios[5] are key indicators relied upon by US regulators to monitor the effectiveness of health insurance related products. While loss ratios are more relevant in a one year span, which eliminates the impact interest rate assumptions, they are widely used to assess the long-term behavior of LTCI products. The difference between the US and France is striking. In 2014 the US loss ratio was 76%[4] while the loss ratio in France was 42%[35] in 2016 for stand-alone Long Term Care insurance products. While there is a 10-year lag in the evolution of the market between the US and France and the 2006 US Loss Ratio was 45%, the French trend is not likely to resemble the US trend for several reasons.

- Few French companies have withdrawn from the market, which is very varied. The 42% Loss Ratio comes from 1.6 million policyholders who bought policies from insurance companies. But an additional 5 million policyholders are covered by combination products through insurers such as *mutuelles*, described above, or companies whose primary function is not insurance such as the Post Office or banks (Bancassurance). That is, marketing is not as concentrated in a distribution channel or outlet as in the US.

- Benefits and premiums are much lower than the ones that were found in the initial US products, although their durations were all lifetime. The average annual premium corresponding to 42% Loss Ratio, with €1 = $1.18, is $422, and the average benefit is $176. In the US the 76% Loss Ratio average premium is $1,590 and the average claim is $1,204.
- Claim thresholds are much higher in France (see Section 3). In the US, only 2 ADLs are required to receive benefits, while 4 are required under a Total Dependence policy. Just about all contracts incorporate a three-year Waiting Period for cognitive impairment and a one-year Waiting Period for most others, except accidents, which have none. If the incidence date is within the Waiting Period, the premium is reimbursed, and no benefit is paid. In addition, a three month Elimination Period applies before benefits are paid.
- Few policies offer an annual Cost of Living Adjustment.
- Premiums can be revised through the revalorization provision.
- Many benefits, especially through *mutuelles*, take the form of services.

2.6.12 Marketing Distribution

Distribution channels are strikingly different. As we have seen earlier, the regulatory separation of banking and insurance operations in the US preclude the development of bancassurance. In France about 60% of Life insurance products (including annuities, which are the preferred form of Long Term Care products) is sold through bancassurance.

In the US about half of insurance products are sold through commission compensated independent agents. In France, only 10% of the products are sold through non-salaried agents.

Direct marketing, especially on the internet, also occupies a larger proportion of sales in France, 15%, versus 10% in the US.

These differences have an impact beyond the front-loaded, higher distribution costs of Long Term Care policies. One issue that arose during the rate increase wave in the US was the representation of lifetime level premiums, where the guaranteed renewability of the policy was not clearly disclosed at the time of the sale by the agent.

Bancassurance, through its branch advertising, also brings Long Term Care Insurance in the daily lives of the population.

In France the multiplicity of distribution channels is not confined to banks and the internet, but to operations that would not be expected in the United States, like the national postal service, which offers a wide array of insurance products, including

LTC, advertised and offered in many of its 16,000 offices. The post office has its own actuarial department which designs products that fit its other products and its customers' needs.

2.7 Conclusion

This chapter attempts to give a very brief overview of the LTC risk environment in the US.

Insurance companies have given several reasons for exiting the LTCI market[36]:

1. Poor performance, as exemplified in the three examples cited at the beginning of the chapter.
2. High capital requirement, which Solvency 2 is raising.
3. Ability to raise premium.
4. Worsening assessment of the risk.
5. Lack of risk management expertise.
6. Lack of reinsurers.
7. Reputation risk.
8. Risk of downgrade from rating agencies.
9. Too difficult to market.

Whether these concerns can be lifted over time remains to be seen, but this book attempts to address concerns 4 and 5.

The many factors that influence current and future LTSS costs call for actuarial tools which need to incorporate parametric distributions, and combinations of such distributions, not previously considered, with tools that build upon current developments in technologies such as data, computers, the internet, Artificial Intelligence; and actuarial research. Most of the book addresses the biometric aspect of the insurance risk, which deals with the benefit side of the equation, be it cash or cost reimbursement. But, even if benefits are becoming more limited, many premiums are still on a lifetime basis, which means exposure to non-biometric volatility. And that volatility is not only found in quantifiable factors such as interest rates, assets, persistency, or expenses, but also in less quantifiable social, political and geographical factors. For insurance companies, the Long Term Care risk is also more akin to disability insurance than life insurance, in that it is exposed to a high occurrence of fraudulent claims. While it may be expedient to ignore these non-biometric risks at the time of setting premium rates, when premiums are spread over an extended but receding period and are not easily modified, then it is likely that the current US experience will be repeated. Beyond the insurance sphere, the impact of the Long Term Care risk is not only defined by quantifications and projections, but no adequate approach can be reached without its adequate measurement, which requires new methodologies.

This book concentrates on two markets, France and the United States, and this chapter delves into the legal and regulatory environment of the United States. The legal and regulatory environment of France, now part of the relatively new European Union, will increasingly resemble that of the US.

Looking back at the three cases described at the beginning of this chapter, they illustrate several facets of the Long Term Care risk.

- Global risk: one stated reason for a Chinese insurance company's interest in purchasing Genworth is the LTCI experience Genworth has generated and how it could be beneficial for insuring the risk in China. In 1979, China instituted a one-child policy, which was phased out in 2016. The policy has produced the 'four-two-one' phenomenon where an adult may be faced with the care of two parents and four grandparents.
- Jurisdiction: Penn Treaty's policyholders will not be treated equally through its liquidation process, as each state has its own Guaranty Fund statutes which govern how, and to what extent, their residents insured by the Pennsylvania company can receive their benefits.
- Reinsurance and reserves: General Electric's decision to substantially increase the reserves of its reinsurer has had a major impact on their client ceding companies.

End Notes

1. http://www.penntreaty.com/Liquidation/CourtDocuments.aspx. http://www.penntreaty.com/Portals/0/PDFs/PTNA/penn_treaty_nework_america_ins_company_rehabilitation_order_jan_6,_2009[1].pdf.
2. Statutory reserves regulated by the jurisdiction where the insurance is sold. In this chapter, statutory reserves refer to Active Life Reserves (ALR) and Claim Reserves (CR). http://us.milliman.com/uploadedFiles/insight/2016/long-term-care-insurance-valuation.pdf.
3. https://www.irs.gov/pub/irs-regs/td8792.pdf.
4. http://www.naic.org/documents/cipr_current_study_160519_ltc_insurance.pdf.
   NAIC: National Association of Insurance Commissioners (1871). A nongovernmental organization governed by the chief insurance regulators from the 50 states, the District of Columbia and five U.S. territories. The NAIC assists state insurance regulators in establishing standards and best practices, conduct peer reviews, and coordinate regulatory oversight. Its organization is divided into four zones: Northeastern, Southeastern, Midwestern and Western. The NAIC acts as a forum for the creation of model laws and regulations. Each state decides whether to pass each NAIC model law or regulation, and each state may make changes in the enactment process. The NAIC also acts at the national level to advance laws and policies supported by state insurance regulators. The NAIC is also responsible for creating the statutory accounting principles (SAP) upon which insurance accounting is based and is notable for its very conservative valuation methods. The NAIC promulgates the annual statement which incorporates SAP and must be filed with the department of insurance in every state in which an insurance company conducts business.

5. Loss Ratio.

   In Health insurance products, a Loss Ratio (LR) is the ratio of the annual medical costs and the annual premium. In Long Term Care Insurance pricing and regulatory pricing, this ratio is the present value, or the accumulated value, of the benefits and the present value of premiums. For projections the numerator and denominator can take several forms:

   Benefits can be Paid or Incurred, and may include Active Life Reserve.

   The relationship between Paid (PC) and Incurred (IC) claims is IC = PC + change in IBNR, where IBNR is Incurred But Not Reported Reserve.

   Premium can be on a Paid or Earned basis.

   The relationship between Paid (PP) and Earned (EP) premium is EP = PP + change in UR, where UR is Unearned Premium Reserve. A common Loss Ratio measure is the cumulative ratio, where past paid claims are divided by past paid premiums, accumulated at the applicable interest rates.

6. Genworth Cost of Care Survey.

   https://www.genworth.com/about-us/industry-expertise/cost-of-care.html.

7. Mobility rates

   http://libertystreeteconomics.newyorkfed.org/2016/10/what-caused-the-decline-in-interstate-migration-in-the-united-states.html.

8. LTSS costs, Private Insurance and Gross Domestic Product.

   Source: National Health Expenditure Accounts, Office of the Actuary, Center for Medicare and Medicaid Services. LTSS is the sum of Total Nursing Care Facilities and Continuing Care Retirement Communities and Total Home Health Care Expenditures.

9. Riders are optional benefits that can be added to a policy for an additional premium. A Cost of Living Adjustment is required to be offered for a Long Term Care insurance policy to be considered Tax Qualified, but an applicant can opt not to add it.

10. Medicare (1966) is a single-payer federal social medical insurance program, funded through federal payroll tax and general revenue. It provides health insurance for Americans aged 65 and older who have worked and paid into the system through the payroll tax. In 2016, it provided health insurance for over 48 million people age 65 and older.

    https://www.cms.gov/Research-Statistics-Data-and-Systems/Statistics-Trends-and-Reports/ReportsTrustFunds/Downloads/TR2017.pdf.

11. Activities of Daily Living, as defined in the Internal Revenue Code:

    (1) Eating
    (2) Toileting
    (3) Transferring
    (4) Bathing
    (5) Dressing
    (6) Continence.

12. ASOP (Actuarial Standards of Practice)

LTC: http://www.actuarialstandardsboard.org/wp-content/uploads/2014/02/
asop018_136.pdf.
Data: http://www.actuarialstandardsboard.org/wp-content/uploads/2017/01/
asop023_185.pdf.

13. Partnership: these plans are designed to help an individual to offset their Long
    Term Care insurance benefits from their Medicaid asset limit eligibility require-
    ments. https://www.ltcfeds.com/help/faq/miscellaneous_partnership.html.

14. Medicaid (1966) is social insurance program funded jointly by the national
    (federal) government and each state. It is administered by individual sates but
    subject to federal minimum requirements to receive funding. Medicaid is a Pay
    as You Go program, annually funded. As of this writing Medicaid is the largest
    institutional payor of LTSS. Eligibility to Medicaid benefits depends on need
    and financial resources. As of 2017, 74 million people have received Medicaid
    benefits.

15. https://www.hhs.gov/hipaa/for-professionals/privacy/laws-regulations/
    combined-regulation-text/index.html.

16. CLASS
    file:///C:/A/textbook/Chapters/US/CLASS/Class-Act-Legislation.pdf.
    http://www.ncsl.org/documents/statefed/health/CLASSOvrview21313.pdf.
    https://www.actuary.org/files/publications/class_july09_0.pdf.

17. Economic dependency
    https://www.bls.gov/emp/tables/economic-dependency-ratio.htm.

18. https://www.populationpyramid.net/united-states-of-america/2017/.

19. Current Population Survey: Bureau of Labor Statistics and Census Bureau.

20. https://www.census.gov/data/tables/2013/demo/wealth/wealth-asset-
    ownership.html.

21. 'How are income and wealth linked to health and longevity?', Urban Institute
    https://www.urban.org/research/publication/how-are-income-and-wealth-
    linked-health-and-longevity.

22. https://aspe.hhs.gov/reports.

23. A Single Index of Mortality and Morbidity by Daniel F. Sullivan *HSMHA Health
    Reports*, Vol. 86, No. 4 (Apr., 1971), pp. 347–354: http://dx.doi.org/10.2307/
    4594169.

24. The National Nursing Home Survey (NNHS) is a series of nationally represen-
    tative sample surveys of United States nursing homes, their services, their staff,
    and their residents. The NNHS was first conducted in 1973–1974 and repeated
    in 1977, 1985, 1995, 1997, 1999, and most recently in 2004. Although each of
    these surveys emphasized different topics, they all provided some common basic
    information about nursing homes, their residents, and their staff. All nursing
    homes included had at least three beds and were either certified (by Medicare or
    Medicaid) or had a state license to operate as a nursing home. In 2012, NCHS
    initiated the National Study of Long Term Care Providers (NSLTCP)—a bien-
    nial study of adult day services centers, residential care communities, nursing
    homes, home health agencies, and hospice agencies. NSLTCP uses administra-

tive data for the nursing home sector obtained from the Centers for Medicare and Medicaid Services (CMS).

25. Transactions of the Society of Actuaries 1988–1990 Reports, Report of the Long Term Care Experience Committee, 1985 National Nursing Home Survey, Utilization data. The survey was based on 1982 data collected by the National Center for Health Facilities from which 1,079 facilities were sampled producing a sample of 5,243 residents and 6,023 discharged residents.
https://www.soa.org/Library/Research/Transactions-Reports-Of-Mortality-Moribidity-And-Experience/1980-89/1988/January/TSR886.aspx.

26. SOA Long Term Care Intercompany Experience Study Intercompany study
1984–91: https://www.soa.org/library/research/transactions-reports-of-mortality-moribidity-and-experience/1990-99/1993/january/TSR934.pdf.
1984–93: https://www.soa.org/experience-studies/2000-2004/ltc-84-93-insurance-experience-study/.
1984–2000: https://www.soa.org/experience-studies/2000-2004/hlth-1984-2001-long-term-care-experience-committees-intercompany-study/.
1984–2007: https://www.soa.org/experience-studies/2005-2009/research-ltc-study-1984/.
2000–2011: https://www.soa.org/experience-studies/2015/research-ltc-study-2000-11-aggregrated/.
2015 models: https://www.soa.org/experience-studies/2015/2000-2011-ltc-experience-basic-table-dev/.

27. Policy persistency:
2002–04: https://www.soa.org/experience-studies/2005-2009/ltci-ins-persistency/.
2005–07: https://www.soa.org/experience-studies/2011/05-07-ltc-ins-persistency-report/.
2008–11: https://www.soa.org/experience-studies/2016/research-ltc-insurance/.

28. Waiting Period: The number of years during which the insurer will reimburse premium but not pay any benefit in case of a claim.
Elimination Period: The number of months after which benefit payments start after the incidence date. The EP can vary from one month to 18 months. Some policies will retroactively pay benefits accrued during the Elimination Period. Regulations may limit the maximum Elimination Period.

29. https://mort.soa.org/?_ga=2.170063268.587304493.1514832902-422425784.1465397146.

30. Actuarial Memorandum (from NAIC Model Law 641)
Model 641 s 10
Initial filing requirements
A. This section applies to any long term care policy issued in this state on or after [insert date that is 6 months after adoption of the amended regulation].

B. An insurer shall provide the information listed in this subsection to the commissioner [30 days] prior to making a long term care insurance form available for sale.

Drafting Note: States should consider whether a time period other than 30 days is desirable. An alternative time period would be the time period required for policy form approval in the applicable state regulation or law.

(1) A copy of the disclosure documents required in Section 9; and

(2) An actuarial certification consisting of at least the following:

(a) A statement that the initial premium rate schedule is sufficient to cover anticipated costs under moderately adverse experience and that the premium rate schedule is reasonably expected to be sustainable over the life of the form with no future premium increases anticipated;

(b) A statement that the policy design and coverage provided have been reviewed and taken into consideration;

(c) A statement that the underwriting and claims adjudication processes have been reviewed and taken into consideration;

(d) A complete description of the basis for contract reserves that are anticipated to be held under the form, to include:

(i) Sufficient detail or sample calculations provided so as to have a complete depiction of the reserve amounts to be held;

(ii) A statement that the assumptions used for reserves contain reasonable margins for adverse experience;

(iii) A statement that the net valuation premium for renewal years does not increase (except for attained-age rating where permitted); and

(iv) A statement that the difference between the gross premium and the net valuation premium for renewal years is sufficient to cover expected renewal expenses; or if such a statement cannot be made, a complete description of the situations where this does not occur;

(I) An aggregate distribution of anticipated issues may be used as long as the underlying gross premiums maintain a reasonably consistent relationship;

(II) If the gross premiums for certain age groups appear to be inconsistent with this requirement, the commissioner may request a demonstration under Subsection C based on a standard age distribution; and

(e) (i) A statement that the premium rate schedule is not less than the premium rate schedule for existing similar policy forms also available from the insurer except for reasonable differences attributable to benefits; or

(ii) A comparison of the premium schedules for similar policy forms that are currently available from the insurer with an explanation of the differences.

C. (1) The commissioner may request an actuarial demonstration that benefits are reasonable in relation to premiums. The actuarial demonstration shall include either premium and claim experience on similar policy forms, adjusted for any premium or benefit differences, relevant and credible data from other studies, or both.

(2) In the event the commissioner asks for additional information under this provision, the period in Subsection B does not include the period during which the insurer is preparing the requested information.

31. Insurance Compact:
   http://www.insurancecompact.org/about.htm.
32. Principle-Based Reserving:
   https://www.actuary.org/files/LTC_PBR_Report_012116_0.pdf.
33. Commissioner's reserve valuation method.
   (1) Reserves according to the commissioner's reserve valuation method, for the life insurance and endowment benefits of policies providing for a uniform amount of insurance and requiring the payment of uniform premiums, must be the excess, if any, of the present value, at the date of valuation, of future guaranteed benefits provided for by the policies, over the then present value of any future modified net premiums. The modified net premiums for any policy must be the uniform percentage of the respective contract premiums for the benefits that the present value, at the date of issue of the policy, of all modified net premiums must be equal to the sum of the then present value of the benefits provided for by the policy and the excess of Subsection (1)(a) over Subsection (1)(b), as follows:
   (a) a net level annual premium equal to the present value, at the date of issue, of benefits provided for after the first policy year, divided by the present value, at the date of issue of an annuity of one per annum payable on the first and each subsequent anniversary of the policy on which a premium falls due. However, the net level annual premium may not exceed the net level annual premium on the 19-year premium whole life plan for insurance of the same amount at an age 1 year higher than the age at issue of the policy.
   (b) a net 1-year term premium for benefits provided for in the first policy year.
   (2) (a) For each life insurance policy issued on or after January 1, 1987, for which the contract premium in the first policy year exceeds that of the second year, for which a comparable additional benefit is not provided in the first year for the excess, and that provides an endowment benefit, a cash surrender value, or a combination of both in an amount greater than the excess premium, the reserve according to the commissioner's reserve valuation method, as of any policy anniversary occurring on or before the assumed ending date as the first policy anniversary on which the sum of any endowment benefit and any cash surrender value then available is greater than the excess premium, is the greater of the reserve as of the policy anniversary calculated as described in Subsection (1) or the reserve as of the policy anniversary calculated as described in Subsection (1) with the following exceptions:
   (i) the value defined in Subsection (1)(a) is reduced by 15% of the amount of the excess first-year premium;
   (ii) all present values of benefits and premiums are determined without reference to premiums or benefits provided for in the policy after the assumed ending date;
   (iii) the policy is assumed to mature on the assumed ending date as an endowment; and

(iv) the cash surrender value provided on the assumed ending date is considered an endowment benefit.

(b) In making the comparisons in Subsection (2)(a), the mortality and interest bases [stated in another section] must be used.

(3) Reserves according to the commissioner's reserve valuation method for the following must be calculated by a method consistent with the principles of this section, except that any extra premiums charged because of impairments or special hazards must be disregarded in the determination of modified net premiums:

(a) life insurance policies providing for a varying amount of insurance or requiring the payment of varying premiums;

(b) group annuity and pure endowment contracts purchased under a retirement plan or plan of deferred compensation, established or maintained by an employer, including a partnership or sole proprietorship, or by an employee organization, or by both, other than a plan providing individual retirement accounts or individual retirement annuities under Section 408 of the Internal Revenue Code, as amended;

(c) disability and accidental death benefits in all policies and contracts; and

(d) all other benefits, except life insurance and endowment benefits in life insurance policies and benefits provided by all other annuity and pure endowment contracts.

(4) (a) Subsection (4)(b) applies to any annuity and pure endowment contracts other than group annuity and pure endowment contracts purchased under a retirement plan or plan of deferred compensation established or maintained by an employer, including a partnership or sole proprietorship, or by an employee organization, or by both, other than a plan providing individual retirement accounts or individual retirement annuities under Section 408 of the Internal Revenue Code, as amended.

(b) Reserves according to the commissioner's annuity reserve method for benefits under annuity or pure endowment contracts, excluding any disability and accidental death benefits in the contracts, must be the greatest of the respective excesses of the present values, at the date of valuation, of the future guaranteed benefits, including guaranteed nonforfeiture benefits, provided for by the contracts at the end of each respective contract year, over the present value, at the date of valuation, of any future valuation considerations derived from future gross considerations required by the terms of the contract that become payable prior to the end of the respective contract year. The future guaranteed benefits must be determined by using the mortality table, if any, and the interest rate or rates specified in the contracts for determining guaranteed benefits. The valuation considerations are the portions of the respective gross considerations applied under the terms of the contracts to determine nonforfeiture values.

(c) The commissioner's reserve valuation method provided by this section is subject to the provisions of the valuation manual as adopted by the commissioner.

34. Mandatory insurance:

http://www.argusdelassurance.com/mediatheque/9/5/4/000013459.pdf.
35. Long Term Care Insurance in France
    https://www.ffa-assurance.fr/content/assurance-dependance-68-millions-de-personnes-couvertes-la-fin-de-annee-2015.
36. "Exiting the Market: Understanding the Factors behind Carriers' Decision to Leave the LTC Insurance Market" Lifeplans 2013.

Bibliography

1. Stallard, E.: Compression of morbidity and mortality: new perspectives. N. Am Actuar. J. **20**(4) (2016)
2. Lally, N.R., Hartman, B.M.: Predictive modeling in long-term care insurance. N. Am Actuar. J. **20**(2) (2016)
3. Report of the American Academy of Actuaries' Long Term Care Risk Based Capital Work Group to the NAIC Capital Adequacy Task Force
4. Report on Principle-Based Reserve Modeling for Long-Term Care (LTC) Insurance, American Academy of Actuaries

Chapter 3
Long Term Care in France

Fabio Castaneda and François Lusson

3.1 Introduction

As of 2018, the Long Term Care market in France has close to thirty years of history, which is long compared to other countries in this market but is short compared to life expectancy. This Long Term Care insurance ("Loss of Autonomy" would probably be more judicious), now covers millions of individuals but policies in the market are very heterogeneous, using many types of guaranties, many types of benefits, and heterogeneous underwriting modes.

A positive message: the market has had a tendency to converge with respect to Total Dependence,[1] so that some credible actuarial knowledge is now available. With caution, however, as the risk is by nature very volatile.

What event are we talking about?

We consider a dependent state when a person needs the help of a third party to carry out one or several Activities of Daily Living (ADL). The cause of this state is either physical or psychological, which necessitates the inclusion of both components in its definition.

The provider, be it a public or private insurer, is faced with a continuum of situations. Hence, the number of different cases appears considerable: how to compare an individual who can move without assistance and needs help only for preparing a meal or housekeeping, and a bedridden individual, or one whose mental functions are severely impaired and necessitate 24/7 assistance?

This continuum is categorized by the public sector, administrators having defined 6 levels of autonomy, the Iso-Resource Groups (IRG),[2] through a grid. These IRGs

F. Castaneda · F. Lusson (✉)
Actense, Paris, France
e-mail: francois.lusson@actense.fr

F. Castaneda
e-mail: fabio.castaneda@actense.fr

© Springer Nature Switzerland AG 2019
E. Dupourqué et al. (eds.), *Actuarial Aspects of Long Term Care*,
Springer Actuarial, https://doi.org/10.1007/978-3-030-05660-5_3

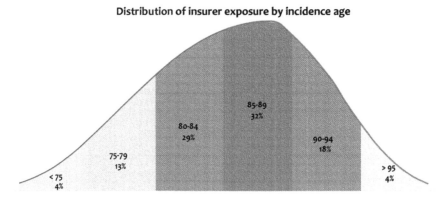

Fig. 3.1 Distribution of insurer exposure by incidence age

first distinguish two categories: Total Dependence (DT)[2] and Partial Dependence (PD).[1]

So, what risk are we talking about?

It is clearly a risk increasing with age, and even "strongly increasing" (compared to death rates) and which affects the insured, on average, several months or even several years before their "probable death".

The graph (Fig. 3.1) "distributes" probable benefits for an insurer insuring a cohort of policyholders with a monthly lifetime annuity contingent on entering a dependent status. This a qualitative presentation for the Total Dependence risks. The same insurer having to face prefunding, and hence reserving,[3] for this risk requires a premium which must be set carefully and for which the present value will precisely anticipate probable future benefits.

It is by nature a long risk, even very long (from the issue age horizon, spanning roughly a life expectancy of at least 2 years). This is a very volatile risk since it depends on numerous biometric hazards, but also numerous financial hazards.

Other guaranties were added to this basic Total Dependence[1] lifetime cash payment. Over several years the following benefits have appeared,

- Partial Dependence[1] annuities,
- Payment of a lump sum at the time of the first annuity payment (Total Dependence[1] or Partial Dependence),
- Funeral cost benefits,
- Also, service benefits for the various levels of risks identified.

Recently "Caregiver Assistance" benefits have also appeared.

Finally we should also note that, fundamentally, Loss of Autonomy coverage concerns the few years before probable death and therefore belongs to cohorts of retired people: group retirement income policies frequently cover the case of total disability with an anticipated lump sum death benefit, but the frequencies encountered (of the order of 'per 1000', maybe less) turn out to be much lower compared to those dealing with the Total Dependence[1] risk ('per 100') or the Partial Dependence[1] risk ('per 10' at the very high ages).

3.2 Social Stakes

3.2.1 Demography: Aging of the Population

It is difficult to find one's way among the diverse forecasts from the numerous organizations following and anticipating demographic evolutions, but concerning the Loss of Autonomy risk, their conclusions agree and finally appear clear.

Beyond their annual evaluations, public organizations now undertake to publish quarterly reports on current events (heat wave, flu epidemic, …). These follow-ups, which are better approached with a long-term logic, were started because statisticians, demographers, and actuaries struggled for decades to fully understand and describe the increase of life expectancy of the population.[4] Insurers kept correcting their mortality tables by differentiating by gender and introducing trends on one hand, and on the other hand differentiating so-called "best-estimate" approaches or more prudent approaches of the "lifetime annuities" type, among others. The subject of life expectancy appears to be better and better understood by all the operators (Fig. 3.2, Table 3.1).

The 2017 picture has been the object of many commentaries. In particular it has been accompanied by a number of very long-term projections (2080), and it is possible to see the changing shape, by gender, of the different age bands in the population: it now appears possible to distinguish the "payors" (age 20–64 band) and the beneficiaries of social protections (starting with retirement income after 65).

Salient points:

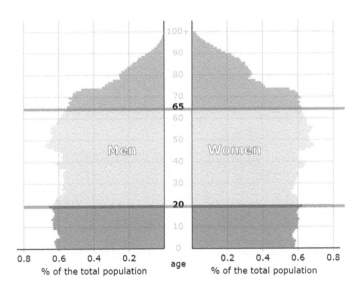

Fig. 3.2 2020 age pyramid [12]

Table 3.1 2020 population distribution by age group

Age	Million	%	% Women
65+	12.85	19.2	57.1
20–64	37.73	56.3	50.8
<20	16.41	24.5	48.8
Total	66.99	100	51.6

- Life expectancy at birth doubled since 1900 (also true in many neighboring countries[5] [8]).
- It is customary to forecast an annual increase for life expectancy[6] [10] of about three months per year (which "sounds" good); upon analysis this phenomenon appears somewhat overestimated and, moreover, it tends to wear off. No matter, the trend remains strong and constant.
- Life expectancy at birth in 2014[7] [8]:

  – close to 86 years, for women,
  – close to 80 years, for men.

  Beyond these facts, with respect to the Loss of Autonomy risk within the framework of a national policy, the most important consists in displaying the different proportions by age classes and their evolution over time (an insurer using a prefunding approach will concentrate more on mastering and anticipating life expectancy increases and continuance in a dependent state than the relative evolution of cohorts through time).

  INSEE[6] brings interesting information on the matter as within 70 years (1990–2060) the proportion of "75 year and older" would go from 6 to 17% of the national population, almost a three-fold increase (Fig. 3.3).

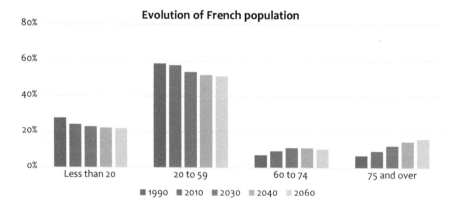

Fig. 3.3 Age distribution projection

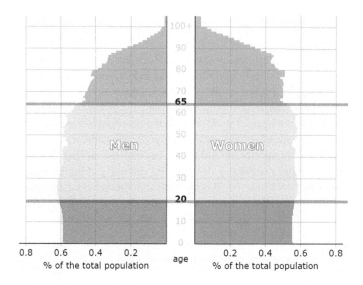

Fig. 3.4 2060 age pyramid [12]

Table 3.2 2060 population distribution by age bands [12]

Age	Million	%	% Women
65+	21	27.9	53.7
20-64	37.85	50.9	50.3
<20	16.36	21.7	48.8
Total	75.21	100.5	50.9

This phenomenon directly ensues from the deformation of the age pyramid in 2060 presented by INSEE.[6] These pyramids are presented as an animated series starting in 2015 with projections going to 2080 (Fig. 3.4, Table 3.2).

Globally, even if to this day no focus has been brought at the national level to the Loss of Autonomy risk itself, by documenting Total Dependents and Partial Dependents on one hand and population age classes on the other hand, it is possible, without taking too much risk, to clearly conclude that no matter what source of information is used, the aged dependent population will at least double in about fifty years. Rare are the markets that cover a doubling of required resources over a two-generation horizon!

3.2.2 Nature of the Risks

The nature of the risks depends on the incidence age. Pathologies at the origin of loss of autonomy are strongly correlated to age. Likewise, life expectancy while dependent (continuance) relies on the pathology, hence on this incidence age.[7]

Tumors and cancers are over-represented among the youngest ages in the observed portfolios, namely from ages 75 to 80. Hence, the resulting life expectancy observed is shorter than the one observed, year after year, for neurological and dementia ailments affecting the majority of "old ages", beyond age 80 (Fig. 3.5).

Extrapolation of life expectancy or increased mortality at higher ages appears to be particularly delicate. Moreover, it is possible that this schematic breakdown, after a full cohort has evolved in the insurance portfolio (beyond age 95), changes slightly.

Using the usual definitions, the following points are worth emphasizing:

- Incidence rates increase with age at about 15–20% per year until at least age 90;
- Incidence rates evolve in significantly separate ways between males and females;
- High age risks strongly evolve:

  - The nature of pathologies evolves with age,
  - Multiple pathologies and neurological disorders become overwhelmingly predominant,
  - But the risk is still insufficiently known after age 90. How does the annual incidence rate evolve? To date, different models exist.

Concurrently, in 2008 INED[5] published a study on Life Expectancy without Disability at 65 and without dependence, distinguishing male and female lives [9] (Fig. 3.6).

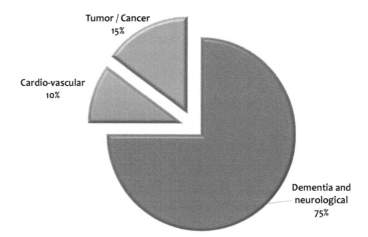

Fig. 3.5 Repartition of Total Dependence incidence by diagnostic

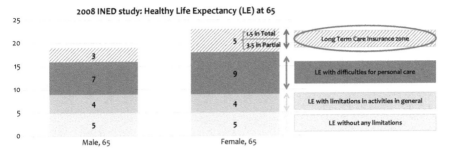

Fig. 3.6 Disability-free life expectancy at 65

It is interesting to note that the excess life expectancy at 65 for women, over men, occurs exclusively in the life expectancy zone where difficulties in undertaking personal care occur and in the Long Term Care insurance zone, Partial or Total.

Secondly, the 5 years of female life expectancy in dependence correspond to a mix of about 1.5 years in Total Dependence[1] and 3.5 years in Partial Dependence for policies covering women, this total being 3 years for men with a comparable repartition for the two Dependence states[1] (note that we are looking at life expectancy at 65 and not from the time of incidence).

The last enlightening element of this risk from the viewpoint of insurers is that between 2/3 and 3/4 of Individual contracts are issued to women, showing their greater level of awareness… or of their well understood interest; but current contracts all offer coverage at unisex premiums, although the equilibrium price should be sharply higher for women than for men.[8]

3.2.3 Sizable Out-of-Pocket Costs

3.2.3.1 Lifetime Annuity

A person taking steps to provide long-term insurance aimed at being compensated in case of loss of autonomy will seek an appropriate coverage, which would be on a lifetime annuity basis. This undertaking consists in thinking ahead of time in case of the occurrence of an adverse event from the issue date until the end of benefit payments due to that event. The benefit level will need to appear satisfactory and meet the anticipated need, but also evolve with time through a revalorization of benefits.

This expectation appears legitimate during the benefit period: it amounts on average to about 4 years in Total Dependency,[1] combining extra mortality during the first 2 years and life expectancies linked to neurological disorders which can go from 8 to 10 years. It is useful to point out that this revalorization also applies during the deferred period.

It is important that the policyholder invests in a plan of action which, between policy issue and the probable occurrence of the risk, sees that benefit levels and premiums evolve by comparable levels to assure a satisfactory (at claim date) and permanent (during claim duration) response.

3.2.3.2 Benefit Levels

Several studies exist on the question, based on differential calculations, of estimating aggregate monthly requirements,

- From which must be deducted public assistance;
- From which must be finally subtracted average retirement income.

  In practice, monthly aggregate benefits are conditioned on whether:

- The insured can stay at home (based on the nature of his or her state and/or the proximity of a caregiver);
- The insured must enter a residential care facility.

According to studies, costs of monthly home care are estimated at between €300 and €500 (in practice the amount that must be financed varies with average monthly retirement income levels).

In the case of a residential care facility, monthly costs approach rather rapidly, and almost systematically, €2,000, which assumes, when looking at current benefit levels offered by insurers, that supplemental sources of income are available.

Today's market naturally displays two exclusion phenomena:

- Individuals with modest income and/or assets forgo insurance;
- At the other end, individuals with the means to self-insure thanks to their income and/or assets prefer to avoid an insurance solution, by nature insufficient when residence in a residential care facility is required.

3.2.3.3 The Revalorization Mechanism

The differential calculation outlined above integrates several elements for which clearly all the "warning signals are red":

- Most Home Care benefits correspond to expenses for which long-term trends grow rather faster than CPI: energy costs, food, medical care, services.
- Public benefits bear heavily on national and regional finances and it is likely that future increases will be modest.
- Social security retirement income, based on periodically updated projections, seem to struggle to keep up with long-term inflation trends.

The remaining gap (that is, the unfunded costs in the case of loss of autonomy) will mechanically increase much faster than the inflation rate.

Concurrently, insureds who have or are considering Long Term Care Insurance will need to carefully pay attention to revalorization provisions. However, from a contractual point of view, those are often fragmented, if not obscure, ... even non-existent. It is worthwhile to recall that Long Term Care contracts are usually, and more and more frequently, approached in logics or cultures close to non-life insurance; whereas by definition, the lifetime nature of Long Term Care coverage belongs to life insurance with a mechanism of policy accounts, with both underwriting and investment gains participation, which would be logical to find in such a contract. In practice, this wish goes against an economic reality: any lifetime guaranties will bear heavily on Solvency 2 capital requirements.

However, the fact is that since introducing Long Term Care coverage, insurers:

- did not experience underwriting gains (insureds live longer, and are expected to have a higher incidence, requiring a strengthening of reserves due to an increasing risk),
- have included premium expense loads that are rarely sufficient to manage, monitor, and control these contract provisions,
- these last few years have experienced lower investment income and lower underwriting gains than expected, which bore heavily on the financial soundness of this coverage.

Overall, revalorization activity for this coverage remains modest; the market has regularly, and sometimes significantly, revised premium upward for existing contracts to follow the lifetime logic of long-term guarantees needed for Total Dependence[1] and Partial Dependence[1] payments.

3.3 Definitions

The following are three different practical solutions to measure the risk, defined by three types of definitions or tests privileged by the different players:

- IRG grid,[2]
- ADL,
- Different cognitive tests implemented to incorporate the notion of Loss of Autonomy due to neuropsychological disorders.

3.3.1 The Iso-Resource Groups Grid

This grid, introduced in 1997,[2] became an instrument shared by social insurance providers (regional funds, retirement funds). Important training efforts were undertaken to improve the skills needed by these different users of this tool [17]. It has two functions:

NATIONAL ISO-RESOURCE GROUP GRID
IDENTIFICATION OF PERSON BEING EXAMINED
Last Name: .... First Name: ....
Identification Number: |__|___|___|___|___|___|___|___|___|___|___|___|___|___|
Address: ....
Zip code: |__|___|___|___|___| City: ....

SITUATION WITH RESPECT TO ESSENTIAL AND ORDINARY DAILY ACTIVITIES		
DISCRIMINATING VARIABLES - PHYSICAL AND PSYCHOLOGICAL AUTONOMY		
1	COHERENCE: to communicate and/or behave in a sensible manner	
2	ORIENTATION: orientation in time, through the day, and on the premises	
3	TOILETTING: concerns personal hygiene	High
		Low
4	DRESSING: to dress, undress, and be presentable	High
		Average
		Low
5	FEEDING: to eat prepared food	Serve
		Eat
6	ELIMINATION: to assume hygienic urination and fecal elimination	Urinate
		Defecate
7	TRANSFER: to get up, lie down, and sit down	
8	INDOOR MOBILITY: with or without a cane, a walker, or a wheelchair	
9	OUTDOOR MOBILITY: from the front door, without transportation means	
10	LONG DISTANCE COMMUNICATION: to use communication tools, telephone, bell, alarm ...	
	ILLUSTRATIVE VARIABLES: DOMESTIC AND SOCIAL AUTONOMY	
11	MANAGEMENT: to manage one's individual business, budget, and assets	
12	COOKING: to prepare one's meal and serve them	
13	HOUSEKEEPING: to be able to do all housekeeping tasks	
14	TRANSPORTATION: to take and / or order a means of transportation	
15	SHOPPING: to be able to shop directly or by order	
16	DOCTOR TREATMENT: to be able to follow a doctor's prescription	
17	LEISURE TIME ACTIVITIES: sport, cultural, social activities, for leisure or hobbies	

A: does alone, completely, habitually, correctly
B: does partially, not habitually, not correctly
C: does not do

Fig. 3.7 Example of an IRG grid

- Evaluate the degree of dependence or autonomy of individuals in the performance of Activities of Daily Living,
- Assess the eligibility to and the level of a monthly cash benefit[8] by an algorithmic translation of certain items of the grid in one synthetic indicator "Iso-Resource Group" (IRG) (Fig. 3.7).

A national grid, a sample of which is shown above, allows regional medico-social teams or primary care physicians to classify the degree of dependence of an elderly person in six Iso-Resource groups[2] ("IRG"). These groups are detailed in the Table 3.3:

Table 3.3 Iso-Resource Group definitions

Group	Definition
IRG 1	• A person confined in bed or armchair, whose mental functions are severely altered and whose state requires an indispensable and continuous presence of caretakers, • Or a person in an end-of-life state
IRG 2	• A person confined to bed or an armchair, whose mental functions are not totally altered and whose state requires assistance for most activities of daily living, • Or a person whose mental functions are altered, but who is able to move and who needs permanent supervision
IRG 3	A person keeping his or her mental autonomy, and partial autonomy of motion, but who needs assistance for they bodily functions, daily and several times each day
IRG 4	• A person who cannot assume alone his or her transfers but who, once up, can move around indoors, and who needs assistance for toileting and dressing, • Or a person who, while having no motion impairment must be assisted for bodily care and meals
IRG 5	A person needing only timely assistance for toileting, meal preparation, and housekeeping
IRG 6	A person autonomous for essential activities of daily living

An IRG score between 1 and 4 allows one to determine the level of cash benefit paid through the national social insurance program (persons classified between IRG 5 and 6 can nevertheless receive housekeeping assistance through regional programs), in a certain number of cases (the minority), it is retained by insurers as a trigger for the benefit stipulated in the contract.

In practice insurers, however, through their experience but also with the support of reinsurers and consultants, design contractual terms relying on ADLs (sometimes on ADLs coupled with IRG scores), notably in order not to rely on decisions stemming from the application of the IRG grid, which could be adverse. From experience, Total Dependence issuers are practically in sync with IRG 1 or IRG 2 scores; but when doctors are responsible for evaluating Partial Dependence, more significant deviations are observed, especially at the IRG 3 level (even more for IRG 4 if that score is retained for "light" dependence benefits).

3.3.2 Activities of Daily Living

Insurers in the Long Term Care market have proposed their own Activities of Daily Living (ADL) definitions and also their own number of activities. Policies issued in 1990–2000 are more often based on six ADLs, or on four by grouping two of them. These ADLs are used to measure the level of an individual's physical dependence. ADL grids offer an alternative to the IRG grid, which can be used as a complement to determine the level of dependence of an individual.

Through the years, definitions used by insurers have become more and more homogenous (for Total Dependence). The Table 3.4 clarifies several definitions, showing the existence of zones of interpretation of some ADLs.

Arguably, the table shows that these definitions appear very limited with respect to Total Dependence. Each insurer can use its own claim evaluation system if this system meets contractual terms (through a combination of notation system for each ADL according to the ability to perform or not perform an activity); for each ADL it should be clear (for example) that:

- The individual is totally and permanently unable to perform this act knowingly in a spontaneous, habitual manner, without the help of a third party, or
- all the above-mentioned definitions in the activity description cannot be met irreversibly at the time of evaluation.

3.3.3 Cognitive Tests

Just as the IRG grid is not sufficient to define mental dependence, ADLs are not sufficient to satisfactorily define Total Dependence and Partial Dependence once an ailment of neuropsychic origin appears. It became necessary to have recourse to cognitive tests. They are used to evaluate loss of psychological autonomy.

A study of contracts in the market reveals that the most used tests are[11] [4]:

- *The Mini-Mental State Examination (MMSE)* (Fédération pour la Recherche sur le Cerveau), or Folstein test (the most widely used), which explores orientation to time and space, learning, memory, … For instance, the following questions could be asked: "what is today's date?", "Which floor are we on?", "Can you name the object in front of you?"…
- The 'Blessed' test, which seeks to evaluate behavior after questioning people close to the applicant.

Over the years the market has integrated these cognitive tests to precisely identify the risks by combining their results with physical indicators (ADLs).

3.4 Public Sector

3.4.1 Social Insurance

Long Term Care social insurance programs were introduced in the mid-1990s, and their financial responses have evolved to meet needs linked to dependence; this is how a program to compensate caregivers, ACTP[12] (1975), was progressively replaced with other programs as the time-line below shows (Fig. 3.8).

Table 3.4 Total Dependence definitions: LTCIBG[10] Label and three insurers

Activity	Definition of total dependence[a]			
	FFSA 2013 LTCIBG[10] label definition	Insurer 1	Insurer 2	Insurer 3
State identified if	Total Dependence	Total Dependence	Total Dependence	Total Dependence
	4ADL/5	4ADL/6	IRG 1, 2 + 3 ADL/4	IRG 4 + 3 ADL/6
Toiletting	Wash one's entire body. Be able to insure elimination hygiene. Satisfy daily corporal hygiene level conforming to usual norms	Ability to achieve a level of corporal hygiene level conforming to usual norms	Ability to wash the entire upper body (front, side, and back of head and torso) and ability to wash the entire lower body (legs and intimates parts)	Ability to achieve a level of corporal hygiene level conforming to usual norms
Dressing	Put on usual clothing, eventually adapted to one's disability. Remove usual clothing, eventually adapted to one's disability	Ability to dress and undress	Ability to correctly put on and remove one's clothing (except shoes)	Ability to dress and undress
Eating	Handle previously served and cut food. Drink	Ability to feed oneself and eat prepared food which has been served	Ability to cut one's food or to pour a drink and ability to bring food to one's mouth and swallow it	Ability to feed oneself and eat prepared food which has been served
Elimination	Included in "toiletting"	Ability to insure urinal and fecal elimination hygienically, including using protective or surgical equipment	Not mentioned	Ability to insure urinal and fecal elimination hygienically, including using protective or surgical equipment
Transferring	Transition from each of 3 standing, lying down, sitting to the other, both ways: to get up from a bed of from a chair, to lie down, to sit down	Ability to transfer from a bed to a chair or armchair or inversely	Not mentioned	Ability to transfer from a bed to a chair or armchair or inversely
Indoor mobility	Move indoor on a flat surface	Ability to move on a flat surface inside one's home or to leave the premises in case of danger	Once standing, ability to move inside one's home on a plane surface, after using adapted equipment	Ability to move on a flat surface inside one's home or to leave the premises in case of danger

[a]Exclude neuropsychological tests

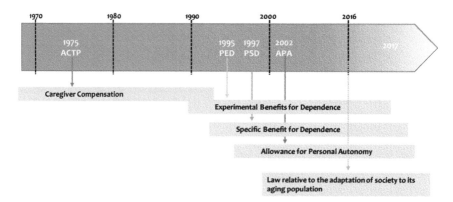

Fig. 3.8 Timeline of national insurance Long Term Care programs to address the risk of dependence[10]

These successive mechanisms brought piecemeal responses which were partially criticized: Experimental Benefits for Dependence (PED)[12] was paid only for Home Care and was implemented as an experiment in a few regions. It prefigured the implementation of the Specific Benefit for Dependence (PSD)[12] which had a limited success due to its claim on a person's estate. Moreover, benefit amounts were highly unequal in different regions.

Lastly, Allowance for Personal Autonomy (APA),[12] started in 2002, brought stabilizing responses over fifteen years until an expected law, which never arrived. The law was intended to reinforce the national responses as well as to create favorable conditions for a complementary system within the second pillar of group coverage and, more importantly, within the third pillar framework of individual insurance offerings.

The law relative to the adaptation of society to its aging population[12] has brought responses privileging prevention principles and responses addressing the availability of public financing. But it did not address insurance market expectations and for ten years the market has remained anemic. However:

- The needs are destined to explode (in both resources and amount),
- The earlier the financing occurs in the insured's life, the more moderate and affordable the premium would be.

3.4.2 Allowance for Personal Autonomy

The Allowance for Personal Autonomy[9] (APA) went into effect January 1, 2002 [13][13]. It eliminated the recourse to the beneficiary's estate and covers individuals with 1–4 IRG scores. Maximum income level requirements for eligibility are eliminated, but the benefit amount varies based on income levels. Benefit levels are set nationally, but payments are implemented in a regional management framework (hence benefit payments weigh heavily on regional finances), depending on:

- Beneficiary's IRG score,
- Beneficiary's income,
- Beneficiary's needs according to the nature of assistance which appear necessary (notably compensation for home caregivers, payments for licensed family caregivers, eventual transportation costs, …),
- Beneficiary's residence (home or institutional).

APA is principally financed regionally, and nationally through a dedicated administration,[14] at an annual cost of €5,5 billion benefiting 1.2 million people. Between 2002 and 2009, the number of beneficiaries increased from 700,000 to 1.13 million, or a 7% annual rate [3]. 60% of beneficiaries reside in their home. 2040 projections of APA beneficiaries are estimated between 1.6 million [16] and 2 million [11] (Fig. 3.9).

The law relative to the adaptation of society to its aging population, which went into effect on January 1, 2016 [14], integrates a partial reform of the Long Term Care national insurance program by modifying the maximum benefits for Home Care (regulation relative to residential care remains unchanged) as well as improving the recognition of the needs and expectations of beneficiaries. This law also introduces the recognition of the "benevolent caregiver" of an aged dependent person. This provision principally considers the "right of respite", consisting in improving caregiver assistance by making available one or more temporary care programs (Portail national d'information pour l'autonomie des personnes âgées et l'accompagnement de leurs proches).

Fig. 3.9 APA beneficiaries

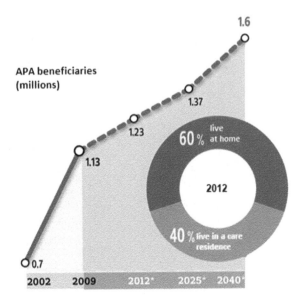

3.5 The Insurance Market

3.5.1 By the Numbers

To take an inventory of the activities of the Long Term Care insurers is difficult as just about all insurance organizations have a presence in the market and some reinsure between themselves. As in tallying business for Group Health and Disability coverage, some transactions are double counted. However, it is possible to establish a few quantitative elements, as the Fédération Française d'Assurance (FFA)[15] points out, on classifications relative to "contracts with principal and unique Long Term Care benefits".

In other words, corporate group insurance coverage as well as hybrid offerings incorporating Long Term Care coverage are difficult to integrate in such a total. In general, the LTC benefit amount turns out to be much smaller in the last case; moreover, these contracts have much lower premiums relative to contracts recorded by the FFA (Table 3.5, Fig. 3.10).

Important points to note:

Table 3.5 Long Term Care insurance top 10 [2]

Rank	Company	2016 direct business premium[a] (M€)	Variation 2016/2015 (%)
1	Crédit Agricole Assurance	91.0	3.4
2	Groupama	78.7	1.4
3	La Banque Postale Prévoyance	70.0	2.6
4	Mutex	67.8	0.7
5	Axa France	65.0	16.1
6	Groupe CNP Assurances	57.0	−19.1
7	Sgam AG2R La Mondiale	49.0	−5.8
8	Groupe des Assurances du Crédit Mutuel	33.0	3.1
9	Groupe Istya	27.7	−8.2
10	La Mutuelle Générale	27.2	−0.7

[a]Does not include reinsurance

Fig. 3.10 Long Term Care insurance key numbers [7]

2015 Long Term Care Incurance key numbers for contracts with principal and unique LTC coverage

Covered persons: 1.6 million
Premium: €551.1 million
Claims paid: €219.8 million
Average monthly payments: €582 (Individual contracts)
Statutory reserves: €4.8 billion

- The presence of all classes of insurers (bancassurance, traditional insurers, group mutual, group health, …),
- The very high concentration of the market: for several years the highest four companies have captured more than half of the Long Term Care insurance market.

The average monthly annuity is the norm and ranges between €550 and €600 for Individual policies; however, the number of insureds is likely to cover disparate situations and without doubt is not technically comparable, as it incorporates voluntary Individual membership and mandatory Group participation (their monthly benefits can hardly be compared).

Issue age is close to 60, if not slightly higher; it is close to average retirement age, when insureds are planning their long term financial contingencies by integrating in their young retired budget the monthly cost of Long Term Care.

At this average issue age, the monthly premium amounts to about €30–€40 for Total Dependence depending to the richness of benefits. Today, at age 60, at a 1% annual interest rate, and a premium rate reflecting reasonable loads, the actuarial premium for a Total Dependence monthly annuity of €600 appears to be about €30 (without lump sum payment, service benefit, nor nonforfeiture benefit).

The FFA average premium is €28 per month, but benefits include a first-expense lump sum benefit, a reduced Paid Up benefit, and service benefits. Moreover, this market statistic includes premiums reflecting both Individual and Group rates. It is a good bet that many people covered as categorized by FFA numbers have premium levels lower than the actuarial equilibrium premium if coverage were on a lifetime basis: either they subscribed at a younger age than 60, or they were issued a policy before the strong decline in interest rates.

Strong and very contrasting major trends should be pointed out:

- Convergent phenomena: Total Dependence rates through the years (as well as benefit definitions) tend to converge, leading to a more uniform market;
- Divergent phenomena: most insurers have offered or are introducing Partial Dependence benefits, which is a rather new risk and for which rate setting is not yet well defined. It is not well defined from an actuarial standpoint because the long-term risk is still not well known (incidence and continuance), and poorly measured from a statistical standpoint since, to this day, the largest portfolios lack a level of credible experience which makes it difficult to extrapolate sufficiently ahead to know the actual cost of the risk.

Another divergent element worth mentioning is coverage definitions. There remain in the market plans that do not cover the same contingencies but which use the same names. This is regrettable and potentially dangerous for the whole market, thus we have:

- Pre-funding premium for Individual policies covering lifetime benefits (these policies are filed either as "Individual contract" or "Open Group contract";
- Annually cancellable group contracts which only cover open claims in case of cancellation.

Clearly this second item, especially regarding a strongly increasing risk, cannot respond to the long-term expectations of the insureds who in good faith believe that they will keep their lifetime coverage if they pay their premium. The "truth serum", in this case, lies in the answer to the simple question: what happens in case of cancellation by the insurer? Statistics from professional insurance organizations should clearly distinguish pre-funded contracts with a lifetime coverage objective, and the others, annually cancellable. The whole market would benefit.

To conclude, the French market is strongly concentrated on Individual insurance ("Third Pillar"), and seems even to be a world leader. It is still finding its way in Group insurance, premium income remains marginal compared to Individual insurance (Fig. 3.11).

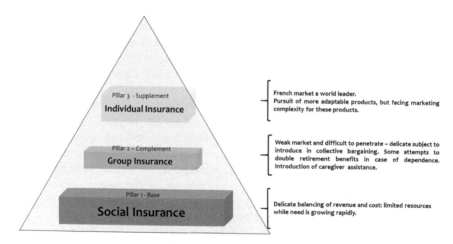

Fig. 3.11 The three pillars of Long Term Care insurance

3.5.2   Individual Insurance

3.5.2.1   Total Dependence

The French market has always been organized around the payments of monthly lifetime cash annuities in case of Total Dependence: when a claim is approved, payments are disbursed independently of costs incurred by the beneficiary. It usually consists of a basic contract benefit, and the amount, in €100 steps, is chosen at issue, between €300 and €4,000 per month. The average monthly amount issued has been stable for an extended period in the French market, slightly under €600.

These contracts are no longer subject to "non-Life" taxes, and premiums integrate in their calculation a load close to 20%. These load levels can seem rather high, but they must be compared to the monthly premium amount which remains low (since monthly premiums are close to €40.) What is at stake is to consider contract duration, and hence based on a logic of "contingent present value", all aspects of administrative costs linked to premiums and claims, medical expertise (whether at the underwriting or claim stages), marketing, regulatory, and legal, as well as technical support and development.

Contracts offer more and more frequently a lump sum benefit not directly related to actual expenses, generally as part of the basic coverage. This lump sum is paid once a claim is approved, specifically to help finance the remodeling of a claimant's residence. This amount is often close to €3,000 and is uniform across the market, with few contracts offering various levels. It is usually triggered by a Total Dependence claim, but can be paid for Partial Dependence.

3.5.2.2   Industry Standards

The Long Term Care Insurance Basic Guarantee label (LTCIBG)[10] arose from a mid-2003 insurance industry initiative which proposed to identify standard conditions for Total Dependence benefits (level, minimum, eligibility, access to information, ...): it was meant to "offer the insured more understandable contract provisions, more protection, and be accessible to more people".

Several technical features were integrated into this label:

- Minimum monthly benefit established at €500;
- Elimination Period no longer than 3 months from claim incurred date;
- Limited maximum Waiting Periods: none in case of accident, 1 year in case of illness, and 3 years in case of neurodegenerative or psychiatric ailment;
- Maximum issue age cannot be younger than 70.

Lastly the Total Dependence definition kept by the label was reached after discussions within the profession, considering the experience of insurance organizations over the years. Notably, this synthesis bears on an objective assessment for the cognitive risk in Total Dependence, an assessment which combines ADL and tests. To receive benefits one of the following three conditions must be met:

- Cannot complete 4 ADLs out of 5; or
- 3 ADLs out of 5 and an MMSE score less than or equal to 15; or
- 2 ADLs out of 5 and an MMSE score less than or equal to 10.

Several products followed these guidelines to obtain the label, but the introduction occurred during an unresponsive market, waiting for political decisions, on one hand, and increasing Partial Dependence benefit offerings on the other hand (these are not subject to the label, which could only apply to part of a policy with Partial Dependence benefits).

3.5.2.3   Partial Dependence

Monthly lifetime cash benefits for Partial Dependence were introduced in some contracts, and now in the majority of contracts since 2000. It usually consists of an option from the basic benefit, the policyholder being able to choose or decline this coverage as long as he or she is covered for Total Dependence. The benefit is also fixed as a function of the Total Dependence benefit level (a 50% level is most often observed).

The policy usually stipulates that the full premium is waived once Partial Dependence benefits begins: it would be difficult to explain to the policyholder with the two guaranties (Total Dependence and Partial Dependence) that he or she would receive on one hand a Partial Dependence monthly benefit but must continue to pay a Total Dependence premium. This premium waiver is not free, since it adds from 5 to 10% of total premium.

Frequently, some policies with Reduced Paid Up nonforfeiture benefit clauses offer reduced Total Dependence benefits but no Partial Dependence benefit, upon non-payment of premiums.

The expertise for the risk of Total Dependence seems more and more relevant at increasingly older ages. About thirty years were needed to reach a situation where the principal insurers had at their disposal a converging assessment of the guaranties to insure, and hence a converging view of proposed rates. To date, several public statistical studies, as well as regular monitoring by insurance organizations, seem to show that the Total Dependence risk is insurable; forecasting active life expectancy, increasingly robust knowledge incidence distributions, and more refined knowledge of continuance even at the oldest ages and longest durations (which now remains the main area to perfect).

Meanwhile "morbidity compression" over a long period allows for the Total Dependence risk to consider that active lives file claims at later policy durations and remain in claim status for a duration that does not lengthen (even at times decreasing).

But all these rather reassuring indicators for Total Dependence come from several trends, which, for Partial Dependence, raise some caution: insurers offering Partial Dependence coverage observe significantly lower actual-to-expected ratios (insureds should have more frequent claims and longer claim continuances than for Total Dependence, which remains to be observed); premium rates seem to vary widely in the market, from one insurer to another. The risk is also more subjective and litigious than Total Dependence, which makes it costlier to approve a claim and more difficult to manage medically and legally.

Moreover, medical research tends to bear on neurodegenerative disease treatments which will increase Partial Dependence continuance and delay Total Dependence incidence as well as shorten its continuance.

To date, most global indicators appear economically favorable for Total Dependence, and symmetrically unfavorable to Partial Dependence. But these two categories are contractually bounded at the bottom line level.

3.5.2.4 Premium, Underwriting and Other Characteristics

- **Premium**

  The most widely used form of premium for Individual Long Term Care insurance policies is a level premium, monthly or quarterly, calculated based on issue age.

  An ambiguity for this type of insurance must be pointed out as it is most often sold commercially within a framework of non-cancellability: premiums cannot be cancelled or revised as long as the policyholder pays them. But from a regulatory standpoint, premiums are guaranteed renewable and can be revised up or down. Moreover, Solvency 2 capital calculation guidelines encourages that flexibility, as an assumption of lifetime non-cancellability turns out to bring very onerous capital requirements and results in very high premiums. Hence, premium rates appear adjustable, but reserves and experience monitoring are undertaken in a logic of "personal insurance" by continuously evaluating the insurer's current liabilities and the policyholder's ability to continue to pay his or her premium.[16]

  These calculations are usually done on an inventory basis with a logic of premium and benefits at the "attained level", the reserve being set for increasing risks. Each upward revision of the premium rate will decrease the reserve, everything else being equal. Reciprocally, any decrease of the non-life interest assumption will instantly increase, and sometimes significantly, the amount of reserve required[17] (ACTENSE).

- **Reduced Paid Up nonforfeiture benefit**

  Most often, Individual policies incorporate a reduced paid up benefit under the provision of a minimum number of premium paying years (often fixed at 8 years in the French market). The policy specifies that the policyholder is eligible for a reduced annuity based on a pre-defined schedule depending on the date premiums cease to be paid. This schedule can be revised and communicated at the request of the policyholder.

  The reduced benefits turn out to be relatively small compared to the need of a dependent person. Moreover, contacts between the insurer having ceased, many dependents and their caregivers do not necessarily have sufficient information to activate their claim.

- *Underwriting*

Most insurers fix the maximum issue age between 70 and 75 (nowadays the risk is virtually not insurable in the market after 75).

However, the trend is toward shortened medical questionnaires for underwriting, adverse selection not being at the forefront of insurers' (and reinsurers') concerns since incidence occurs on average twenty years after issue. Depending on the benefit level and the answers to the shortened health form, insurers reserve the right to decline issuing the policy, or to apply higher rates for the substandard risk.

- *Couple discount*

Many contracts offer "couple discounts" allowing one of the two policyholders to lower his or her premium if the policy is issued to both individuals. Technical motivations are diverse:

  - Lower cost of issuing two policies instead of one,
  - The cost of the risk for a male and a female may be lower than for two individuals on average in a portfolio (the high majority being female),
  - In the event of occurrence of dependence for one of the couple, the probability that the individual remains at home, and in good health for a longer time, tends to lower the price of the risk.

- *Waiting Period*

Policies incorporate Waiting Period clauses which are converging over time and are logically reflected in the LTCIBG[10] label.

In practice the usual provisions are the following:

  - Loss of autonomy due to somatic ailment: 1 year
  - Loss of autonomy due to cognitive ailment, such as Alzheimer's: 3 years
  - Loss of autonomy due to accident: 0 year

For Total Dependence, these clauses do not appear to be determinant on rates. They may reveal themselves to be more important for Partial Dependence.

- *Elimination Period*

The lifetime annuity payment is often contingent on an Elimination Period. The Elimination Period can be absolute (no annuity payment if death occurs during the period) or retroactive (payment from the incurred date if the claimant survives the period). The introduction of an Elimination Period leads to several non-negligible implications for:

  - Risk and premium: substantial first year mortality increases after a claim: (annual mortality rates are about 30% depending on the cause of dependence), and an Elimination Period has an important impact on the premium (near 10% for an absolute Elimination Period of 3 months).

- Claim administration: an Elimination Period complicates claim approval (systems, dates for incurred, reception, approval, deferral, and first benefit payment). Besides, average processing time (around three months on average) are lengthened significantly: receipt of complete information, administrative processing, medical reviews ....
- Insurer's image and risk of lawsuits: a claimant who dies during the Elimination Period would not be eligible to any payment, even after paying premiums for twenty years.

3.5.2.5 Caregiver Assistance

This is a new benefit belonging to "Service Benefits" provisions. Most often, these benefits are optional and allow relatives or friends of the insured to gain access to social or legal information (internet sites or applications). These benefits, which are usually considered services, were developed over time and can also cover three types of innovative benefits:

- Help in undertaking administrative procedures (for instance, help in looking for a nursing home or finding a home care solution),
- Psychological support and hiring a person in case of the unavailability of the caregiver, consultations,
- Help in housekeeping and emergencies, and tele-assistance or advice.

Lastly, but this benefit entails substantially higher cost, more ambitious assistance can now be contemplated for caregivers who have a dependent relative, to give them rest time by discharging them of all or part of their professional or domestic tasks.

These caregiver assistance benefits present two characteristics:

- They add more appeal and strengthen service benefits which have existed since the origin of Long Term Care insurance contracts;
- They will have a stronger resonance in the case of collective contract negotiations for companies, allowing employees to receive supplemental benefits in line with savings/retirement benefits. These benefits can be negotiated in the case of collective agreements, by analogy with savings/retirement contracts, they would be cancellable annually and would allow responses to be brought to employees who would co-finance with the employer, the "insured" being a family member (generally the spouse or parents) and the beneficiary being the employee. To reach an optimal plan, one should remain vigilant on the formulation of the benefits and their implementation relative to other complementary social insurance programs so that they can safely be within the conditions for social and tax exemptions.

This type of benefit can rekindle the interest in Group Long Term Care insurance. Human Resource professionals see more advantages to offering an immediately measurable service for current employees rather than co-financing a Long Term Care annuity which could occur on average twenty years after the employee's retirement.

3.5.3 Group Versus Individual Insurance

3.5.3.1 Mandatory Group Insurance

These contracts developed through different channels, but it is clear that the market for mandatory Group insurance is still finding its way because it is trying to resolve obstacles arising from collective bargaining, which is by nature difficult to overcome. At the end of 2016, for contracts where the main and only guarantee is dependence, 80% of the insured persons fall under the third pillar "Individual procedures" against only 20% under pillar 2 "Companies and Branches" [7].

Collective bargaining negotiations concerning all employees must be able to reach a plan, with its contributions independent of employees' age, either fixed, or proportional to salary, which would allow the funding of future claims (the present value of future benefits calculated at the time of incidence). But this prefunding cannot stop at retirement, it therefore assumes a choice by the retiring employee whether or not to continue premium payments necessary to reach satisfactory benefit levels.

The earlier the pre-funding occurs, the lower the post retirement premium becomes, which would ease the adoption of such plans. Three obstacles remain when, on their last day of employment, young retirees see:

- their income decrease,
- employer contributions cease,
- social and fiscal benefits are no longer available.

Insurers have designed different provisions around the funds collected through payroll deductions. There are questions during the employment phase which to this date remain delicate to resolve and have not found satisfactory solutions:

- What happens when an employee terminates before retirement?
- How does one take seniority into account upon cessation of employment? A function of participation years?
- What happens when a young retiree ceases to pay individual premiums?

3.5.3.2 Another Group

Contracts between the two extremes described above are not easily classifiable. At one end of the market spectrum are individual policies and facultative group insurance; they rely on reserves with a rate structure, a culture, and management akin to health insurance, taking into consideration probable future benefit and premium payments. At the other extreme are group mandatory contracts also using reserves, but they are much rarer.

Between these extremes, contracts were marketed to offer Long Term Care benefits, often with lower benefits than Individual policies, but incorporating the concept of repartition, or generational funding. These types of plans were able to exist or continue to be issued for benefits such as burial expenses or Long-Term Disability, for instance. For a strongly increasing risk, reaching affordable prices seems even more possible and attractive.

These 'repartition' contracts usually carry a guaranty to pay benefits until the end of the coverage period, as stipulated in the contract. However, nothing indicates that the increasing risk reserves, which involve a rigorous actuarial approach (considering probable future benefits and probable future premium at the time of valuation), are in balance with available funds. In this case the contractual mechanism relies partly on repartition, that is, it relies on the "demographic engine", in other words on the perpetual arrival of an increasing number of new participants (by assuming that the plan was designed at a lower price than one using a pre-funding approach).

In this scheme, there is a real risk that the contract-holder will have the plan cancelled by an insurer that considers it to be too exposed in its long-term actuarial balance. The commercial and image risk are also real since, at the cancellation date, participants who have paid for many years could find themselves without any right, and may even find that they are unable to buy Long Term Care coverage, due to their age. These plans remain in the minority but their rate calculations and the assumptions which underlie their long-term solvency deserve a very careful study.

3.5.3.3 A Particular Case

A Group insurance organization offering Health, Disability, Long Term Care, and Life insurance plans[18] launched a product based on a logic of Long Term Care benefit point purchases with each annual premium. This plan has evolved, integrating for instance payment floors or mutual acquisition mechanisms within a group based on its average age. But fundamentally, it remains original in the field of Long Term Care plans since:

- The plan relies on a deferred lifetime annuity expressed in points,
- The plan uses a pre-funding mechanism.

By design, the monthly annuity amount at incidence is not known at issue, it will depend on the willingness of the participant to keep paying his or her premium before a claim (for instance at retirement or later). Reciprocally, if the annuity amount is not known at issue, as long as the insured keeps paying premiums, the minimum annuity payment keeps increasing.

Hence, like point-based retirement plans, the plan is based on a mechanism of regular acquisition rights; but the difference with common retirement plans is that the contract functions with a pre-funding mechanism inherent to all personal insurance plans with lifetime benefits.

For the employer or the employee, such a mechanism is akin to a defined contribution plan, but for the insurer, the participant is the beneficiary of a deferred lifetime annuity; the mechanism is close to defined benefit plans used in some corporate retirement plans[19] in which the premium, when received, is directly converted in a deferred lifetime annuity with tabular and rate guarantees. The main difference comes from the fact that the retirement age is known and considered, and, moreover, a benefit payment is quasi certain; whereas the incidence age, even that the risk will materialize, and lastly the duration of the lifetime annuity, are by nature very random.

The plan is offered on an Individual basis but also added to Group contracts which offer Social Insurance Supplemental Life and Health coverage. Purchase points appear by design more sensitive to lower interest rates than level lifetime premium plans (more so because participation ages are much lower than Individual policy issue ages).

3.5.4 Reinsurance

The market and the number of reinsurers being very limited and the risk being so volatile, some thought was given at one time to non-proportional treaties covering claim payments after a contractually defined continuance, or generational stop-loss, but the reinsurance market is now concentrated on co-insurance treaties.

These treaties organize the proportional provisions for all aspects of policy issuance (except loads and expenses), with the reinsurer paying a ceding commission. From an economic standpoint, these treaties achieve a dilution of production in exchange for a technical, medical, and financial partnership with an operator which has access to a larger market.

Besides sharing outcome, the added value offered by reinsurers comes from:

- Technical partnership: rate setting and statistical studies of the risk, hence developing biometric distributions for ages and continuance beyond the experience of the insurer;
- Medical partnership: the reinsurer having the role of a market observer but also, when exposures are significant, auditing underwriting procedures, especially during claim application reviews.

For the ceding insurer, the co-insurance approach presupposes from the reinsurer:

- Rate making transparency,
- Transparency in the principles underlying annual reports and a true partnership which allows one to allocate the respective responsibilities and the manner of communicating potential adverse experience,
- Ideally, an initial cessation clause in the treaty allowing one to define the conditions of the eventual recapture of the portfolio by the ceding company.

The treatment of reinsurance in Solvency 2 calculations has been analyzed [5]. Reinsurance is now becoming a potential tool to manage and monitor capital which, above all, helps to reduce its need.[20]

Globally, reinsurance in the framework of Solvency 2 leads to:

- an increase in the need for capital for the counterparty default risk (this aggregate risk being a function of the rating of one or both counterparties), which is usually compensated by

- a lower underwriting risk capital requirement (cession of risk and potential adverse deviation.)

3.5.5 Solvency 2

Solvency 2 did not integrate Long Term Care specificity[21] into its development or its introduction. If Long Term Care insurance, combining numerous risks, could be burdened by minimum surplus (Solvency Capital Requirement or SCR) at such a level, then the product price would be considerably impacted.

Long Term Care assurance was already difficult to sell without this innovation (insurers point out that two or three face-to-face meetings are necessary to sell a Long Term Care contract).

For its "standard formula", the current directive requires specific tests of a certain number of shocks to measure the SCR linked only to the issue risk; shocks whose principal characteristics are the following:

- Longevity risk: lower active and claimant mortality rates by 20%,
- Morbidity risk: increase incidence by 35% in the first year, then by 25% in the following years,
- Severity risk: increase severity by 10%.

Beyond these specific shocks on each of these biometric events, regarding a very long-term risk, the "interest rate" dimension appears equally determinant during the product design; it remains so year after year through the interest curve evolution retained to calculate the minimum surplus necessary.

Overall, and after all the calculations are done, one shows that SCR is a function of the insurer guaranties and for which the result appears much steeper than the results reached in a Solvency 1 logic; moreover, if it was necessary to integrate in all the premiums a load to fund the required surplus, it is a safe bet that rates would just about double. Enough to kill the patient in one blow instead of curing the disease!

Fortunately, "management actions" were integrated in the Solvency 2 standards. Principally, they allow one to identify beforehand all the levers that the insurer can and must "pull" in order to balance its Long Term Care insurance portfolio in the long term. In principle, these "management actions" bear on the revalorization granted (through result reportings, which are usually explained by general conditions) and on rate revisions decided upon annually by the insurer, within the constraints of contractual terms, if those exist.

This action concretely assumes an effort of inquiry by management and preliminary reflections on rate setting to itemize and formulate measures to undertake against problems arising in future years, in statistical studies, and by monitoring claim trends.

Globally, Solvency 2 makes management of Long Term Care insurance more complex; but practically, the standard is not of a nature to discourage offering these benefits; on the contrary, it points to and reinforces the necessity to manage this type of insurance as a "state within a state", that is, through an organism incorporating reflexes of the ORSA type (Own Risk and Solvency Assessment) allowing one to organize all operational contributions to manage the risk. Only a few did not wait for ORSA to incorporate its principles; as soon as the volume allowed it, the quality step taken by utilizing all sources of knowledge within the insurance organism to manage the Long Term Care risk made complete sense, even before Solvency 2, given the nature and the inherent risks of Long Term Care insurance.

3.6 Conclusion

3.6.1 A Market at the Crossroad

Year after year, technical departments will tend to privilege Total Dependence coverage, financial managers will concentrate on integrating the effect of long lasting low interest rates, marketers will of course privilege the introduction of benefits for which the availability is frequent, immediate, and conspicuous (Partial Dependence, Services, Caregiver Assistance, …), all of this under the scrutiny of risk managers who will need to consolidate the entire system to manage profitability, integrate in their assessment "management actions", the effect of more extensive coverage, … and the necessity of rate revisions.

Also, for 10 years, the Long Term Care insurance market has waited for encouraging signals from the government, without success; now it knows that it must rely only on itself.

3.6.2 Potential Innovations

As a basis for future reflections and eventual innovations, there are several characteristics of "bricks" that are now increasingly observed, in "mixed products":

1. Total Dependence (annuity and lump sum)
2. Partial Dependence (annuity and lump sum)
3. Service
4. Caregiver assistance.

The first is a "true risk" insurance, with low incidence rates combined with continuance rates integrating very high mortality rates (the "half-life", that is, the duration at which 50% of deaths are observed, occurring less than 24 months after the beginning of a Total Dependence[1] claim.) This continuance risk still remains partially controlled, being opposite to the expected healthy lifetime of premium paying policyholders (mortality table of active lives), for which the cost turns out to increase year after year.

The second risk occurs more frequently as the contractual definition would cover the very partial risk which would lead to incapacity or the need for regular assistance. INED[5] and DREES[11] released edifying studies (by gender, in particular) showing the very strong occurrence after 75 of the need for punctual and chronic help.

The third seems more classical but completely leaves the logic of insuring an individual on a lifetime basis, targeted by the first two risks above.

Lastly the fourth leads to a set of practical observations which all lead us to believe that it will continue to grow.

But what are the common points between these different benefits in terms of management and risk taking? What could be common management and risk cultures between them:

- Lifetime financing of long-term protection with low frequency (Total Dependence annuity),
- Prefunding a very likely risk (Partial Dependence annuity or lump sum),
- Cost and accounting, on an annual basis, of service benefits,
- Monitoring caregiver assistance benefits, a short/medium term risk?

The first risk has by nature a lifetime horizon, the second appears, at this date, very clearly under-estimated in insurance portfolios, and finally the third and fourth are legally cancellable annually; the first two, with their lifetime approach, are not, nor should they be.

3.6.3 Insurable, Yes ... But Not at Any Cost

To conclude, with twenty years of experience managing such guaranties, two beliefs, two messages of hope:

- The Total Dependence risk should be manageable; the risk can now be approached with solid statistical bases (given that the chain of risk assessment remains sustainable and thus claim approval closely follows contract terms, which remains to be demonstrated),
- Long Term Care requires, by nature, exacting skills of all the professions within the insurance organizations and heightened attention; ORSA requirements under Solvency 2 find natural grounds for application here, still requiring, at the scale of insurance organization portfolios, a Long Term Care portfolio with a significant number of insureds, premiums, and reserves.

No sooner had the Total Dependence risk been better understood than many products were distributed in the market for which there is not yet robust knowledge and of which a part (Partial Dependence benefits) are naturally combined with Total Dependence results. There is enough to remain for many years, enduring, humble, determined, and vigilant. Enough to keep a sharp eye on Long Term Care insurance. An actuarial glance.

Notes

1. Total Dependence: 3 ADL out of 5 or Level 1 or 2 for IRG
   Partial Dependence: 2 ADL out of 5 or Level 3 or 4 for IRG.
2. IRG, Iso-Resource Groups, translation of GIR, Groupes Iso-Resources
   AGGIR: Autonomie Gérontologique Groupes Iso-Ressources
   A claim evaluation algorithm designed by doctors at the national public insurance administration (Sécurité Sociale) and the Société française de gérontologie
   See Sect. 3.3.1 for definitions and "AGGIR, The Work of Grids", Dupourqué, E., Long Term Care News, Issue 32 September 2012.
3. See Chap. 2.d: Reserving.
4. See Chap. 1.a: Disability and Life Expectancy.
5. INED: Institut National d'études Démographiques, a public research institute specialized in population studies that works in partnership with the academic and research communities at national and international levels.
6. INSEE: Institut National de la Statistique et des Études Économiques collects, analyses and disseminates information on the French economy and society.
7. See Chap. 4.
8. See Chap. 5.
9. APA, Allocation Personnalisée d'Autonomie
   Translated as Allowance for Personal Autonomy.
10. LTCIBG, Long Term Care Insurance Basic Guarantees, translation of GAD, Garantie Assurance Dépendance.
11. DREES: Direction de la recherche, des études, de l'évaluation et des statistiques
    Part of the Health and Social ministry of the national government. It was created with the mission to provide guidance to governmental departments and agencies, with a better capacity of observation, expertise and evaluation on their actions and their environment.
12. ACP: Allocation Compensatrice pour Tierce Personne
    PED: Prestation Expérimentale Dépendance
    PSD: Prestation Spécifique Dépendance
    APA: Allocation Personnalisée d'Autonomie
    Loi relative à l'adaptation de la société au vieillissement.
13. Law # 2001-647 of July 20, 2001 relating to the care for the loss of autonomy of the aged and personalized assistance for autonomy.
14. CNSA: Caisse Nationale de Solidarité pour l'Autonomie
    Established in 2004 to administer national programs against loss of autonomy, including Long Term Care.
    It oversees Personalized Allocation for Autonomy (APA) benefit payments, issues guidelines, collects revenues and funds Long Term Care residential establishments.
    It also publishes periodical studies on Long Term Care developments.
15. FFA: Fédération Française d'Assurance

Created in July 2016, it gathers the French Féderation des Sociétés d'Assurance (FFSA) and the Groupement des Entreprises Mutuelles d'Assurance (GEMA) under a single umbrella.

FFA represents 280 insurance and reinsurance companies operating in France, covering over 99% of the market.

16. Premium rate increases, depending on the policies, can be subject to a maximum over the policy duration.
17. Reporting on December 2016 non-life valuation interest rate and regulation summary: http://actenseactualites.typepad.com/actu/tme/.
18. OCIRP, Organisme Commun des Institutions de Rente et de Prévoyance, (or Union of Annuity, Health, and Disability Insurers).
19. Plan d'Epargne Retraite Entreprise (old article 83 of Code Général des Impôts).
20. See Chaps. 8 and 9.
21. See Chap. 6.

References

1. ACTENSE: Récupéré sur (s.d.) http://actenseactualites.typepad.com/actu/tme/
2. Argus de l'Assurance N°7508 26 mai 2017. *Top 10 Dépendance* (2017)
3. CNSA: *2016: Les chiffres clés de l'aide à l'autonomie*. Récupéré sur (2016). http://www.cnsa.fr/documentation/cnsa_chiffrescles2016-web.pdf
4. DREES: *Les contrats d'assurance dépendance sur le marché français en 2006*. Récupéré sur (2008). http://drees.solidarites-sante.gouv.fr/IMG/pdf/serieetud84.pdf
5. Dubois, D., Frédéric, P.: *Quel impact de Solvabilité 2 sur les dispositifs de réassurance?* (2015)
6. Fédération pour la Recherche sur le Cerveau: Récupéré sur (s.d.). https://www.frcneurodon.org/wp-content/uploads/2016/09/Mini-Mental-State.pdf
7. FFA: *Assurance dépendance: 6,8 millions de personnes couvertes à la fin de l'année 2015*. Récupéré sur (2017). https://www.ffa-assurance.fr/content/assurance-dependance-68-millions-de-personnes-couvertes-la-fin-de-annee-2015
8. INED: *France 2004: l'espérance de vie franchit le seuil de 80 ans*. Récupéré sur (2005). https://www.ined.fr/fr/tout-savoir-population/graphiques-cartes/graphiques-interpretes/esperance-vie-france/
9. INED: *La dépendance: aujourd'hui l'affaire des femmes, demain davantage celle des hommes?* Récupéré sur (2011). https://www.ined.fr/fichier/s_rubrique/19151/483.fr.pdf
10. INSEE: *Tableau de l'économie Française*. Récupéré sur (2012). http://www.insee.fr/fr/ffc/tef/tef2012/T12F036/T12F036.pdf
11. INSEE: *Insee Analyses: L'allocation personnalisée d'autonomie à l'horizon 2040*. Récupéré sur (2013). https://www.insee.fr/fr/statistiques/1521329
12. INSEE: *Insee, Age Pyramids*. Récupéré sur (2015). https://www.insee.fr/en/statistiques/2418104
13. LegiFrance; Récupéré sur (2001). https://www.legifrance.gouv.fr/affichTexte.do?cidTexte=JORFTEXT000000406361
14. LegiFrance: Récupéré sur (2015). https://www.legifrance.gouv.fr/affichTexte.do?cidTexte=JORFTEXT000031700731
15. Portail national d'information pour l'autonomie des personnes âgées et l'accompagnement de leurs proches. Récupéré sur (s.d.). http://www.pour-les-personnes-agees.gouv.fr/dossiers/les-nouvelles-mesures-de-la-loi

16. SÉNAT: *Rapport d'Iinformation: prise en charge de la dépendance et la création du cinquième risque*. Récupéré sur (2011). https://www.senat.fr/rap/r10-263/r10-2631.pdf
17. Service-Public.fr. Récupéré sur (2016). https://www.service-public.fr/particuliers/vosdroits/F1229

Part II
Liabilities Measurement

Bob Yee

Introduction

There are three components in developing any actuarial liabilities, namely, data, assumptions, and methodology. Long term care insurance liabilities pose special challenges in that scarcity and complexity of data lead to tenuous assumptions and intricate models. The likelihood of needing long term care services and supports increases imperceptibly before retirement age but quite rapidly in later ages beyond 70. As such, longitudinal experience data takes time to develop. Within each country, the types of services, severity, intensity, and care settings are dependent on various economic, cultural, and geographical factors. The benefit structure of the insurance coverage adds to complexity in estimating future events. Other insurance product risks, namely, mortality, voluntary lapse, investment returns, and expenses, also have a significant impact on liabilities.

The task of setting assumptions involves transforming useful experience data into suitable forms as inputs to the modeling of future liabilities. In practice, the process is often reversed in that a model is postulated, then data is sourced and manipulated to fit the model.

Since long term care risks are uncertain, the traditional deterministic method of measuring liabilities is generally inadequate to assess the range of possible outcomes. Premium rate increases on in force business in the United States are rampant and caused loss of consumer confidence. One of the culprits can be traced to the traditional pricing method that failed to produce sufficient contingency margins in the pricing of long term care insurance policies. Increasingly actuaries are employing stochastic modeling techniques to understand the variability of long term care risks.

However, stochastic methods need to be sufficiently robust in order to capture the underlying events that drive future liabilities. At any time, a policyholder can be healthy, in a number of disabled states (with insurance benefits), dead or lapsed. All such statuses are determined by a set of probability distributions that represent reality. The models need to account for both parameter and volatility risks. That is, the risk that the underlying probability distributions are inaccurate and the risk that actual experience fluctuates from expectation based on the distribution.

Given the data as described in Chaps. 4, 5 starts with an exploration of the issues with determining expected values for certain assumptions such as mortality, voluntary lapse, incidence of claim, and severity of claim. Then projection models for long term care insurance are introduced for pricing and then extended for valuation of liabilities. Models of various insurance designs are discussed, the traditional policy that insures against long term care risk only as well as those designs that combine long term care with other forms of insurance. The models illustrated in this chapter suggests that long term care insurance is one of the most challenging insurance products for actuarial professionals.

Chapter 4
Measuring Long-Term Insurance Contract Biometric Risks

Quentin Guibert and Frédéric Planchet

4.1 Introduction

Long Term Care insurance contracts are complex insurance products covering an individual for which pricing and reserving issues are traditionally addressed by the introduction of multi-state models. This type of model allows one to describe the transitions of each insured through different states that correspond to events determining, under the terms of the contract, the respective commitments of the parties. The description of insurance contracts through multi-state models is the subject of several studies in the actuarial literature (cf. [21, 31] or [14]). To implement this approach to pricing or reserving, actuaries need to establish suitable statistical bases.

In most cases, the estimation of such models in Long Term Care insurance is addressed in a Markovian framework, where the transition from one state to another depends on the current state and the age of the insured (see [20, 53, 57, 43] or [24]). However, the use of the Markov assumption is unrealistic since the insured dependent mortality is both a function of age and continuance. Using this simplified framework is generally justified by the use of public and aggregate data that do not follow the length of stay in a state where the health of the insured is degraded. In such a case, the number of transitions and the exposure at risk for each age are available,

Q. Guibert (✉)
CEREMADE, UMR 7534, Université Paris-Dauphine, PSL Research University, 75016 Paris, France
e-mail: q.guibert@hotmail.fr

Q. Guibert · F. Planchet
Univ Lyon - Université Claude Bernard Lyon 1, ISFA, Laboratoire SAF EA2429, 69366 Lyon, France
e-mail: frederic@planchet.net

Q. Guibert · F. Planchet
Prim'Act, 42 avenue de la Grande Armée, 75017 Paris, France

© Springer Nature Switzerland AG 2019
E. Dupourqué et al. (eds.), *Actuarial Aspects of Long Term Care*,
Springer Actuarial, https://doi.org/10.1007/978-3-030-05660-5_4

which allows one to fit transition intensities by means of a Poisson regression model or the usual parametric model, e.g. Weilbull's law. Using institutional or insurance databases with an individual follow up, some recent works alternatively use multi-state semi-Markov models to understand the length of stay in dependence (see [7, 18, 25]).

In Long Term Care insurance, the construction of bases of experience adapted to these models turns out to be a delicate exercise on the one hand because of the low volumes observed until today by insurers and, on the other hand, because of the evolving nature of this phenomenon. The construction process becomes more complex when the number of considered states is high, and in the presence of reversible states. These limitations are pushing practitioners to turn to approaches using survival models for the sake of efficiency and to favor the robustness of the model [23, 47].

In this context, the aim of this chapter is to present common tools used for the construction of multi-state tables in Long Term Care insurance. It addresses in detail corresponding practical aspects, both before and after entry into dependence. For this, we illustrate our approach with the generic framework of a multi-state *illness-death* model described in Fig. 4.1. It is based on three non-reversible states: an "active" or "autonomous" state corresponds to the status of an inforce and non-dependent insured, the state "invalid" or "disabled" is as defined in the contract for claim eligibility, and "death" is the terminal state. This formalism allows us to approach the construction of tables for:

- Active life survival,
- Incidence in disability,
- Disabled survival.

Considering the data formats that the insurer can collect, with this example in mind we illustrate step-by-step an approach to constructing and validating multi-state tables, which reproduces accurately Long Term Care risk features in France. Solutions for estimating more complex multi-state schemes are also discussed. In addition, our approach is designed to consider these risks realistically, in accordance with the current accounting and solvency standards, and not just from a perspective of pure risk rating.

This chapter is organized as follows. Section 4.2 describes the common framework of observation data for an insurer and discusses the implications in terms of estimation techniques. Section 4.3 introduces the main notations and concepts for

Fig. 4.1 Illness-death model

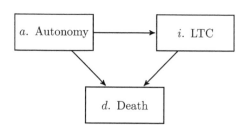

handling Markov and semi-Markov models. Section 4.4 explains how to estimate the transition intensities and probabilities of these models, with a special focus on approaches developed for survival data. Section 4.5 focuses on the practical construction's process of each multi-state table of an illness-death model by using several smoothing and relational techniques. Section 4.6 presents the tools that allow us to control the accuracy and consistency of the results, as well as risk monitoring tools. Section 4.7 deals with the issues raised when taking into account the systemic risk related to the construction's process. It introduces a basic framework that allows one to measure this risk for risk management purposes. Finally, Sect. 4.8 discusses some relevant points and future issues for the estimation of multi-state tables.

4.2 Outline of Data Collection

Data gathered by insurers at issue, upon collection of premiums and claims payments, generally allow individual insured information. Monitoring steps during the coverage period of a contract allows the collection of:

- Insured date of birth,
- Coverage starting date,
- Transition times from one state to another,
- Ending date of observation.

Data is collected during a specific observation period, based on the reliability of information systems and management processes. Figure 4.2 describes the observed information collected by the insurer with the state space of the illness-death model.

For the insured 1, the insurer can observe the entry date t_{10} in the portfolio and the exit due to death at time t_{11}. The time spent in the autonomous state is then precisely measured and no transition to the disability state occurs. The insured 2 experiences the disability state, which triggered the payment of benefits since the date t_{21}. Before that, the insurer collects premiums in the autonomous state between times t_{20} and t_{21}. The sojourn time in the disability state corresponds to the time $t_{22} - t_{21}$.

The history of transition times is the required material for estimating the multi-state models with the techniques introduced in this chapter. This information is collected in continuous time during the coverage period if exact dates are available.

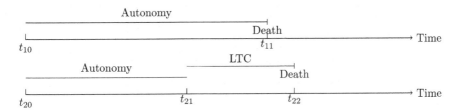

Fig. 4.2 Example of follow-up based on individual data

However, this observation mode can lead to loss of information resulting from censoring and truncation mechanisms [41, 51]. Indeed, the issue date of the original active insured appears in survival analysis as a left truncated event. In addition, the ending date may correspond to the death of the individual or the event of surrender of the contract. The latter case corresponds to right censoring which can be random, for example if the individual performs a nonforfeiture option or terminates the contract, or fixed at the end of the study. In most situations in insurance, these two phenomena can be regarded as independent of the survival in each state. This working hypothesis is usually satisfied and is held in the rest of this chapter. In addition, to simplify the notation and without loss of generality, we only consider the right censoring as a source of missing data. Left truncation can easily be added.

More complex situations can appear in practice, as the manner of observing the data is dependent on the claims management system. Thus, the terms of an insurance contract may affect the available information. For instance, deaths before entry into disability can be poorly known if no benefit is payable for these events. Other policy conditions may increase the missing information,[1] such as:

- a Waiting Period for the subscription period may cause left-truncation for the incidence in dependence, but can induce left-censoring for the duration time in the Disability state,
- an Elimination Period at incidence induces a left-truncation for the continuance.

These characteristics should be considered in the construction of the tables, which is not a major problem, as we shall see in the next section. However, they can be restrictive over time if contractual changes are being considered: for example, the reduction of an Elimination Period may involve a specific approach if past data provide no information on this point. This case is outside the scope of this chapter.

In this chapter, a single irreversible state of disability is considered. Note, however, that specific contractual provisions can rise, in principle, additional states such as:

- the adjustment of premiums based on the level of dependence,
- premium reduction clauses (e.g. Reduced Paid Up, regulatory denial of a rate increase).

Contracts are most often of the periodic premium type, with rate increase clauses varying from one insurer to another, from the insurer being free to evaluate the premium change to the premium being based on a predetermined index subject, for instance, to a maximum annual level of increase triggered by adverse experience. Such provisions have an impact on behaviors in case of lapse, nonforfeiture options, and benefit reductions.

The multi-state scheme can therefore quickly become complex. It should be noted that the constraints of data availability also severely limit the ability to use refined descriptions, with the prospect of not being able to accurately estimate the probabilities of transition. Some policyholders may stop paying premiums and are, as a result,

---

[1]On this point, see Chaps. 1–3 by F. Castaneda and F. Lusson.

beneficiaries with reduced benefits in the event of dependence: a table for this state of "Reduced Paid Up" should sometimes also be built. This topic is not covered in this chapter.

Moreover, factors capturing the heterogeneity of a portfolio such as sex, region, income, professional categories, can be observed, and thus integrated into the construction method.

Establishing a reliable database on a sufficiently long history containing the information above is a prerequisite and requires that the information system and the management of the insurer processes have integrated the needs of actuaries in their design. The rest of this chapter assumes that the required data are available and reliable, which in practice is always a sensitive issue requiring specific controls.

4.3 Notations and Setting

In this section, we introduce the main concepts needed for a semi-Markov model as introduced by [33]. This model represents the current state of an insured, which depends on the previously visited state and the time in this state. The Markov model [34] is seen as a particular case in this presentation. In this context, two representations are generally introduced in the actuarial literature. The first one is based on the joint process with the current state and the duration time in that state, whereas the second considers the jump times and the state between each jump. We refer to [21] for a detailed presentation.

4.3.1 Two Representations of a Semi-markov Model

We introduce a continuous time multi-state stochastic process $(X_t, t \in T)$ càd-làg (right continuous with left limit) taking its values in a space of m states, $S = \{e_1, e_2, \ldots, e_m\}$. $\mathcal{F}_t$ is the observed history arising from this process. To describe it, we introduce *transition probabilities* from a state h to a state j with $s, t \in T$ and $s \le t$

$$p_{hj}(s, t) = \Pr(X_t = j | X_s = h, \mathcal{F}_{t-}).$$

In particular, for an *illness-death* model, the information $\mathcal{F}_{t-}$ stands for the duration within the current state at time s. This duration time is denoted Z_t.

To be general, a semi-Markov model, defined by the process (X_t, Z_t), is introduced for the insured's situation (see [32, 33] or [30], Chap. 1). We focus then on the transition probability

$$p_{hj}(s, t, z, v) = \Pr(X_t = j, Z_t \le v | X_s = h, Z_s = z).$$

If the transition probabilities do not depend on the time spent in a state, we refer to a Markov model [34], which can be described by

$$p_{hj}(s, t) = \Pr(X_t = j | X_s = h).$$

Alternatively, one may focus on the market point process (T, J) where $T_n, n \in \mathbb{N}$, is the nth transition from the state J_n to the state J_{n+1}. The state J_n corresponds to the value of the process X between the times T_n and T_{n+1}. In that case, the market point process is defined from the inhomogeneous semi-Markov kernel [36, 37].

$$
\begin{aligned}
Q_{hj}(t, z) &= \Pr\left(\Delta T_{N(t)+1} \leq z, J_{N(t)+1} = j \middle| J_{N(t)} = h, T_{N(t)} = t\right) \\
&= \Pr\left(J_{N(t)+1} = j \middle| J_{N(t)} = h, T_{N(t)} = t\right) \\
&\quad \Pr\left(\Delta T_{N(t)+1} \leq z \middle| J_{N(t)+1} = j, J_{N(t)} = h, T_{N(t)} = t\right) \\
&= \phi_{hj}(t) K_{hj}(t, z),
\end{aligned}
$$

where $\phi_{hj}(t)$ is the transition probability at time t from the state h to j, $N(t)$ is the number of jumps at time t, and $K_{hj}(t, z)$ represents the duration law for this transition. If the process depends only on the sojourn time, the process is called a homogenous semi-Markov model and the semi-Markov kernel is expressed as

$$Q_{hj}(t) = \Pr(\Delta T_{n+1} \leq z, J_{n+1} = j | J_n = h) = \phi_{hj} K_{hj}(t, z).$$

These two alternatives are not equivalent, but if the process (T, J) is Markovian then the process X is semi-Markov. Hence, for modeling a semi-Markov model, they required different quantities of interest.

4.3.2 Transition Intensities

This section describes the transition intensities for both introduced representations. For the sake of simplicity, all functions are assumed to be derivable.

In the context of a multi-state model X, we naturally introduce transitions intensities

$$\mu_{hj}(t) = \lim_{\Delta t \to 0} \frac{p_{hj}(t, t + \Delta t)}{\Delta t}, \text{ for the Markov case,}$$

or

$$\mu_{hj}(t, z) = \lim_{\Delta t \to 0} \frac{p_{hj}(t, t + \Delta t, z, \infty)}{\Delta t}, \text{ for the semi-Markov case.}$$

Regarding the marked point process (T, J), we introduce the following density

$$k_{hj}(t, z) = \lim_{\Delta z \to 0} \frac{K_{hj}(t, z)}{\Delta z}.$$

As a result, the key quantity of interest for deriving the transition probability is the transition intensity

$$\lambda_{hj}(t, z) = \frac{\phi_{hj}(t)k_{hj}(t, z)}{\sum_{j=1}^{e_m} \phi_{hj}(1 - K_{hj}(t, z))}.$$

For a homogeneous semi-Markov model, the same functions can be defined by removing the time variable t.

4.3.3   The Link Between Transition Intensities And Transition Probabilities

Transition probabilities can be obtained by solving systems of integro-differential equations.

4.3.3.1   The Markov Model

Let's begin with a general Markov model. Denoting by M the transition intensity matrix, the transition probability matrix P can be obtained as the solution of the Kolmogorov differential equations

$$\frac{\partial}{\partial t}P(s, t) = P(s, t)M(t) \text{ and } P(s, s) = I,$$

where I is the identity matrix. For example, the transition probabilities of an illness-death model are

$$p_{ai}(s, t) = \int_s^t p_{aa}(s, \tau)\mu_{ae}(\tau)p_{ee}(\tau, t)d\tau,$$

$$p_{ad}(s, t) = \int_s^t p_{aa}(s, \tau)\mu_{ae}(\tau)p_{ed}(\tau, t)d\tau + \int_s^t p_{aa}(s, \tau)\mu_{ad}(\tau)d\tau,$$

$$p_{ed}(s, t) = \int_s^t p_{ee}(s, \tau)\mu_{ed}(\tau)d\tau,$$

$$p_{aa}(s, t) = 1 - p_{ai}(s, t) - p_{ad}(s, t).$$

This model also satisfies the Chapman–Kolmogorov equations for $0 \leq s \leq u \leq t$

$$P(s, t) = P(s, u)P(u, t).$$

The solution of the Kolmogorov differential equations is unique and could be expressed as a matrix product integral, see [5]

$$P(s, t) = \mathcal{P}_{\tau \in]s,t]} \, (1 + M(\tau)d\tau)$$

If the transition intensity matrix is constant over time, the solution takes the form of a matrix exponential

$$P(s, t) = P(0, t - s) = \exp(M(t - s)),$$

which is interesting in terms of implementation if the matrix M is diagonalizable.

4.3.3.2   The Semi-markov Model

For a semi-Markov model, the integro-differential equation systems satisfied by the transition probabilities are more complex and are solved numerically. The degree of complexity depends upon the size of the state space and the presence of loops in the model.

Where we are interested in the process (X, Z), the transition probability is described by Kolmogorov forward and Kolmogorov backward differential equations for $0 \leq z \leq s \leq t$ and $v \geq 0$

$$\frac{\partial}{\partial t} p_{hj}(s, t, z, v) = \int_0^v \mu_{hh}(t, \tau) p_{hj}(s, t, z, d\tau)$$

$$+ \sum_{l \neq j} \int_0^\infty \mu_{hl}(t, \tau) p_{lh}(s, t, z, d\tau) - \frac{\partial}{\partial v} p_{hj}(s, t, z, v),$$

and

$$\frac{\partial}{\partial s} p_{hj}(s, t, z, v) = - \sum_{l \in S} \mu_{hl}(t, z) p_{lj}(s, t, 0, v) - \frac{\partial}{\partial z} p_{hj}(s, t, z, v).$$

The backward equations are generally simpler to resolve knowing the transition intensities.

For the model based on the process (T, J), the transition probabilities $p_{hj}(s, t)$ are given by Janssen and Manca [37]

$$p_{hj}(s, t) = \delta_{hj} \left(1 - \sum_{l \neq h} Q_{hl}(s, t - s) \right)$$

$$+ \sum_{l \in \mathcal{S}} \sum_{\tau=s+1}^{t} (K_{hl}(s, t) - K_{hl}(s, \tau - 1)) p_{hj}(\tau, t),$$

with δ_{hj} the Kronecker symbol.

4.4 Estimation Methods

The methodologies implemented for the construction of multi-state tables essentially depend on how the data are observed. The key quantities to estimate and the complexity of the underlying multi-state model (number of states, the presence of recursive transitions) should also be considered. In this section, we discuss the usual approaches used for estimating a Markov or a semi-Markov model in the literature. For both types, they consist in first estimating the transition intensities, and then simply plugging these estimates into the transition probabilities formula. Alternative, we briefly introduce direct approaches for the transition probabilities in Sect. 4.4.3.

With the observation scheme introduced in Sect. 4.2 with independent right censoring, we observe in continuous time for a censoring event C

$$(X_t, Z_t, 0 \leq t \leq C) \text{ and } \left(J_{N(t)}, T_{N(t)}, 0 \leq t \leq C \right).$$

For each individual of a sample, the current state and the transition times are observed until the censoring event. For the sake of simplicity, we assume to have a homogeneous sample with independent individuals.

This section is partially taken from [30].

4.4.1 Estimation of Transition Intensities

We present different parametric and non-parametric techniques for estimating transition intensities. An interesting feature of Markov models is that they can be estimated using approaches very close to models for survival analysis. Non-parametric techniques are used for estimating crude transition intensities or crude transition probabilities with the aim of validating the accuracy of fitted quantities or as an input for a relational model.

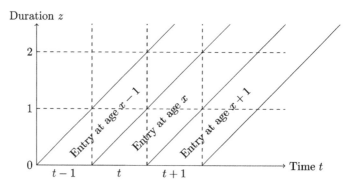

Fig. 4.3 A representation of a Lexis diagram with the time and duration dimensions

4.4.1.1 Crude Intensities

For a Markov model, the integrated intensity for transition from state h to state j

$$M_{hj}(t) = \int_0^t \mu_{hj}(\tau)d\tau,$$

can be estimated non-parametrically with the Nelson–Aalen estimator [5]

$$\hat{M}_{hj}(t) = \sum_{\{k:t_k \le t\}} \frac{d_{hj}(t_k)}{r_h(t_k)},$$

where t_k is the kth jump time of the process, $d_{hj}(t_k)$ is the number of transition from state h to state j at time t_k, and $r_{hj}(t_k)$ is the number of individuals exposed at risk in state h at time t_k. This estimator is quite flexible and verifies asymptotic properties, but requires a sufficient amount of data to be robust. The increment of the sum for this estimator provides an estimate of the transition intensities, which are required to be smoothed thereafter, e.g. using a kernel function [5].

This type of estimator is not available for a general (non-homogeneous) semi-Markov model. However, simplifications appear in the framework of an illness-death model. For transitions from state a, the Markov assumption is generally satisfied, and we can consider transition intensities with a unique time dimension $\mu_{ai}(t)$ and $\mu_{ad}(t)$. Hence, it can be used for estimating the crude intensities $\hat{\mu}_{12}(t)$ and $\hat{\mu}_{13}(t)$.

A second approach, valid for Markov and semi-Markov models, is based on a piecewise constant hypothesis for transition intensities. This consists in defining a discretization grid for the time in one dimension for states where the Markov assumption is satisfied, and a Lexis diagram [40] for states satisfying the semi-Markov assumption, as displayed in Fig. 4.3. Individuals with the same age at entry in disability in the illness-death model appear on the same diagonal.

In general, the transition intensities are assumed to be constant:

- for each integer date $t \in T$ for a Markov model,
- within each rectangle (t, z) of the Lexis diagram on an annual basis for age t and on a monthly basis for the duration time z.

For this parametric model, the maximum likelihood estimator for transition intensity of the model is directly obtained as a ratio of *the number of transitions/the exposure at risk*. This result is known as the Poisson regression model. For instance, the transition intensities for the illness death model are derived by

$$\hat{\mu}_{12}(t) = \frac{N_{12}(t)}{E_1(t)}, \hat{\mu}_{13}(t) = \frac{N_{13}(t)}{E_1(t)} \text{ and } \hat{\mu}_{23}(t, z) = \frac{N_{23}(t, z)}{E_2(t, z)},$$

where:

- $N_{12}(t)$ corresponds to the number of transitions from state 1 to state 2, $N_{13}(t)$ from state 1 to state 3, and $E_1(t)$ corresponds to the risk exposure in state 1 at integer time t,
- $N_{23}(t, z)$ corresponds to the number of dependent deaths occurring after spending z months in the disability state at integer time t, and $E_2(t, z)$ corresponds to the risk exposure in state 2 on the same rectangle (t, z) of the Lexis diagram.

These estimators are asymptotically normal with the following standard errors

$$\sigma_{12}(t) = \frac{\sqrt{N_{12}(t)}}{E_1(t)}, \sigma_{13}(t) = \frac{\sqrt{N_{13}(t)}}{E_1(t)} \text{ and } \sigma_{23}(t, z) = \frac{\sqrt{N_{23}(t, z)}}{E_2(t, z)}.$$

The link between these estimators and regression techniques is detailed in Sect. 4.5. Note that for practical reasons, it is often more convenient to keep a uniform distribution of dependent deaths for each segment of the (x, z) parallelograms, where x represents the entry age in the dependent state [29].

4.4.1.2 Parametric Models

The model can be inferred parametrically by specifying intensity functions. We consider the transition intensity to be a fixed function which depends on a vector of parameters $\boldsymbol{\theta}$

$$\mu_{hj}(t, z) = f(t, z; \boldsymbol{\theta}),$$

with f a parametric function. In one dimension, for a Markov model, one has the usual laws such as the Weibull distribution

$$\mu_{hj}(t) = \lambda_{hj} \alpha_{hj} t^{\alpha_{hj}-1}, \quad \lambda_{hj} > 0, \alpha_{hj} > 0,$$

or the Gompertz distribution

$$\mu_{hj}(t) = b_{hj} c_{hj}^t, \quad b_{hj} > 0, \, c_{hj} > 0,$$

where $\lambda_{hj}, \alpha_{hj}, b_{hj}$ and c_{hj} are positive parameters. For a homogeneous semi-Markov model, a similar approach can be implemented by considering a parametric function for the density of the duration time $k_{hj}(z)$. In this framework, the parameters are computed numerically based on a maximum likelihood estimation.

For an inhomogeneous semi-Markov model where the intensity or the density depends on both times variables, it is not easy to find a specific function which fits well. For feasibility, some authors assume that the density function $k_{hj}(t, z)$ is free of the time t, see e.g. [45]. Another approach, valid for a non-recurrent model, consists in fixing one of the time dimensions as a time-dependent covariate and defining the other one as the main time variable. In this situation, the Cox semi-Markov model (see [17] or [4]), if the time t is treated as a covariate, is defined as

$$\mu_{hj}(t, z) = \mu_{hj,0}(z)\mathbf{exp}(\theta f(t)),$$

with $\mu_{hj,0}(t)$ a baseline intensity function, θ a parameter to estimate and f a fixed function. If the amount of data is enough, it is also possible to stratify on the time t, having already transformed it into a discrete variable beforehand.

4.4.1.3   Taking into Account Heterogeneity

The Cox model [15] is often used to measure the impact of segmentation variables on the risk. This model assumes the effect of covariates is proportional with respect to a baseline function, according to the following specification

$$\mu_{hj}\left(t \mid \mathbf{U}_{hj}; \boldsymbol{\beta}_{hj}\right) = \mu_{hj,0}(t)\mathbf{exp}\left(\boldsymbol{\beta}_{hj}^T \mathbf{U}_{hj}\right),$$

with $\mathbf{U}_{hj} = \left(U_{1,hj}, \ldots, U_{p,hj}\right)^T$ a vector of explanatory variables which varies with transition from state h to state j, and $\boldsymbol{\beta}_{hj} = \left(\beta_{1,hj}, \ldots, \beta_{p,hj}\right)$ a vector of parameters to estimate. For this presentation, we consider only one time dimension, as a semi-Markov model can be handled by a Cox semi-Markov specification. More parsimonious, it is possible to assume that the regression coefficients are the same for several transitions. Time-dependent covariates and time-dependent coefficients are also allowed. In such a model, the baseline intensity function is from the sub-population for which $\mathbf{U}_{hj} = 0$. For a Markov model, the integrated intensity can be estimated as a semi-parametric model by the Breslow estimator [5]

$$\widehat{M}_{hj,0}(t) = \int_0^t \frac{1_{\{L_h(\tau)>0\}}}{S_{hj}^{(0)}(\tau, \boldsymbol{\beta})} dN_{hj}(\tau),$$

where $S_{hj}^{(0)}(t, \boldsymbol{\beta}) = \sum_{i=1}^{n} \exp(\boldsymbol{\beta}_{hj}^T \mathbf{U}_{hj}) L_{h,i}(t)$, $L_{h,i}(t)$ is the exposure in time t in state h for individual i, $L_h(t) = \sum_{i=1}^{n} L_{h,i}(t)$, and $N_{hj}(t)$ is the number of transitions from state h to state j at time t.

Using methods derived from survival analysis, many alternatives can be envisaged, e.g. the additive hazard model of Aalen [44]. Latent heterogeneity can be handled by using the frailty model [35]

$$\mu_{hj}(t \mid \mathbf{U}_{hj}; \boldsymbol{\beta}_{hj}) = \mu_{hj,0}(t) \omega \exp(\boldsymbol{\beta}_{hj}^T \mathbf{U}_{hj}),$$

where ω is a frailty positive random variable.

4.4.2 Estimation of Transition Probabilities Based on Transition Intensities

The relationships presented in Sect. 4.3.3 provide a simple way to estimate the transition probabilities from the estimated intensities. For a general semi-Markov model, transition probabilities are carried out by numerically solving Kolmogorov equations. A presentation of these algorithms is outside the scope of this chapter, and we refer to Janssen and Manca [37] for a detailed presentation. In addition, [8, 25] use numerical integration techniques to compute the transition probabilities of a homogeneous semi-Markov model.

In this section, we focus on situations where an explicit solution is available. For that purpose, we present the usual methods for Markov models and explains how methods from survival analysis can be used for a progressive illness-death model. Indeed, this model responds to many situations in practice and can be easily extended by splitting the cause of entry into the disability state, e.g. the severity of disability or the type of disease (dementia, cancers, ...).

4.4.2.1 Solutions for a Markov Model

The product integral formula

$$P(s, t) = \mathop{\mathcal{P}}_{\tau \in]s,t]} (I + M(\tau) d\tau),$$

admits an explicit solution for a parametric model with exponential distributions. In that case, the transition intensity functions are constant over time, and the transition probability matrix is an exponential matrix. Similarly, the piecewise constant intensities model admits an exponential representation for each interval where the transition intensity is assumed to be constant. The estimated transition probabilities are obtained for all dates $s \leq t$ using the Chapman–Kolmogorov equations. For other parametric models, the product integral formula is resolved numerically.

A general solution is also available using non-parametric estimation techniques. With independent right censoring, the Aalen–Johansen estimator [1] is obtained by plugging the Nelson–Aalen estimator into the product integral formula. As the latter are step-functions, the product-integral formula for $\widehat{\mathbf{P}}(s, t)$ can be expressed as a finite product of matrices

$$\widehat{\mathbf{P}}(s, t) = \prod_{k=1}^{n} \left(\mathbf{I} + \Delta \widehat{\mathbf{M}}(t_k) \right),$$

where $s < t_1 < t_2 < \cdots < t_n \leq t$ are the observed jump times.

This non-parametric formulation has important implications for estimating survival and competing risks models, which are both models of interest for the construction of multi-state tables. For an illness-death model, incidence and active life deaths can indeed be studied through concurrent risk models, whereas continuance is based on a survival model. Observed covariates can be easily integrated by substituting the Breslow estimators in place of the Nelson–Aalen estimators.

4.4.2.2 Survival Models for the Death Probabilities in Disability State

A survival model is a multi-state model with only two states: alive and dead. Within the framework of a homogenous population, the Aalen–Johansen estimator is exactly the Kaplan–Meier estimator [39], allowing one to estimate the survival function in an efficient way. Let consider an i.i.d. sample of failure times $(T_1, \ldots, T_n)$ not completely observable and a second sample, assumed to be independent, composed of right-censoring variables $(C_1, \ldots, C_n)$. The available observations are then $(\widetilde{T}_1, D_1), \ldots, (\widetilde{T}_n, D_n)$ with

$$\widetilde{T}_i = T_i \wedge C \text{ and } D_i = \begin{cases} 1 \text{ if } T_i \leq C_i \\ 0 \text{ if } T_i > C_i \end{cases}.$$

The Kaplan–Meier estimator for the survival function of T is simply calculated by

$$\widehat{S}_{KM}(t) = \prod_{\widetilde{T}_i \leq t} \left(1 - \frac{d_i}{r_i} \right),$$

where d_i represents the number of non-censored exits at time $\widetilde{T}_i$ and r_i the number of individuals exposed to exit risk just before $\widetilde{T}_i$. Conditional probabilities or death rates are then easily derived from

$$\widehat{q}(t) = \frac{\widehat{S}_{KM}(t) - \widehat{S}_{KM}(t + 1)}{\widehat{S}_{KM}(t)}.$$

Using these quantities in discrete time to derive transition probabilities for an illness-death model is relevant. For instance, the conditional probability for monthly deaths is defined as $\hat{q}_{id}(t, z) = \hat{p}_{id}\left(t + z, t + z + \frac{1}{12}, z, \infty\right)$ and the estimated hazard function $\hat{\mu}_{id}(t, z)$ is linked by the simple relation

$$\hat{q}_{id}(t, z) = 1 - e^{-\hat{\mu}_{id}(t,z)},$$

which is a direct consequence of the general relation $q(t) = 1 - \exp\left(-\int_t^{t+1} \mu(\tau)d\tau\right)$ when the hazard function is assumed to be constant over $[t, t + 1]$, see [51] for basic properties of survival models.

As the hazard function should be a two time dimensions function, we can apply the Cox semi-Markov specification or stratify the Kaplan–Meier estimator on age intervals.

When the effect of the observed covariable is modeled using a Cox multiplicative model, the effects on death-rates can simply be interpreted by comparison with a baseline sub-population

$$q(t|U; \beta) = 1 - \left(\frac{S(t + 1|U; \beta)}{S(t||U; \beta)}\right) = 1 - \left(\frac{S_0(t + 1)}{S_0(t)}\right)^{\exp\left(\beta_{hj}^T U_{hj}\right)} = 1 - (1 - q_0(t))^{\exp\left(\beta_{hj}^T U_{hj}\right)}.$$

When $q_0(t)$ is small, we get the approximation $q\left(t|(\beta_{hj}^T U_{hj})\right) \approx q_0(t) \times \exp\left(\beta_{hj}^T U_{hj}\right)$. However, this proportionality is false in the general case and, when the rate $q_0(t)$ becomes non-negligible, it leads to a wrong assumption of proportionality of hazard functions.

4.4.2.3 The Case of Multiple Causes of Exit

In the presence of several causes of exit for state a, the Kaplan–Meier estimator can be applied on each of the causes, by assimilating the other causes to one censoring variable. However, the assumption of independence that the Kaplan–Meier estimator requires cannot be controlled between causes because only those that occur are observable [26]. Hence, one cannot study the structure of dependence between causes [3, 59]. It is appropriate to turn toward the Aalen–Johansen estimator, as a competing risks model can also be viewed as a multi-state model.

Noting

$$\begin{cases} \tilde{T} = T \wedge C \\ D = 1_{\{T \leq C\}} \end{cases},$$

where T corresponds to active life duration and V is the cause of the exit from that state. We estimate $q_{aj}(t) = P(T \leq t + 1, V = j | T > t), j = i, d,$ by

$$\hat{q}_{aj}(t) = \sum_{t < \tilde{T}_i \le t+1} \frac{S_{KM}\left(\tilde{T}_i-\right)}{S_{KM}(t)} \frac{D_i \times 1_{\{V_i=j\}}}{\sum_{k=1}^n 1_{\left\{\tilde{T}_k \ge \tilde{T}_i\right\}}},$$

on the sample base $\left(\tilde{T}_i, D_i\right)_{1 \le i \le n}$ which arises from the $\left(\tilde{T}, D\right)$ vector. In practice this estimator is close to the preceding one in its definition.[2] Hence, the Kaplan–Meier estimator is valued marginally, i.e. by ignoring one of the two exit causes from state a

$$\hat{q}_{aj}^*(t) = \hat{P}\left(T_{aj} \le t + 1 \big| T_{aj} > t\right),$$

where T_{aj} is the unobserved failure time due to cause j. One approach used by practitioners consists in giving priority to one cause over the other, which translates for an order of fixed priority $(j_1, j_2) \in \mathbb{N}^2$ by the formula

$$\begin{cases} \breve{q}_{aj_1}(t) = \hat{q}_{aj_1}^*(t) \\ \breve{q}_{aj_2}(t) = \hat{q}_{aj_2}^*(t) \times \left(1 - \hat{q}_{aj_1}^*(t)\right) \end{cases}.$$

The incidence rate estimator $\breve{q}_{aj}$ is bounded, whatever the order of priority retained, between the two boundaries b_j^- and b_j^+ defined such that for all $t \ge 0$

$$b_j^-(t) = \hat{q}_{aj_1}^*(t) \times \left(1 - \hat{q}_{aj_2}^*(t)\right) \le \breve{q}_j(t) \le \hat{q}_j^*(t) = b_j^+(t).$$

The boundary b_j^+ is sometimes called the partial transition probability (Example IV.4.1 in [5]) and corresponds to the incidence rate estimated if only one exit cause is considered.

4.4.3 Estimation Using Direct Approaches

This section describes an alternative approach to directly estimate probability rates without going through transition intensities. These estimation tools do not require the introduction of a Markov assumption and are built according to a non-parametric approach which appears to be relatively robust. We illustrate this approach for the illness-death model.

Given T_1, the duration in state a, and T_2, the total life span of an insured, and the cumulative distribution functions F_1 and F_2, Meira et al. [46] proposed to express the probabilities $p_{aa}(s, t)$, $p_{ai}(s, t)$ and $p_{ii}(s, t)$ as ratios of Kaplan–Meier integrals

$$p_{aa}(s, t) = \frac{1 - F_1(t)}{1 - F_1(s)}, \quad p_{ai}(s, t) = \frac{E\left[\varphi_{s,t}(T_1, T_2)\right]}{1 - F_1(s)} \text{ and } p_{ii}(s, t) = \frac{E\left[\phi_{s,t}(T_1, T_2)\right]}{E\left[\phi_{s,s}(T_1, T_2)\right]},$$

---

[2] A comparison is proposed by [27].

with $\varphi_{s,t}(u, v) = I(s < u \leq t, v > t)$ and $\phi_{s,t}(u, v) = I(u \leq s, v > t)$. More efficient alternatives are proposed by Uña-Álvarez and Meira-Machado [22]. Titman [58] and Putter and Spitoni [54] also introduce estimators valid for non-Markov models with loops.

As noted by [28], who propose similar adaptations to derive probabilities of interest directly used in reserve calculations, the main merit of this approach is to not require the Markov assumption. Indeed, the semi-Markov specification might introduce bias in practical situations as the state space defined by the contract is not enough to accurately model the dynamic followed by the process. For example, the evolution of progressive diseases, such as the Alzheimer's disease, is influenced by intermediary states which are not observed by the insurer.

4.4.4 Time-Dependent Death Probabilities

In general, the totality of biometric phenomena associated with a Long Term Care contract depend on time; however, taking into account the difficulties arising from building incidence and continuance longitudinal exit probabilities are usually only modeled for Active lives (see Sect. 4.5.1).

In this situation, the estimators of choice are the Kaplan–Meier estimator by generation or the Hoem approximation, also by generation. Such a practical approach is contemplated by Levantesi and Menzietti [43]. This point is discussed in more detail by [29].

4.5 Regression and Smoothing Techniques for Multi-state Tables

This section focuses again on the regulation of crude transition probabilities. In a similar manner as models for mortality tables, the estimation approaches are generally addressed by means of parametric techniques, relational models, see e.g. Brass [11] or regression models, see [56] or [12].

This section focuses on a progressive illness-death model. In particular, the use of estimation methods from survival analysis can be justified. We outline here how to implement this in a framework with which actuaries are familiar. According to the usual actuarial operational approach, table construction can be broken down into three steps:

1. estimate crude transition intensities ($\hat{\mu}_{ai}(t)$, $\hat{\mu}_{ad}(t)$ and $\hat{\mu}_{id}(t, z)$), or estimate crude transition probabilities, ($\hat{p}_{ai}(s, t)$, $\hat{p}_{ad}(s, t)$ and $\hat{p}_{id}(s, t, z, \infty)$),
2. select a set of parametric or regression models,
3. analyze the goodness-of-fit and the performance of the models.

The techniques described in this section can be implemented with the R software [55]. Practical explanations concerning this implementation are available in Chaps. 9 and 10 of [13].

4.5.1   Active Lives Mortality

First, we consider the measurement of the death rates of Active lives covered by a Long Term Care insurance contract. When the risk undertaken by the insurer is long, it can be useful to consider mortality rate trends over time in order not to risk underestimating the number of Active lives, and hence the resulting number of dependants.

The literature on mortality table construction is vast, but we stress that most of the usual models, such as Lee–Carter [42] and its various generalizations (see [6, 10] for examples of reviews), are questionable here. Due to the lack of historical data, the suitability of these models to derive reliable trends by time series techniques cannot be granted, both for direct mortality rates and incidence rates. Then, it is often preferable to favor positioning approaches in relation to prospective reference tables, see e.g. [2] or [50].[3] The contemplated positioning technique can be applied to death rates estimated from aggregate incidence functions or to rates marginally estimated after prioritizing, as described in Sect. 4.4.2.3.

Obtaining reliable data on Active lives is not always easy, especially with brokerage marketing. Depending on the guaranties offered (for instance death and non-forfeiture benefits), it is not always possible to distinguish among Active lives who are no longer paying premiums, those that have a reduced paid up coverage, and those who died. Attention must be paid on this point to avoid the risk of underestimating the Active lives mortality.

In the presence of a small volume of data and when one seeks to position estimated death rates versus a table, a familiar practical example consists in applying a linear regression on the *logits of* observed rates compared to that extracted from a reference table. Following [11], we have the following relationship for the age x and the calendar time t

$$\ln(q(x,t)/(1 - q(x,t))) = a\ln\left(q_{xt}^{ref}/\left(1 - q_{xt}^{ref}\right)\right) + b,$$

with $q(x,t) = q(x,t|a;b)$ the experience mortality rate, where a and b are two regression coefficients to estimate. The estimates can be achieved by minimizing a loss function

$$\varphi(a,b) = \sum_{x,t} E_{x,t} \times (q(x,t|a;b) - q(x,t))^2,$$

---

[3]See also http://www.ressources-actuarielles.net/gtmortalite, which proposes a complete analysis framework for the construction of such tables.

which is a natural generalization of the maximum likelihood criterion in a discrete model. The choice of the loss function is important and for this reason several variants should be tested. In addition, this model allows an easy extrapolation of the logit of observed rates for the old ages.

The choice of the exogenous reference for mortality may have an important effect on the premium calculation and should be carefully analyzed. In fact, reference mortality tables usually used by actuaries include all causes of mortality, reflecting insured mortality or public mortality, whatever their state of health. By contrast, Active lives tables for a Long Term Care insurance contract only measure mortality without disability (without passing through a dependent state as defined in the contract). Indeed, estimated death rates do not consider individuals who have become dependent, and then died within a year. In other words, this amounts to breaking down an insured's life expectancy to an autonomous life expectancy, measured by the Active lives mortality table, and a dependant life expectancy, which ensue from incidence tables and dependence continuance. This observation will be used in Sect. 4.6 to build consistency tests.

4.5.2 Entry into Disability

Incidence rate estimates, from one or several causes, can be achieved as described in Sect. 4.4.2.3. Several limitations from the way data are collected can arise and should be considered during the estimation process. Indeed, Long Term Care insurance contract clauses generally include Elimination Periods and Waiting Periods, the presence of which can lead to overlooking some events. For instance, dependent individuals whose continuance is smaller than the Elimination Period may not be observed if the claim is not recorded, which corresponds to a left-censoring phenomenon. Thus, it is possible to determine incidence rates only for disabled insured with duration greater than the smallest Elimination Period of the portfolio. This is a source of selection bias that the insurer should consider in its premium and reserve settings.

In practice, it is usual to observe an exponential growth of incidence rates until ages 90 or 95. Hence the observed incidence rates, calculated on an annual basis, can be smoothed using the usual parametric laws (e.g. exponential, Weibull, Gompertz–Makeham distributions). Figure 4.4 compares several parametric smoothings of the incidence rates for a contract without Elimination Period, as well as the relational Brass model, estimated from the prescribed French table TH 00-02.

One can typically observe annual increases of incidence rates around 15 to 20% and total dependence incidence rates around age 85 are at the 5 to 6% level. In this example, the mortality rate increases by about 8% per year on average.

Reliable data from an insurance portfolio being generally limited after ages 90 or 95, it is necessary to extend incidence rules beyond such ages. A parametric extension based on an exponential growth can lead rapidly to rates higher than 100% and thus must be limited, as illustrated in Fig. 4.5, based on the adjustment of Fig. 4.4.

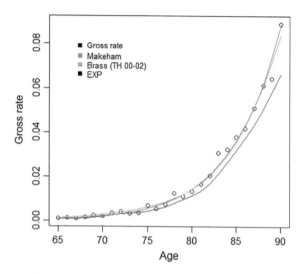

Fig. 4.4 Crude and adjusted incidence rates (total dependence without elimination period)

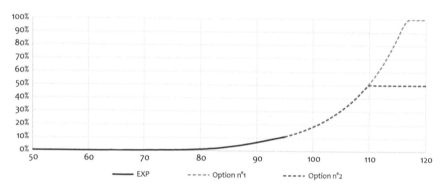

Fig. 4.5 Extrapolation of incidence rates at older ages

Several common incidence rate extrapolation approaches are considered. Option#1 assumes that an individual who reaches high ages will necessarily become dependent, option#2 limits the incidence rate to a level strictly lower than 100% (here 50% for the illustration). The choice can be motived by advice from medical experts and can be subject to sensitivity analysis.

In addition, when data is lacking, a great difficulty with incidence rates comes from the absence of reference tables to allow for positioning techniques. In this context, it is possible to derive these incidence rates from prevalence data that can be found in epidemiology, see e.g. the PAQUID study of [19].

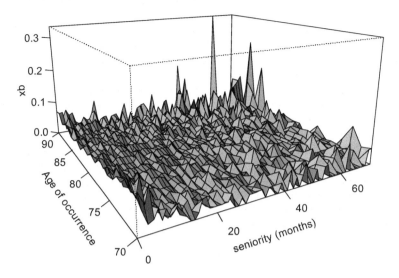

Fig. 4.6 Observed monthly death rates in severe dependancy

4.5.3   Continuance

Although it is certainly possible to consider adjustment to existing mortality tables by worsening them, for instance (see [48]), these techniques do not always lead to satisfying results in terms of goodness-of-fit. The survival law in disability generally requires two time variables: age and duration time. This framework, which can be handled by a semi-Markov multi-state model, can also be considered in practice in several ways:

- using a regression model with one of the two variables as the reference time variable and the second as a covariable, as seen in Sect. 4.4,
- estimating death rates for each integer incidence age by stratifying the sample, and then smoothing them through a parametric or non-parametric approach.

In the following we concentrate on the second approach. Kaplan–Meier estimators of corresponding survival rates or hazard rates can be estimated from individual observations. Figure 4.6 illustrates the shape of observed rates estimated with the Kaplan–Meier approach.

Due to its irregular appearance, it is a matter of smoothing the surface to bring out its trends. This global structure results from the aggregation of subsets of the population presenting mortality levels which are potentially very heterogeneous. As an example, pathologies responsible for incidence have a strong influence on the residual survival of the insured [8]. Figure 4.7 illustrates this heterogeneity in monthly death rates for major types of pathologies in an insurer's portfolio. Cancers in a terminal stage (pathology 3) generally lead to a significant change in the physical status of an individual and are associated to a rapid death after incidence.

The volume of information available for insurers rarely allows one to correctly capture this heterogeneity and notably its interaction with incidence age. Also, this aspect is virtually never considered during table construction, which partly leads to the need for smoothing the surface presented in Fig. 4.6.

Of course, classical smoothing methods can be applied, such as the Whittaker–Henderson or splines methods.[4] However, the complex structure of the mortality surface is more often poorly represented through such approaches. Effectively, this complexity arises from a bias-variance trade-off not only for the overall surface, but also on local areas. Indeed, the usual approaches such as that of Whittaker–Henderson can offset a too regular surface via a smoothness constant but is not able to simultaneously resolve several local issues, leading to some areas being over-smooth and some under-smooth.

Below we present two methods allowing us to overcome this problem, one requiring some statistical tools (the local likelihood approach) and the second very pragmatic, derived from observations of the evolution of mortality rates between the first continuance year and the following ones (the positioning approach).

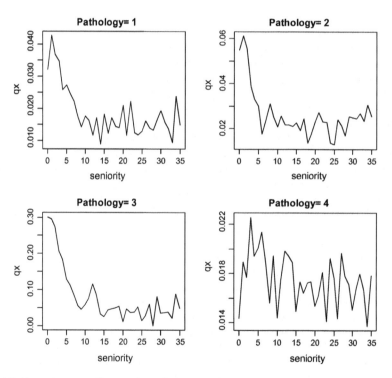

Fig. 4.7 Dependent mortality rates by pathology

---

[4]See, for instance, [51] for a presentation of these techniques or [9] for a very detailed description.

4.5.3.1 Local Likelihood Approaches

The local likelihood approaches presented by Planchet and Tomas [52] provide a framework well adapted to this type of problem. For this one, observe that, for the hazard function estimate, the distribution of non-censored exits for an incidence age x and duration z can be represented by a Poisson distribution

$$D(x, z) \sim Poi(E(x, z) \times \mu_{id}(x, z)),$$

with $E(x, z)$ the risk exposure on the rectangle (x, z). This assimilation is valid under the not very restrictive assumption that the hazard function is constant in each interval $[x, x + 1[\times [z, z + \frac{1}{12}[$.

In the framework of a Poisson Generalized Linear Model (GLM) with the log-linked function, we specify for one observation i

$$\psi_i = \ln(\mu_i) = \beta_0 + \beta_1 z_i + \beta_2 z_i + \beta_3 z_i^2 + \beta_4 z_i v_i + \beta_5 v_i^2 = \mathbf{x}^T \boldsymbol{\beta},$$

with z_i continuance (monthly) and v_i incidence age. By locating this rigid parametric form in the neighborhood of the point x_i, we obtain a local likelihood model. This is equivalent to assuming that the function Ψ is regular, followed by a limited development around x_i, each observation j in the neighborhood of i having weight w_j. The weight is built using the length between the points and a moving average parameter, with

$$w_j = W\left(\frac{d(x_i, x_j)}{h}\right),$$

with $\frac{d(x_i, x_j)}{h} \leq 1$, or zero elsewhere, and W a positive kernel function with mean equal to 1. We can then improve the performance of the model by locally picking the optimal bands to have a so-called adaptive model. Figure 4.8 displays four adjustment examples derived from these techniques. In particular, we can focus on (e) the local likelihood, (f) a P-splines, (g) a confidence interval intersection, and (h) a local adaptive approach.

Examination of the above surfaces points to the importance of the local adaptive approach, which allows one to capture the complexity of the death rates surface, which the P-splines approach does not achieve.

4.5.3.2 Relational Models

In practice, considering the difficulty of extracting a priori a structure allowing an extrapolation beyond age 90 from the data given above, we choose to study the mortality rates beyond one year with a relational model. For that, we consider the excess of mortality with respect to a reference table. These surplus rates are adjusted

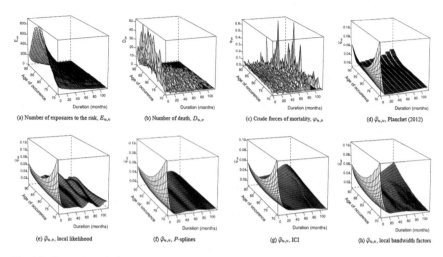

(a) Number of exposures to the risk, $E_{u,v}$ (b) Number of death, $D_{u,v}$ (c) Crude forces of mortality, $\varphi_{u,v}$ (d) $\hat{\varphi}_{u,v}$, Planchet (2012)

(e) $\hat{\varphi}_{u,v}$, local likelihood   (f) $\hat{\varphi}_{u,v}$, P-splines   (g) $\hat{\varphi}_{u,v}$, ICI   (h) $\hat{\varphi}_{u,v}$, local bandwidth factors

Fig. 4.8 Examples of adjustments to dependent mortality surface [52]

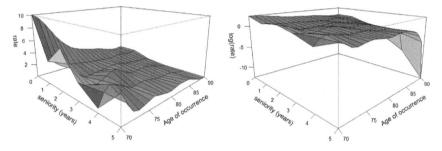

Fig. 4.9 Excess mortality in relation to TD 88-90

from a parametric model, which allows simple extrapolations. Considering its leading role in France for dependent mortality analysis purposes, we retain the table TD 88/90 as a reference. First year death mortality rates are carried out by a specific treatment and a direct adjustment through a parametric model for which the principal objective is to eliminate sampling fluctuations.

Figure 4.9 displays the smoothed excess mortality ratio compared to the mortality rates from table TD 88–90, as well as their logarithm.

It remains to adjust a parametric function to this surface. The rather flat shape[5] of the logarithm of excess mortality rates suggests retaining the function of the form

$$f(x, z) = \exp(a + b \times x + c \times z),$$

---

[5]The convexity found at age 90 beyond the fifth year of continuance is likely a consequence of a lack of data and could be reduced.

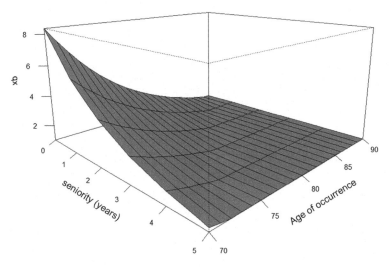

Fig. 4.10 Excess mortality rates adjusted from TD 88–90

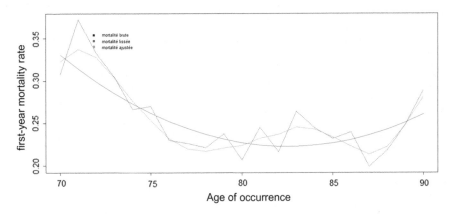

Fig. 4.11 Fitted rates for the first-year excess mortality

where (a, b, c) are parameters to be determined. This calibration is achieved by considering exposure to risk in the loss function and leads to the surface in Fig. 4.10.

We keep adjusting first year mortality rates by the function $f(x) = a + b \times x + c \times x^2$ along with the incidence age, as presented in Fig. 4.11.

This rather robust adjustment then allows us to build a complete table. Of course, this simplest method does not allow us to finely reproduce the structure of the mortality surface, but it greatly facilitates its use. The principal characteristics illustrated of the survival surface in graph (h) of Fig. 4.8 are correctly reproduced. Graph (h) is presented below using a different point of view (Fig. 4.12).

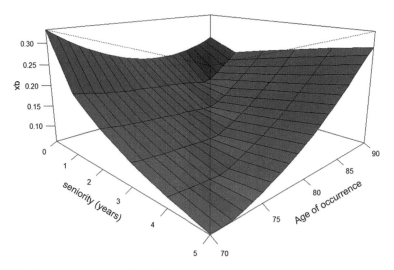

Fig. 4.12 Continuance surface adjusted by the relational approach

With the method proposed above, yearly continuance rates are determined. However, it is necessary, at least for the first year of continuance, to use smaller intervals, usually monthly, to consider the strong convexity of death rates.

In practice, a method often used consists in making, for the first year of continuance, locally linear interpolations of the *logit* of the adjusted rates. We define $q_{id}(x, z)$ as the annual mortality rate at incidence age x and duration z which is derived from the process described above. With age x fixed and for $h = 0, \ldots, 12$ we have the equation

$$\mathbf{logit}\left(q_{id,m}\left(x, z + \frac{h}{12}\right)\right) = \alpha_x + \beta_x \times h,$$

and we seek to estimate (α_x, β_x). We first have

$$1 - q_{id}(x, z) = \prod_{h=0}^{11}\left(1 - q_{id,m}\left(x, z + \frac{h}{12}\right)\right).$$

We introduce as a second constraint the fact that the monthly rates be continuous by assuming that $q_{id,m}(x, z + 1) = 1 - (1 - q_{id}(x, z + 1))^{1/12}$. This equality allows us to find $\alpha_x + \beta_x \times 12 = \mathrm{logit}\left(1 - (1 - q_{id}(x, z + 1))^{1/12}\right)$. We finally solve the equation in β_x

$$(1 - q_{id}(x, z))^{-1} = \prod_{h=0}^{11}\left(1 + \mathbf{exp}\left(\mathbf{logit}\left(1 - (1 - q_{id}(x, z + 1))^{1/12}\right) + \beta_x \times (h - 12)\right)\right).$$

Figure 4.13 displays the smoothed surface that we obtain.

We note the similarity of this outcome with the surface obtained with a local adaptive approach. This approach relying on simple techniques supplies coherent results and hence can be a credible alternative to more sophisticated approaches such as the ones presented in Sect. 4.5.3.1. However, this simplified framework does not allow us to build more refined death rate surfaces, as in illustration (h) of Fig. 4.8, which enables us to better reproduce some local irregularities of dependent mortality and to better capture the interactions between incidence age and duration.

4.6   Consistency Checks

No matter what method is retained to produce the different tables, it is essential to check the consistency of results.

4.6.1   Initial Checks

Standard validation techniques allows to assess the relevance of the outcomes, typically:

- statistical tests, e.g. residual analysis or qualitative comparison of models, see [51] or [50],

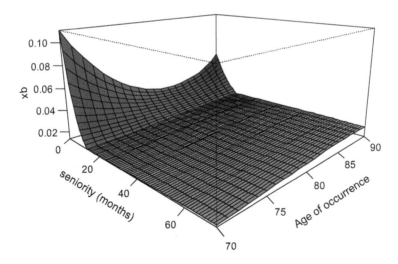

Fig. 4.13 Final continuance rate surface

Table 4.1 Components of life expectancies

Scénario		0	1	2
Biometric Laws	Mortality	(1/3TGH + 2/3TGF)	(1/3TGH + 2/3TGF) – 40%	
	Incidence DT	–	Ref n°1	*EXP*
	Longevity DT	–	Ref n°1	Ref n°1
Life expectancies	EV VV	26.1	24.5	24.1
	EV DT	0.0	1.6	1.7
	TOTAL	**26.1**	**26.1**	**25.8**
Scenario 1 base 100	EV VV	106.6%	100.0%	98.6%
	EV DT	0.0%	100.0%	106.2%
	Total	**100.1%**	**100.0%**	**99.1%**

Note "DT" designates Total Dependence and "VV" Active. "EXP" designates an incidence table built from data, "réf. n°1", a normative table used as a comparison base

- comparison, on the observation range, between exits due to the model and observed exits from different segmentation levels.

Beyond these checks, it is essential to verify the overall consistency of the tables. Hence, by combining Active lives survival (Actives), incidence, and continuance, it is possible to reconstruct the aggregate mortality and therefore to compare the overall life expectancy and the life expectancies in active and disability states. As an example, let's take an Active life age 65 using the following assumptions:

- annual linear distribution calculation (hence, with the classic actuarial calculations, $e_x = \frac{1}{l_x} \sum_{y \geq x} \frac{l_y + l_{y+1}}{2}$),
- Active life population is composed of one third male and two third female,
- calculation date is 12/31/2014,
- mortality table used is an interpolation of mortality rates of tables[6] TGH 05 and TGF 05, reduced by 40%.

Life expectancy for a 65-year-old is then 26.1 years (23.9 years for TGH 05 and 27.4 for TGF 05 before mortality rate interpolation). We use dependence tables corresponding to a severe disability contract without Elimination Period. We compare three scenarios in Table 4.1, according to the choices made as to Active lives mortality rates, incidence rates, and continuance rates.

For this illustrative example, the use of mortality rates based on the TGH/F 05 tables reduced by 40% leads to a life expectancy of 65 years equivalent to that of TGH/F for an insured in Active state. The breakdown of active and continuance years is also shown.

---

[6]The mortality tables used are accessible on http://www.ressources-actuarielles.net/references.

4.6.2   Monitoring Over Time

All the control indicators described so far are purely retrospective and allow us to assess the consistency of tables constructed over the observation period.

Considering that several sources of error affect these tables (e.g. the estimation risk mentioned in Sect. 4.7, future changes in behavior of the insured or in the nature of the risk) it is essential to follow the validity of the different tables over time.

Such monitoring approaches can notably consist in comparing forecasts based on experience tables with observations, for different key quantities: count of monthly incidence, dependent deaths, active lives deaths, etc.

For instance, to predict the incidence counts in the case where an Active life can either die or become dependent, one could use the following approximation of the probability to observe an entry in dependancy between ages x and $x + 1$ knowing that the individual is active at x

$$p \approx q_{ai}(x) \times \left(1 - \frac{q_{ad}(x)q_{ai}(x)}{q_{ai}(x) + q_{ad}(x)} \right).$$

Hence, we see that, when several exit causes are possible, it is not sufficient to simply apply the conditional exit probability to everyone at risk at the beginning of the period, but we should consider the interaction between all exit causes (here death censures incidence).

Once should note, from a statistical point of view, that the construction of tests to detect an abnormal deviation is not simple, due to the progressive arrival of data which update the initial data over time. This question leads to sequential tests, which will not be covered here. The interested reader could consult the review presented on this subject in [16].

4.7   A Simple Model for the Estimate Risk

The different tables so far estimated may contain several errors:

- Due to limited data volume, the model parameters are estimated with some impre-cision,
- The model itself can be incorrect,
- Insured behavior, or the risk, can evolve in time.

Considering these possible errors is indispensable in order not to under-estimate premiums or reserves.

The risk associated to the limited volume of data can be quantified in a rather straightforward way by re-sampling techniques (*bootstrap*), see e.g. the approach proposed in [38]. Intervals and confidence bands can then be designed using two parametric bootstrap techniques:

- a re-sampling of death and incidence rates by age, by duration, and by gender, these observed rates then being smoothed using the model used for calculating the rates,
- a re-sampling of the residuals of the model used for calculating the rates.

Taking into account the two other uncertainties is more complex. The second can be partially considered by testing different models and measuring the impact on premiums and reserves. However, this approach is cumbersome to implement.

Alternatively, [49] proposed a model adding a random effect on the logit of death rates

$$\ln\left(\frac{q_x^\omega}{1 - q_x^\omega}\right) = \ln\left(\frac{q_x}{1 - q_x}\right) + \varepsilon,$$

with ε a zero-mean random variable, e.g. a Gaussian white noise. Likewise, if $\varepsilon = \ln(\omega)$ we have

$$q_x^\omega = \frac{\omega \times \exp(\text{logit}(q_x))}{1 + \omega \times \exp(\text{logit}(q_x))}.$$

The random effect is controlled by its volatility parameter σ. This approach is notably well adapted to continuance rates as this source of randomness of the failure time can be linked to that on the conditional continuance expected value δ, where T denotes the duration time in disability. We measure the uncertainty on this expected value by calculating the relative spread between the expected value and the empirical quantile at the level of the distribution of the expected durations

$$\delta = \frac{VaR_\alpha(E(T|\omega)) - E(E(T|\omega))}{E(E(T\omega))} = \frac{VaR_\alpha(E(T|\omega)) - E(T)}{E(T)}.$$

The volatility parameter can then be estimated by expert judgment. The expert first supplies an evaluation of the uncertainty δ of the conditional continuance expected value, and then deduces the value of σ.

4.8 Conclusion

We present in this chapter several statistical techniques and practical methods allowing us to build experience tables for the Long Term Care risk in a multi-state framework. These approaches can be applied to numerous contracts. Our undertaking involves first estimating the transition intensities and probabilities. Then, we focus on techniques to derive tables, considering Long Term Care-specific risks. We note that the age-continuance structure death rates can be complex, and must be carefully followed. From this point of view, local regression techniques appear as an interesting tool if the available volume of data are sufficient to obtain accurate estimates of the death rates.

Another subject examined in this chapter relates to checking the consistency of the results with respect to some known references. In fact, this type of analysis allows a critical look with respect to an evolving risk which is today still poorly known. However, this type of analysis is not yet sufficient. The average age at issue is now around 63 in France and the average age at incidence is 75 for men and 80 for women. Hence, an insurer wishing to set a level Long Term Care insurance premium (which could be adjusted based on predetermined conditions) must have a realistic view of the Long Term Care risk for the next 20 years. However, Long Term Care is risk extremely sensitive to medical progress. One can cite as an example Alzheimer's disease. This neurodegenerative disease of the cerebral tissue is the cause of great number of Long Term Care incidences. An insurer belatedly taking medical advance will offer a non-competitive premium, exposing itself to an anti-selection risk.

More generally, one cannot determine what the impact of the increase in life expectancy on the incidence age as well as the continuance will be. Indeed, there are several contradictory theories, some forecasting an increase in age at incidence and others an increase in disabled life expectancy. Hence, all these exogenous variables will strongly impact incidence and continuance rates.

Beyond the construction of experience tables reflecting prior insured data, it is thus critical to establish monitoring processes for the evolution of these quantities and the evolution of the underlying heterogeneous structure to better anticipate possible structural evolutions.

References

1. Aalen, O.O., Johansen, S.: An empirical transition matrix for non-homogeneous Markov chains based on censored observations. Scand. J. Stat. **5**, 141–150 (1978).
2. Ahcan, A., Medved, D., Olivieri, A., Pitacco, E.: Forecasting mortality for small populations by mixing mortality data. Insur. Math. Econ. 54, 12–27 (2014). https://doi.org/10.1016/j.insmatheco.2013.10.013
3. Andersen, P.K., Keiding, N.: Interpretability and importance of functionals in competing risks and multistate models. Stat. Med. **31**(11–12), 1074–1088 (2012). https://doi.org/10.1002/sim.4385
4. Andersen, P.K., Perme, M.P.: Inference for outcome probabilities in multi-state Models. Lifetime Data Anal. **14**(4), 405–431 (2008). https://doi.org/10.1007/s10985-008-9097-x
5. Andersen, P.K., Borgan, Ø., Gill, R.D., Keiding N.: Statistical Models Based on Counting Processes, p. 767. Springer, New York Inc., (Springer Series in Statistics) (1993). ISBN: 0-378-97872-0, https://doi.org/10.1002/sim.4780131711
6. Barrieu, P., Bensusan, H., El Karoui, N., et al.: Understanding, modelling and managing longevity risk: key issues and main challenges. Scand. Actuar. J. **2012**, 203–231 (2012). https://doi.org/10.1080/03461238.2010.511034
7. Biessy, G.: Continuous-time semi-markov inference of biometric laws associated with a long-term care insurance portfolio. ASTIN Bull. J. IAA **47**(2), 527–561 (2017). https://doi.org/10.1017/asb.2016.41
8. Biessy G.: A semi-Markov model with pathologies for Long-Term Care Insurance, Working paper (2016). https://doi.org/10.1080/10920277.2014.978025

9. Boor, De: A Practical Guide to Splines. Springer-Verlag, Berlin (1978)
10. Booth, H., Tickle, L.: Mortality modelling and forecasting: a review of methods. Ann. Actuar. Sci. **3**, 3–43 (2008). https://doi.org/10.1017/S1748499500000440
11. Brass, W. (ed.): On the Scale of Mortality, Biological aspects of demography, pp. 69–110 (1972). London, Taylor and Francis
12. CMIR12.: *The Analysis of Permanent Health Insurance Data, Continuous Mortality Investigation Bureau*, The Institute of Actuaries and the Faculty of Actuaries (1991)
13. Charpentier, A. (ed.): Computational Actuarial Science, with R. Chapman and Hall, The R Series (2014)
14. Christiansen, M.C.: Multistate models in health insurance. Adv. Stat. Anal. **96**, 155–186 (2012). https://doi.org/10.1007/s10182-012-0189-2
15. Cox, D.R.: Regression models and life-tables. J. R. Stat. Soc. Ser. B (Methodological) **34**, 187–220 (1972). https://doi.org/10.1007/978-1-4612-4380-9_37
16. Croix J.C., Planchet F., Thérond P.E.: Mortality: a statistical approach to detect model misspecification. Bull. Français d'Actuariat **15**(29), 115–130 (2015)
17. Czado, C., Rudolph, F.: Application of survival analysis methods to long-term care insurance. *Insur. Math. Econ.* **31**(3), 395–413 (2002). https://doi.org/10.1016/S0167-6687(02)00186-5
18. D'Amico, G., Guillen, M., Manca, R.: Full backward non-homogeneous semi-Markov processes for disability insurance models: A Catalunya real data application. *Insur. Math. Econ.* **45**(2) 173–179 (2009). https://doi.org/10.1016/j.insmatheco.2009.05.010
19. Dartigues, J.-F., Gagnon, M., Barberger-Gateau, P., Letenneur, L., Commenges, D., Sauvel, C., Michel, P., Salamon, R.: The PAQUID epidemiological program on brain ageing. Neuroepidemiology **11**, 14–18 (1992). https://doi.org/10.1159/000110955
20. Deléglise, M.P., Hess, C., Nouet, S.: Tarification, provisionnement et pilotage d'un portefeuille Dépendance. Bull. Français d'Actuariat **9**(17), 70–108 (2009)
21. Denuit, M., Robert, C.: Actuariat des assurances de Personnes—Modélisation, tarification et provisionnement, Paris: Economica, p. 405 (Assurance Audit Actuariat) (2007). ISBN: 978-2-7178-5329-2
22. de Uña-Álvarez, J., Meira-Machado, L.: Nonparametric estimation of transition probabilities in the non-Markov illness-death model: a comparative study. Biometrics **71**(2), 364–75 (2015). https://doi.org/10.1111/biom.12288
23. Ferri, S., Olivieri, A.: Technical bases for LTC covers including mortality and disability projections. In: Proceedings of the 31th International ASTIN Colloquium, Porto Cervo, pp. 135–155 (2000)
24. Fong, J.H., Shao, A.W., Sherris, M.: Multistate Actuarial models of functional disability. North Am. Actuar. J. **19**, 41–59 (2015). https://doi.org/10.1080/10920277.2014.978025
25. Fuino M., Wagner J.: Long-term care models and dependence probability tables by acuity level: new empirical evidence from Switzerland. Insur. Math. Econ. **81**, 51–70 (2018). https://doi.org/10.1016/j.insmatheco.2018.05.002
26. Gooley, T.A., Leisenring, W., Crowley, J., Storer, B.E.: Estimation of failure probabilities in the presence of competing risks: new representations of old estimators. Stat. Med. **18**(6), 695–706 (1999)
27. Guibert, Q., Planchet, F.: Construction de lois d'expérience en présence d'évènements concurrents: Application à l'estimation des lois d'incidence d'un contrat dépendance. Bul. Français d'Actuariat **14**(27), 5–28 (2014)
28. Guibert, Q., Planchet, F.: Utilisation des estimateurs de Kaplan-Meier par génération et de Hoem pour la construction de tables de mortalité prospectives. Bull. Français d'Actuariat **17**(33), 5–24 (2017)
29. Guibert, Q., Planchet, F.: Non-parametric inference of transition probabilities based on Aalen-Johansen integral estimators for acyclic multi-state models: Application to LTC insurance. Math. Econ. **82**, 21–36 (2018). https://doi.org/10.1016/j.insmatheco.2018.05.004
30. Guibert Q.: Sur l'utilisation des modèles multi-états pour la mesure et la gestion des risques d'un contrat d'assurance, Ph.D. thesis, Université Lyon 1 (2015)

31. Haberman, S., Pitacco, E.: Actuarial Models for Disability Insurance 1re édn, p. 280. Chapman and Hall/CRC (1998). ISBN: 0-8493-0389-3
32. Helwich, M.: Duration effects and non-smooth semi-Markov models in life insurance, Ph.D. thesis, University of Rostock (2008)
33. Hoem, J.M.: Inhomogeneous semi-Markov processes, select actuarial tables, and duration-dependence in demography, Population Dynamics, 251–296 (1972). https://doi.org/10.1016/B978-1-4832-2868-6.50013-8
34. Hoem J.M.: Markov chain models in life insurance. Blätter der DGVFM. 9(2), 91–107 (1969)
35. Hougaard, P.: Frailty models for survival data. Lifetime Data Anal. 1(3), 255–273 (1995)
36. Janssen, J., de Dominicis, R.: Finite non-homogeneous semi-Markov processes: theoretical and computational aspects. Insur. Math. Econ. 3(3), 157–165 (1984). https://doi.org/10.1016/0167-6687(84)90057-X
37. Janssen, J., et Manca, R.: Applied Semi-Markov Processes. Springer (2006)
38. Kamega A., Planchet F.: Construction de tables de mortalité prospectives sur un groupe restreint: mesure du risque d'estimation. *Bull. Français d'Actuariat* 13(25), 5–33 (2013)
39. Kaplan, E.L., Meier, P.: Nonparametric Estimation from Incomplete Observations. J. Am. Stat. Assoc. 53(282), 457–481 (1958). https://doi.org/10.1080/01621459.1958.10501452
40. Keiding, N.: Statistical inference in the Lexis diagram. Philos. Trans. R. Soc. Lond. A: Math. Phys. Eng. Sci. 332, 487–509 (1990). https://doi.org/10.1098/rsta.1990.0128
41. Klein J.P., Moeschberger M. L. (2003) *Survival Analysis*, Springer, 560 p., ISBN: 0-387-95399-X
42. Lee, R.D., Carter, L.R.: Modeling and forecasting U.S. mortality. J. Am. Stat. Assoc. 87, 659–671 (1992). https://doi.org/10.2307/2290201
43. Levantesi, S., Menzietti, M.: Managing longevity and disability risks in life annuities with long term care. Insur. Math. Econ. 50, 391–401 (2012). https://doi.org/10.1016/j.insmatheco.2012.01.004
44. Martinussen, T., Scheike, T.H.: Dynamic Regression Models for Survival Data. Springer, Statistics for Biology and Health (2006). https://doi.org/10.1198/jasa.2007.s230
45. Mathieu, E., Foucher, Y., Dellamonica, P., Daures, J.P.: Parametric and Non-Homogeneous Semi-Markov Process for HIV Control. Methodol. Comput. Appl. Probab. 9(3), 389–397 (2007). https://doi.org/10.1007/s11009-007-9033-7
46. Meira-Machado, L., de Uña-Álvarez, J., Cadarso-Suárez, C.: Nonparametric estimation of transition probabilities in a non-Markov illness–death model. Lifetime Data Anal. 12(3), 325–344 (2006)
47. Olivieri, A., Pitacco, E.: Facing LTC risks. In: Proceedings of the 32th International ASTIN Colloquium, Washington (2001)
48. Pitacco, E.: Mortality of disabled people. Technical Report (2012). Available at SSRN: http://ssrn.com/abstract=1992319
49. Planchet, F., Tomas, J.: Constructing Entity Specific Mortality Table: Adjustment to a Reference. Eur. Actuar. J. 4(2), 247–279 (2014). https://doi.org/10.1007/s13385-014-0095-y
50. Planchet, F., Tomas J.: Uncertainty on Survival Probabilities and Solvency Capital Requirement: Application to LTC Insurance. Scand. Actuar. J. (2014) https://doi.org/10.1080/03461238.2014.925496
51. Planchet F., Thérond P. E. (2011) Modélisation statistique des phénomènes de durée—Applications actuarielles, *Paris: Economica*, (Assurance Audit Actuariat), ISBN: 2-7178-5234-4
52. Planchet, F., Tomas J.: Multidimensional smoothing by adaptive local kernel-weighted log-likelihood with application to long-term care insurance. Insur. Math. Econ. 52, 573–589 (2013). http://dx.doi.org/10.1016/j.insmatheco.2013.03.009
53. Pritchard, D.J.: Modeling disability in long-term care insurance. North Am. Actuar. J. 10, 48–75 (2006). https://doi.org/10.1080/10920277.2006.10597413
54. Putter H., Spitoni C.: Non-parametric estimation of transition probabilities in non-Markov multi-state models: The landmark Aalen–Johansen estimator. Stat. Methods Med. Res. 27(7), 2081–2092 (2016). https://doi.org/10.1177/0962280216674497

55. R Development Core Team, *R: A Language and Environment for Statistical Computing*. Vienna, Austria, (R Foundation for Statistical Computing) (2017). ISBN: 3-900051-07-0
56. Renshaw, A.E., Haberman. S.: On the graduations associated with a multiple state model for permanent health insurance. Insur. Math. Econ. **17**(1), 1–17 (1995)
57. Rickayzen, B.D., Walsh, D.E.P.: A Multi-State Model of Disability for the United Kingdom: Implications for Future Need for Long-Term Care for the Elderly. Br. Actuar. J. **8**(2), 341–393 (2002)
58. Titman, A.C.: Transition Probability Estimates for Non-Markov Multi-State Models. Biometrics **71**(4), 1034–1041 (2015)
59. Tsiatis, A.: A non-identifiability aspect of the problem of competing risks. Proc. Natl. Acad. Sci. **72**(1), 20–22 (1975)

Chapter 5
Pricing and Reserving in LTC Insurance

Michel Denuit, Nathalie Lucas and Ermanno Pitacco

5.1 Introduction

The LTC insurance policies, which concern millions of individuals, are at present very heterogeneous, using many types of guarantees and many types of benefits underwriting modes. The French market has offered payments of monthly lifetime cash annuities since the beginning. Yet the growing LTC market is currently proposing some indemnity-based products. In brief, benefits in LTC insurance products can be classified into three main categories:

- Predefined benefits

  Benefits of a predefined amount (usually, a lifelong annuity benefit, for example on a monthly or quarterly basis) can be either a fixed-amount benefit, stated in the policy conditions, or a degree-related (or graded) benefit. A graded benefit is a benefit whose amount is graded according to the degree of disability, that is, the severity of the disability itself; the severity must be assessed relying on a scoring system, for example the Activities of Daily Living (ADL) scale. Benefits of a predefined amount can be provided by a stand-alone LTC cover as well as by several types of combined products.

- Reimbursement benefits

  This category includes LTC insurance products which provide expense reimbursement. Two basic types of products can be recognized. Stand-alone LTC cover

M. Denuit · N. Lucas
Université Catholique de Louvain, Place de l'Université, 1,
B-1348 Louvain-la-Neuve, Belgium

E. Pitacco (✉)
Università di Trieste, p.le Europa, 1, 34127 Trieste, Italy
e-mail: ermanno.pitacco@deams.units.it

© Springer Nature Switzerland AG 2019
E. Dupourqué et al. (eds.), *Actuarial Aspects of Long Term Care*,
Springer Actuarial, https://doi.org/10.1007/978-3-030-05660-5_5

provides the (partial) reimbursement of expenses related to LTC needs, in particular nursery, medical expenses, physiotherapy, etc. Usually, there are limitations on eligible expenses. Further, deductibles (in terms of fixed amount or fixed percentage, or a combination of both) as well as limit values are stated in the policy conditions. LTC benefits can also be provided by an LTC cover as a rider to sickness insurance. The resulting product is a lifelong sickness insurance. However, in order to cover LTC needs, eligible expenses are extended with respect to a usual sickness cover, so to include, for example, nursing home expenses. Further, a fixed-amount daily benefit can be paid to cover expenses without documentary evidence.

- Service benefits

The LTC insurance products providing care service benefits usually rely on an agreement between an insurance company and an institution which acts as the care provider. An interesting alternative is given by the Continuing Care Retirement Communities, briefly CCRCs, which have become established in the US. CCRCs offer housing and a range of other services, including long term care. Costs (in particular related to LTC) are usually met by a combination of entrance charge plus periodic fees (that is, upfront premium plus monthly premiums).

In what follows, we focus on LTC products providing predefined benefits.

This chapter aims to present the actuarial calculation techniques for pricing and reserving in LTC insurance products providing predefined benefits, thus disregarding in particular cost reimbursement benefits. The multistate structure is consistent with such predefined benefits but also allows to quantify the time LTC is needed, i.e. the duration of payment of insurance benefits of any type. The model used here is a hierarchical three-state model, as previously introduced in Chapter "Distribution Construction Methods from Long Term Care Experience". We do not consider temporary loss of autonomy but assume that reactivation is not possible: there is thus a single, irreversible state of dependence. The absence of recovery is justified because only people with a severe disability are eligible for LTC benefits, in general.

The suitable statistical base, experience table and estimation techniques used for pricing and reserving are detailed in Chapter "Distribution Construction Methods from Long Term Care Experience". The Semi-Markov framework in the LTC state relies on two variables: age and continuance or occupation time. As explained in Chapter "Distribution Construction Methods from Long Term Care Experience", there is an important heterogeneity in mortality among LTC beneficiaries, according to major types of pathologies inducing the LTC need, but it will not be considered here.

The remainder of this chapter is organized as follows. Section 5.2 describes the multistate modeling for LTC insurance policies. The quantities entering actuarial calculations (transition probabilities and intensities) are defined in Sects. 5.3 and 5.4. The actuarial equivalence principle is applied to LTC insurance pricing in Sect. 5.5. Sections 5.6–5.9 present analytical expressions for premiums related to some specific LTC insurance products. Combined products are also considered. Section 5.10 discusses the reserving process and provides the reader with some analytical expressions. The final Sect. 5.11 concludes this chapter.

5.2 Multistate Modeling

Multistate models provide a convenient representation for life and health insurance contracts, including LTC. Each state represents a particular status for the policyholder. The benefits comprised in the contract are associated to sojourns in, or transitions between states. See, e.g., Chap. 8 in [6] or [9] for an introduction.

In the remainder of this chapter, we consider a three-state model, and the following notation is adopted. Henceforth, x denotes the policyholder's age at policy issue. We assume that there is an ultimate age $\omega \leq \infty$ and we denote by

$$\omega_x = \omega - x$$

the maximal time until death for an individual aged x. The time t measures time since policy issue and thus corresponds to contract seniority.

The policyholder's history is described by the stochastic process $\{X_t, \ t \geq 0\}$ where X_t gives the state occupied by the individual at time t, with $X_t \in \{a,i,d\}$ as shown in Fig. 5.1, where state a stands for "autonomous" or "active", state i stands for "invalid" or "disabled", and state d stands for "dead". The LTC state where benefits are paid thus corresponds to i. Henceforth, only transitions a→i, a→d and i→d are allowed so that the loss of autonomy is assumed to be permanent (no recovery possible).

This non-reversibility greatly simplifies the calculations (as the 3-state process is hierarchical and trajectories can easily be described in terms of just a few random variables) and appears to be reasonable at older ages (at which the LTC need becomes stronger). In case recoveries are possible, calculations can be carried on using the so-called Waters algorithm based on time discretization. We refer the reader to [12] for further details about the algorithm.

As seen in Chapter "Distribution Construction Methods from Long Term Care Experience", the time spent in the LTC state i influences future transitions. This is

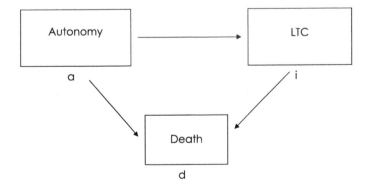

Fig. 5.1 Multistate model for the LTC insurance cover

why we introduce the random variable Z_t, defined as the time spent in the state occupied at time t, i.e.

$$Z_t = \max\{z \leq t \,|\, X_t = X_{t-h} \text{ for all } 0 \leq h \leq z\}.$$

It is assumed that only the current state X_t and the time Z_t spent in the current state influence future transitions so that $\{X_t, t \geq 0\}$ is a Semi-Markov process, i.e. $\{(X_t, Z_t), t \geq 0\}$ is a Markov process. Notice that only the LTC state i requires the Semi-Markov assumption, i.e. probabilities of future transitions from that state also depend on occupation time Z_t.

5.3   Transition Probabilities

We consider a policyholder who is autonomous and aged x at policy issue, i.e. we work conditionally on $X_0 = a$. The following transition probabilities are needed in the three-state LTC model:

$$_u p^{ai}_{x+t} = P[X_{t+u} = i \,|\, X_t = a]$$
$$= \text{probability for an individual in state a at time } t \text{ of being in state i}$$
$$\text{at time } t + u$$

$$_u p^{ad}_{x+t} = P[X_{t+u} = d \,|\, X_t = a]$$
$$= \text{probability for an individual in state a at time } t \text{ of being in state d}$$
$$\text{at time } t + u$$

$$_u p^{id}_{x+t;z} = P[X_{t+u} = d \,|\, X_t = i, Z_t = z]$$
$$= \text{probability for an individual in state i at time } t \text{ since time } t - z$$
$$\text{of being in state d at time } t + u$$

$$_u p^{aa}_{x+t} = P[X_{t+u} = a \,|\, X_t = a]$$
$$= \text{probability for an individual in state a at time } t \text{ of being in state a}$$
$$\text{at time } t + u$$

$$_u p^{ii}_{x+t;z} = P[X_{t+u} = i \,|\, X_t = i, Z_t = z]$$
$$= \text{probability for an individual in state i at time } t \text{ since time } t - z$$
$$\text{of being in state i at time } t + u.$$

The Semi-Markov assumption ensures that these transition probabilities entirely describe the distribution of the stochastic process $\{X_t, t \geq 0\}$ giving the policyholder's individual experience.

By assumption, recovery is not possible. Hence, transition probabilities $_u p_{x+t}^{aa}$ and $_u p_{x+t;z}^{ii}$ are in reality sojourn probabilities, i.e.

$$_u p_{x+t}^{aa} = P[X_{t+h} = a \text{ for all } 0 < h \le u | X_t = a]$$
$$_u p_{x+t;z}^{ii} = P[X_{t+h} = i \text{ for all } 0 < h \le u | X_t = i, Z_t = z].$$

5.4   Transition Intensities

Transition intensities (on transition rates) quantify the instantaneous risk of making a given transition, depending on the state currently occupied. They extend the force of mortality at the heart of life insurance mathematics to more general multistate models describing health insurance products, including LTC products.

From the above transition probabilities, the transition intensities are derived via the following limits:

$$\mu_{x+t}^{ai} = \lim_{h \searrow 0} \frac{_h p_{x+t}^{ai}}{h}$$

$$\mu_{x+t}^{ad} = \lim_{h \searrow 0} \frac{_h p_{x+t}^{ad}}{h}$$

$$\mu_{x+t;z}^{id} = \lim_{h \searrow 0} \frac{_h p_{x+t;z}^{id}}{h}, \qquad z < t.$$

As state a remains Markovian, the transition intensities from that state do not depend on the time spent there, but only on attained age $x + t$. In contrast, there is an effect of the duration of stay in state i so that transition intensities from i depend on both attained age $x + t$ and time z spent in the LTC state.

Transition rates are often assumed to be piecewise constant. This assumption greatly eases the actuarial calculations. There are essentially two approaches to make the Semi-Markov transition intensities $(y, z) \mapsto \mu_{y;z}^{id}$ piecewise constant:

- either transitions intensities vary at integer ages and by sojourn duration in the LTC state, i.e. for every integer y and z,

$$\mu_{y+\xi;z+s}^{id} = \mu_{y;z}^{id} \text{ for all } 0 \le \xi < 1 \text{ and } 0 \le s < 1.$$

Of course, a finer grid can be used (this is often useful for the LTC state, where death rates vary rapidly during the first year after the loss of autonomy);
- or specific, piecewise constant transition rates apply according to the age at entry in the LTC state, i.e.

$$\mu_{y+\xi;z}^{id} = \tilde{\mu}(y + \lfloor \xi - z \rfloor, \lfloor z \rfloor)$$

for some given function $\tilde{\mu}$ defined on $\mathbb{N}^2$, where $\lfloor \cdot \rfloor$ denotes the integer part. The arguments of $\tilde{\mu}(\cdot, \cdot)$ are age at loss of autonomy and time spent in the LTC state, respectively.

The second approximation is very convenient for computations. Rates are displayed in a matrix: the age (last birthday) at loss of autonomy is the first dimension while the time since occurrence is the second dimension (the sum of these two values giving the attained age). We retain the second approximation in this chapter.

Transition intensities are displayed in Figs. 5.2, 5.3 and 5.4. They correspond to values in line with observations made on the French LTC market. We see that μ_y^{ai} and μ_y^{ad} exponentially increase with age y. Considering the death rate in the LTC state, notice that age in the graph corresponds to the age at entry in state i so that individuals are subject to death rates $\mu_{y+z:z}^{id} = \tilde{\mu}(y, z)$ if they lost autonomy at age y. We can see in Fig. 5.4 that mortality is particularly high just after the loss of autonomy (i.e. for small values of z) and then decreases once the individual survived the first years spent in the LTC state.

We also define the exit rate from state a as

$$\mu_y^{a\bullet} = \mu_y^{ai} + \mu_y^{ad}.$$

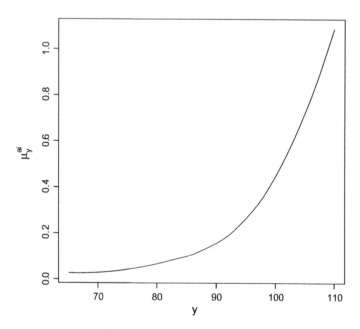

Fig. 5.2 Transition rate $y \mapsto \mu_y^{ai}$

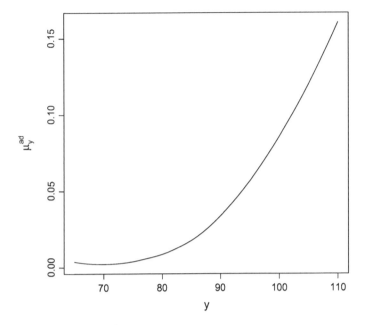

Fig. 5.3 Transition rate $y \mapsto \mu_y^{ad}$

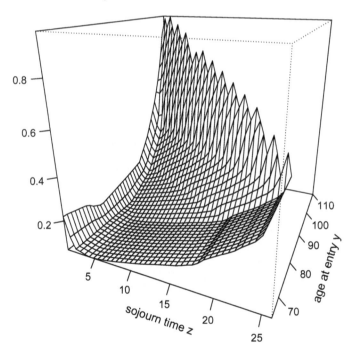

Fig. 5.4 Function $(y, z) \mapsto \tilde{\mu}(y, z)$ defining death rates in the LTC state i

Clearly, the exit rate is also piecewise constant when μ_y^{ai} and μ_y^{ad} both exhibit this feature. The set of transition intensities form a table analogous to a life table in life insurance, allowing the actuary to assign a probability to every event in relation to the LTC cover.

5.5   The Equivalence Principle

This principle used to compute life insurance premiums extends to all health insurance products. It states that at policy issue, the expected present value of the benefits paid to the policyholder is equal to the expected present value of the premiums paid to the insurer. The discount factor $v(s, t)$ is the present value at time s of a unit payment made at time t, $s < t$, with $v(s, s) = 1$. In the numerical illustrations, we assume that the technical interest rate is constant over time, i.e.

$$v(s, t) = \exp\left(-\delta(t - s)\right)$$

for some $\delta > 0$.

The benefits included in LTC policies are as follows:

b_{i} rate of time-continuous benefits paid in the LTC state i;
b_{a} rate of time-continuous benefits paid in the autonomy state a;
c_{ad} benefit paid in case of death if the policyholder occupies the autonomy state a;
c_{id} benefit paid in case of death if the policyholder occupies the LTC state i;
c_{ai} benefit paid when the policyholder enters the LTC state i.

For premiums, we denote by:

π_{i} rate of time-continuous premiums paid in the LTC state i;
π_{a} rate of time-continuous premiums paid in the autonomy state a.

All these quantities may be functions of time, i.e. $\pi_{\mathrm{a}} = \pi_{\mathrm{a}}(t)$ for instance. This allows the actuary to account for periods with no premiums due, for instance. Benefits and premiums in the LTC state may be functions of time and duration of stay in state i, i.e. $b_{\mathrm{i}} = b_{\mathrm{i}}(t, z)$ for instance, because of specific policy conditions.

In general $\pi_{\mathrm{i}} = 0$ but we keep here the possibility of charging premiums in the LTC state, for the sake of completeness. Clearly, in case premiums are charged while benefits are paid, the actuary can always reduce the benefits accordingly so that we assume that the identity

$$b_{\mathrm{i}}(t, z)\pi_{\mathrm{i}}(t, z) = b_{\mathrm{a}}(t)\pi_{\mathrm{a}}(t) = 0$$

holds for all t.

The equivalence principle then states that the expected present value of the premiums paid by the policyholder

$$\Pi = \int_0^{\omega_x} {}_t p_x^{aa} \pi_a(t) v(0, t) dt + \int_0^{\omega_x} {}_t p_x^{aa} \mu_{x+t}^{ai} \left(\int_0^{\omega_x - t} {}_z p_{x+t;0}^{ii} \pi_i(t + z, z) v(0, t + z) dz \right) dt$$

matches the expected present value of the benefits included in the contract

$$B = \int_0^{\omega_x} {}_t p_x^{aa} b_a(t) v(0, t) dt$$

$$+ \int_0^{\omega_x} {}_t p_x^{aa} \mu_{x+t}^{ai} \left(\int_0^{\omega_x - t} {}_z p_{x+t;0}^{ii} b_i(t + z, z) v(0, t + z) dz \right) dt$$

$$+ \int_0^{\omega_x} v(0, t) {}_t p_x^{aa} \mu_{x+t}^{ai} c_{ai}(t) dt + \int_0^{\omega_x} v(0, t) {}_t p_x^{aa} \mu_{x+t}^{ad} c_{ad}(t) dt$$

$$+ \int_0^{\omega_x} {}_t p_x^{aa} \mu_{x+t}^{ai} \left(\int_0^{\omega_x - t} {}_z p_{x+t;0}^{ii} \mu_{x+t+z;z}^{id} c_{id}(t + z, z) v(0, t + z) dz \right) dt,$$

that is, the equality

$$\Pi = B$$

has to hold at policy issue. To make the age x at policy issue visible, we sometimes write Π_x for the single premium $\Pi = B$. The premium rates $\pi_a(\cdot)$ and $\pi_i(\cdot)$ are then set in such a way that the equivalence principle is fulfilled.

5.6 Generalized Annuities

Henceforth, several generalized annuity values will be useful, so we shall give them specific notations. More precisely, we consider actuarial values (i.e. expected present values) of the following time-continuous annuities:

$$\bar{a}_{x+t}^{aa} = \int_0^{\omega_x - t} {}_s p_{x+t}^{aa} v(t, t + s) ds$$

$$\bar{a}_{x+t}^{ai} = \int_0^{\omega_x - t} {}_s p_{x+t}^{ai} v(t, t + s) ds$$

$$\bar{a}_{x+t;z}^{ii} = \int_0^{\omega_x - t} {}_s p_{x+t;z}^{ii} v(t, t + s) ds.$$

In case of temporary annuities, with payments limited to n years, the symbol "$; n\rceil$" is added after age $x + t$, like in

$$\bar{a}_{x+t;n\rceil}^{aa} = \int_0^n {}_s p_{x+t}^{aa} v(t, t + s) ds.$$

Often, policy conditions specify a constant rate of premium payable as long as the insured is in state a. The single premium Π is then easily converted into the constant rate of premium π_a payable continuously in state a:

$$\pi_a = \frac{\Pi}{\overline{a}_x^{aa}}$$

if premium payment is lifelong, or

$$\pi_a = \frac{\Pi}{\overline{a}_{x;\overline{n}|}^{aa}}$$

if premium payment is temporary, limited to n years.

When the transition intensities are piecewise constant, these annuity values can be calculated explicitly because the integrals admit analytical solutions. The idea is to proceed as follows. For integer age x,

$$\overline{a}_x^{aa} = \int_0^{\omega_x} {}_t p_x^{aa} v(0, t) dt$$

$$= \int_0^{\omega_x} \exp\left(-\int_0^t \mu_{x+s}^{a\bullet} ds\right) \exp(-\delta t) dt$$

$$= \int_0^1 \exp(-t\mu_x^{a\bullet} - t\delta) dt + \exp(-\mu_x^{a\bullet}) \int_1^2 \exp(-(t-1)\mu_{x+1}^{a\bullet} - t\delta) dt$$

$$+ \exp(-\mu_x^{a\bullet} - \mu_{x+1}^{a\bullet}) \int_2^3 \exp(-(t-2)\mu_{x+2}^{a\bullet} - t\delta) dt + \dots$$

Each of these integrals now admits an analytical expression so that we obtain

$$\overline{a}_x^{aa} = \frac{1 - \exp\left(-\delta - \mu_x^{a\bullet}\right)}{\delta + \mu_x^{a\bullet}}$$

$$+ \sum_{j=1}^{\omega_x-1} \exp\left(-\sum_{k=0}^{j-1} \mu_{x+k}^{a\bullet} - j\delta\right) \frac{1 - \exp\left(-\delta - \mu_{x+j}^{a\bullet}\right)}{\delta + \mu_{x+j}^{a\bullet}}.$$

Proceeding in a similar way, we get

$$\overline{a}_{x;0}^{ii} = \int_0^{\omega_x} {}_t p_{x;0}^{ii} v(0, t) dt$$

$$= \frac{1 - \exp\left(-\delta - \widetilde{\mu}(x, 0)\right)}{\delta + \widetilde{\mu}(x, 0)}$$

$$+ \sum_{j=1}^{\omega_x-1} \exp\left(-\sum_{k=0}^{j-1} \widetilde{\mu}(x, k) - j\delta\right) \frac{1 - \exp\left(-\delta - \widetilde{\mu}(x, j)\right)}{\delta + \widetilde{\mu}(x, j)}.$$

5.7 Generalized Life Insurances

The actuarial values of a unit lump sum paid in case of a transition, depending on the initial state, are given by

$$\overline{A}_{x+t}^{a;a\to i} = \int_0^{\omega_x-t} v(t,t+s)\, {_sp_{x+t}^{aa}}\, \mu_{x+t+s}^{ai}\, ds$$

$$\overline{A}_{x+t}^{a;a\to d} = \int_0^{\omega_x-t} v(t,t+s)\, {_sp_{x+t}^{aa}}\, \mu_{x+t+s}^{ad}\, ds$$

$$\overline{A}_{x+t}^{a;i\to d} = \int_0^{\omega_x-t} {_sp_{x+t}^{aa}}\, \mu_{x+t+s}^{ai} \left(\int_0^{\omega_x-t-s} {_zp_{x+t+s;0}^{ii}}\, \mu_{x+t+s+z;z}^{id}\, v(t,t+s+z) dz \right) ds$$

$$\overline{A}_{x+t;z}^{i;i\to d} = \int_0^{\omega_x-t} v(t,t+s)\, {_sp_{x+t;z}^{ii}}\, \mu_{x+t+s;z+s}^{id}\, ds.$$

In case of temporary benefits limited to n years, the symbol "; $n\rceil$" is added after age $x+t$, like

$$\overline{A}_{x+t;n\rceil}^{a;a\to i} = \int_0^n v(t,t+s)\, {_sp_{x+t}^{aa}}\, \mu_{x+t+s}^{ai}\, ds.$$

The transition has to occur within the next n years to get the insurance benefit.

When transition intensities are piecewise constant, we get

$$\begin{aligned}
\overline{A}_x^{a;a\to d} &= \int_0^{\omega_x} v(0,t)\, {_tp_x^{aa}}\, \mu_{x+t}^{ad}\, dt \\
&= \mu_x^{ad} \frac{1 - \exp\left(-\delta - \mu_x^{a\bullet}\right)}{\delta + \mu_x^{a\bullet}} \\
&\quad + \sum_{j=1}^{\omega_x-1} \mu_{x+j}^{ad} \exp\left(-\sum_{k=0}^{j-1} \mu_{x+k}^{a\bullet} - j\delta\right) \frac{1 - \exp\left(-\delta - \mu_{x+j}^{a\bullet}\right)}{\mu_{x+j}^{a\bullet} + \delta}
\end{aligned}$$

with a similar expression for $\overline{A}_x^{a;a\to i}$. For a unit death benefit granted to an individual who just entered the LTC state, we have

$$\begin{aligned}
\overline{A}_{x;0}^{i;i\to d} &= \int_0^{\omega_x} {_zp_{x;0}^{ii}}\, \mu_{x+z;z}^{id}\, v(0,z) dz \\
&= \tilde{\mu}(x,0) \frac{1 - \exp\left(-\delta - \tilde{\mu}(x,0)\right)}{\delta + \tilde{\mu}(x,0)} \\
&\quad + \sum_{j=1}^{\omega_x-1} \tilde{\mu}(x,j) \exp\left(-\sum_{k=0}^{j-1} \tilde{\mu}(x,k) - j\delta\right) \frac{1 - \exp\left(-\delta - \tilde{\mu}(x,j)\right)}{\delta + \tilde{\mu}(x,j)}.
\end{aligned}$$

In case of a death benefit in the LTC state, granted to an autonomous individual, we have

$$
\begin{aligned}
\overline{A}_x^{a;i \to d} &= \int_0^{\omega_x} {}_t p_x^{aa} \mu_{x+t}^{ai} \left(\int_0^{\omega_x - t} {}_z p_{x+t;0}^{ii} \mu_{x+t+z;z}^{id} v(0, t+z) \mathrm{d}z \right) \mathrm{d}t \\
&= \mu_x^{ai} \overline{A}_{x;0}^{i;i \to d} \frac{1 - \exp\left(-\delta - \mu_x^{a\bullet}\right)}{\delta + \mu_x^{a\bullet}} \\
&\quad + \sum_{j=1}^{\omega_x - 1} \mu_{x+j}^{ai} \overline{A}_{x+j;0}^{i;i \to d} \exp\left(-\sum_{k=0}^{j-1} \mu_{x+k}^{a\bullet} - j\delta\right) \frac{1 - \exp\left(-\delta - \mu_{x+j}^{a\bullet}\right)}{\delta + \mu_{x+j}^{a\bullet}}.
\end{aligned}
$$

5.8 Some Specific Conditions

Several policy conditions can be included in LTC insurance products. In this section we only address duration-related conditions, i.e. policy conditions which either define the coverage period or the benefit payment period following the claim, that is, the inception of the LTC need.

5.8.1 The Insured Period

The insured period (or "coverage" period) is the time interval during which the insurance cover operates, in the sense that a benefit is payable only if the claim time belongs to this interval. In principle, the insured period begins at policy issue, and ends at policy termination. In LTC policies, given the purpose of the benefits, it is reasonable to assume a lifelong insured period. However some restrictions to the insured period may follow from specific policy conditions.

5.8.2 The Waiting, or Elimination Period

The waiting period (or "elimination" period) is the period following the policy issue during which the insurance cover is not yet operating for sickness-related claims (loss of autonomy due to an accident is generally covered without limitation, from the beginning of the insured period). Different waiting periods can be applied according to the category of sickness. The waiting period aims at limiting the effects of adverse selection, in particular because of any pre-existing health conditions of the insured. It is worth noticing that, although the term waiting period is widely adopted, this time interval is sometimes called the "probationary" period (for instance in the US), while the term "waiting period" is used synonymously with "deferred" period (see below).

In case there is a transition to state i before the end of the waiting period (of duration w, say), the insurer may pay back the premium charged so far, i.e.

$$c_{ai}(t) = \int_0^t \pi_a(s)ds \text{ for } t < w$$

when nominal amounts are reimbursed. We note that $c_{ai}(t)$ constitutes a counter-insurance benefit.

5.8.3 The Deferred Period

In many policies the benefit is not payable until the LTC need has lasted a certain minimum period called the deferred period. This policy condition has a two-fold purpose:

- to reduce the cost and hence the premium of the LTC insurance product; premium reduction can be particularly significant because of the high mortality immediately following the loss of autonomy;
- to ascertain the permanent character of the disease which implies the LTC need (provided that LTC benefits are only paid in the case of permanent disability, as assumed in our model).

5.9 Premium Formulas for Some LTC Insurance Products

A comprehensive set of LTC insurance products have been described in Chapter "The LTC Risk". Here, we only refer to the following products:

1. the stand-alone LTC cover;
2. the enhanced pension, or life care annuity;
3. a package of LTC and lifetime-related benefits;
4. the whole-life insurance with LTC acceleration benefit;
5. an LTC package combining a whole life insurance product comprising a surrender option in case of loss of autonomy, offsetting the financial impact of the deferred period of the LTC annuity.

Formulae for the single premiums of the above products are provided hereafter.

We note that products 2 to 5 constitute special types of insurance packages, or "combined products". From the insurer's perspective, a combined product may be profitable even if one of its components is not profitable. In addition, a combined product may be less risky if it includes some internal hedging mechanism. We refer the reader e.g. to [10] for several examples.

5.9.1 Stand-Alone LTC Cover

The benefit consists in a time-continuous annuity, continuously paid at a constant rate b_i, while the insured is in state i. In the case of no time restriction (that is, in the base case), the single premium is given by:

$$\Pi = b_i \bar{a}_x^{\mathrm{ai}}.$$

Notice that Π can be alternatively rewritten as

$$\Pi = b_i \int_0^{\omega_x} {}_t p_x^{\mathrm{aa}} \mu_{x+t}^{\mathrm{ai}} v(0,t) \bar{a}_{x+t;0}^{\mathrm{ii}} \, dt,$$

which makes explicit the time t of entry into the LTC state. This second formula is useful when policy conditions state some duration-related restrictions, as shown next.

In case policy conditions specify a waiting (or elimination) period w, the single premium becomes

$$\Pi = b_i \int_w^{\omega_x} {}_t p_x^{\mathrm{aa}} \mu_{x+t}^{\mathrm{ai}} v(0,t) \bar{a}_{x+t;0}^{\mathrm{ii}} \, dt.$$

In case policy conditions specify a deferred period d, we get

$$\Pi = b_i \int_0^{\omega_x} {}_t p_x^{\mathrm{aa}} \mu_{x+t}^{\mathrm{ai}} {}_d p_{x+t;0}^{\mathrm{ii}} v(0, t+d) \bar{a}_{x+t+d;d}^{\mathrm{ii}} \, dt.$$

Finally, in case policy conditions specify both a waiting period w and a deferred period d, the single premium is given by

$$\Pi = b_i \int_w^{\omega_x} {}_t p_x^{\mathrm{aa}} \mu_{x+t}^{\mathrm{ai}} {}_d p_{x+t;0}^{\mathrm{ii}} v(0, t+d) \bar{a}_{x+t+d;d}^{\mathrm{ii}} \, dt.$$

If transition intensities are piecewise constant then the single premium of a stand-alone LTC cover can be computed as follows:

$$\begin{aligned}
\Pi &= b_i \bar{a}_x^{\mathrm{ai}} \\
&= b_i \int_0^{\omega_x} \exp\left(-\delta t - \int_0^t \mu_{x+s}^{\mathrm{a\bullet}} ds\right) \mu_{x+t}^{\mathrm{ai}} \bar{a}_{x+t;0}^{\mathrm{ii}} \, dt \\
&= b_i \mu_x^{\mathrm{ai}} \bar{a}_{x;0}^{\mathrm{ii}} \frac{1 - \exp\left(-\delta - \mu_x^{\mathrm{a\bullet}}\right)}{\delta + \mu_x^{\mathrm{a\bullet}}} \\
&\quad + b_i \sum_{j=1}^{\omega_x - 1} \mu_{x+j}^{\mathrm{ai}} \bar{a}_{x+j;0}^{\mathrm{ii}} \exp\left(-\sum_{k=0}^{j-1} \mu_{x+k}^{\mathrm{a\bullet}} - j\delta\right) \frac{1 - \exp\left(-\delta - \mu_{x+j}^{\mathrm{a\bullet}}\right)}{\delta + \mu_{x+j}^{\mathrm{a\bullet}}}.
\end{aligned}$$

The values of Π are displayed as a function of age x at policy issue in Fig. 5.5 with $b_i = 12,000$ (i.e. a monthly payment of 1,000). We can see there that the amount of the single premium increases rapidly until age 80, where it stabilizes due to the effect of high mortality (as this product does not include any benefit in case of death).

Let us now introduce a waiting period and a deferred period. In the case of a waiting period of one year (i.e. $w = 1$), the premium becomes

$$\Pi = b_i \int_1^{\omega_x} \exp\left(-\delta t - \int_0^t \mu_{x+s}^{a\bullet} ds\right) \mu_{x+t}^{ai} \bar{a}_{x+t;0}^{ii} dt$$

$$= b_i \sum_{j=1}^{\omega_x - 1} \mu_{x+j}^{ai} \bar{a}_{x+j;0}^{ii} \exp\left(-\sum_{k=0}^{j-1} \mu_{x+k}^{a\bullet} - j\delta\right) \frac{1 - \exp\left(-\delta - \mu_{x+j}^{a\bullet}\right)}{\delta + \mu_{x+j}^{a\bullet}}.$$

Including a deferred period $d \leq 1$, we get

$$\Pi_x = b_i \int_0^{\omega_x} \exp\left(-\delta(t+d) - \int_0^t \mu_{x+s}^{a\bullet} ds\right) \mu_{x+t}^{ai} \exp\left(-\int_0^d \mu_{x+t+z;z}^{id} dz\right) \bar{a}_{x+t+d;d}^{ii} dt$$

$$= b_i \bar{a}_{x+d;d}^{ii} \exp\left(-d(\delta + \tilde{\mu}(x,0))\right) \mu_x^{ai} \frac{1 - \exp\left(-\delta - \mu_x^{a\bullet}\right)}{\delta + \mu_x^{a\bullet}}$$

$$+ b_i \sum_{j=1}^{\omega_x - 1} \bar{a}_{x+d+j;d}^{ii} \mu_{x+j}^{ai} \exp\left(-d(\delta d + \tilde{\mu}(x+j,0)) - \sum_{k=0}^{j-1} \mu_{x+k}^{a\bullet} - j\delta\right)$$

$$\frac{1 - \exp\left(-\delta - \mu_{x+j}^{a\bullet}\right)}{\delta + \mu_{x+j}^{a\bullet}}.$$

The diminishing effect of the inclusion of a waiting period and of a deferred period in policy conditions is illustrated in Fig. 5.6. Whereas the waiting period moderately decreases the amount of the single premium (the reduction nevertheless getting larger as the age at policy issue increases), the deferred period greatly reduces the single premium because of the high mortality just after the loss of autonomy. The impact of varying the deferred period is illustrated in Fig. 5.7. We can see there that the higher the deferred period, the lower the single premium, as expected.

5.9.2   Enhanced Pension, or Life Care Annuity

Enhanced pensions, or life care annuities, are life annuity products in which the LTC benefit is defined in terms of an uplift with respect to the basic pension. Benefits are then defined as follows. The life annuity, that is, the basic pension, is payable continuously at rate b_a in state a. The LTC annuity is payable at rate b_i in state i, with $b_i > b_a$. The integration of life annuity and LTC cover into a single product

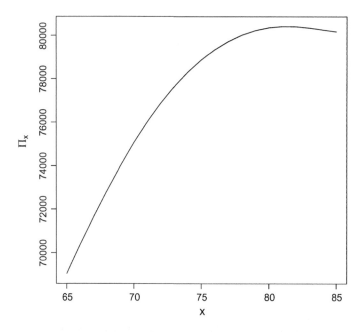

Fig. 5.5 Single premium Π_x as a function of age x at policy issue of a stand-alone LTC cover with $b_i = 12,000$ without limitations

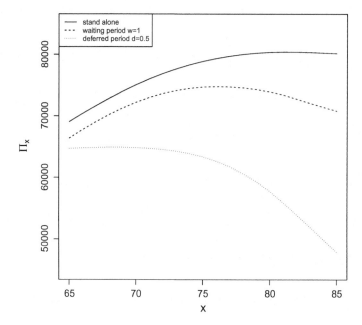

Fig. 5.6 Impact of the waiting period $w = 1$ and of the deferred period $d = 0.5$ on the single premium Π_x of a stand-alone LTC cover, as a function of age x at policy issue with $b_i = 12,000$

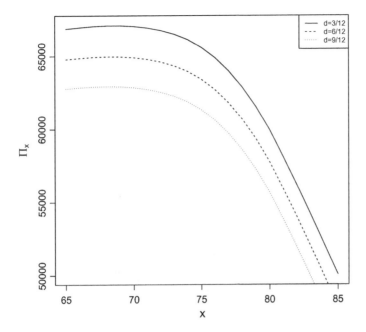

Fig. 5.7 Impact of increasing the deferred period d from 3 to 9 months on the single premium Π_x of a stand-alone LTC cover, as a function of age x at policy issue with $b_i = 12{,}000$

is expected to broaden the population that can be insured, as those individuals with high risk on one component of the package are generally better risks on the other one (see, e.g., [1]).

The single premium is then given by

$$\Pi = b_a \overline{a}_x^{aa} + b_i \overline{a}_x^{ai}$$
$$= b_a \overline{a}_x^a + (b_i - b_a)\overline{a}_x^{ai},$$

where $b_i - b_a$ is the uplift amount, and

$$\overline{a}_x^a = \overline{a}_x^{aa} + \overline{a}_x^{ai}$$

is the price of a regular life annuity sold to an autonomous individual. If $b_i = b_a$, the product comes down to a usual life annuity sold to an autonomous individual. Henceforth, we set $b_i = 2b_a$.

The single premiums of the enhanced pension are displayed in Fig. 5.8 as a function of age at policy issue for $b_a = 12{,}000$. Clearly, Π_x now decreases with age at policy issue. Notice that in this graph, ages at entry up to 85 are considered. If x is greater than 75, an old-age pension (i.e. not a standard one) is involved in the considered package.

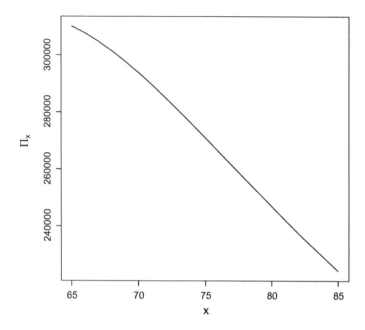

Fig. 5.8 Single premium Π_x of an enhanced pension as a function of age x at policy issue, with $b_i = 2b_a = 24{,}000$

5.9.3 Packages of LTC and Lifetime-Related Benefits

An insurance package can include LTC benefits combined with various lifetime-related benefits, i.e. benefits only depending on the insured's survival and death. We consider the package in which the following benefits are included:

- a deferred life annuity payable at constant rate b_a, paid while the insured is in state a (the deferment period is denoted by n);
- an LTC annuity payable at a rate b_i paid while the insured is in state i;
- death benefits of amount

$$c_{ad} = c_{id} = c_d.$$

The single premium is then given by

$$\Pi = b_a v(0, n)_n p_x^{aa} \bar{a}_{x+n}^{aa} + b_i \bar{a}_x^{ai} + c_d \bar{A}_x^{a},$$

where

$$\bar{A}_x^{a} = \bar{A}_x^{a;a \to d} + \bar{A}_x^{a;i \to d}$$

is the price of a whole life insurance sold to an individual in autonomy and

$$
_nP_x^{aa} = \exp\left(-\int_0^n \mu_{x+t}^{a\bullet}\,dt\right)
$$

$$
= \exp\left(-\sum_{j=0}^{n-1} \mu_{x+j}^{a\bullet}\right).
$$

According to an alternative definition, the death benefits c_{ad} and c_{id} are given by the difference (if positive) between a stated amount c and the amount totally paid as deferred life annuity and/or LTC annuity.

5.9.4 Whole-Life Insurance with LTC Acceleration Benefit

LTC benefits can be added as a rider to a whole-life insurance policy. In particular, the LTC annuity benefit can (totally or partially) be financed by "accelerating" the payment of (part of) the death benefit. Specifically, let c_{ad} be the amount of death benefit for a policyholder in state a. The LTC annuity is payable continuously at rate b_i. The amount of death benefit for an insured in state i dying after having spent a duration z in state i is given by

$$
c_{id}(t,z) = \max\{c_{ad} - b_i z, 0\} = (c_{ad} - b_i z)_+.
$$

In case only the death benefit (or part of the death benefit) is converted into a LTC annuity, the single premium is given by:

$$
\Pi = c_{ad}\overline{A}_x^{a;a\to d} + b_i\overline{a}_{x;\,c_{ad}/b_i\rceil}^{ai}
$$
$$
+ \int_0^{\omega_x} {}_tp_x^{aa}\mu_{x+t}^{ai}\left(\int_0^{c_{ad}/b_i}(c_{ad}-b_i z)\,{}_zp_{x+t;0}^{ii}\mu_{x+t+z;z}^{id}v(0,t+z)dz\right)dt.
$$

In case the LTC annuity is paid until death, with the death benefit decreased accordingly until possible exhaustion, the single premium is given by:

$$
\Pi = c_{ad}\overline{A}_x^{a;a\to d} + b_i\overline{a}_x^{ai}
$$
$$
+ \int_0^{\omega_x} {}_tp_x^{aa}\mu_{x+t}^{ai}\left(\int_0^{c_{ad}/b_i}(c_{ad}-b_i z)\,{}_zp_{x+t;0}^{ii}\mu_{x+t+z;z}^{id}v(0,t+z)dz\right)dt.
$$

Of course, the second arrangement yields a single premium greater than the first one.

Figure 5.9 displays the single premium of this second arrangement. We can see there that the premium increases with age x at policy issue, this increase being steeper for higher amounts of death benefit.

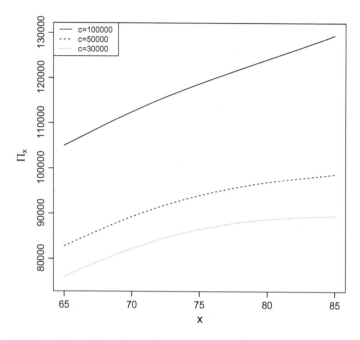

Fig. 5.9 Single premium of the whole-life insurance cover with LTC acceleration benefit, as a function of age x at policy issue, for $b_i = 12{,}000$ and $c_i = c_a = c$

5.9.5 LTC Package with a Whole-Life Insurance Offsetting the Deferred Period

This package consists of a whole-life insurance coverage together with an LTC annuity with monthly payment m, subject to a deferred period. The length of the deferred period depends on policyholder's age x at policy issue; henceforth, we denote it as d_x. The single premium of the package is $m \times d_x$, where the policyholder selects the desired value of m and the insurer's tariff gives d_x corresponding to policyholder's age x. This is also the amount of benefit included in the whole-life insurance cover, which authorizes surrender at entry into the LTC state and thus offsets the financial impact of the deferred period.

Prospects are told that they get their premium $m \times d_x$ back in any case, either at death, at loss of autonomy, or at surrender in state a (but the LTC cover is then automatically cancelled). In case of entry into the LTC state, policyholders can use this amount as benefits during the deferred period (as $m \times d_x$ is precisely the amount needed for the d_x months during which no LTC benefits are paid). Thus, it seems that the LTC cover comes for free, which is particularly attractive from the policyholder's point of view. Let us stress that all benefits are specified in absolute terms (and are not re-evaluated to compensate for inflation) and that the product is sold at relatively young ages (before retirement, in any case).

The product is sold as a combination of a life insurance contract and an LTC cover. Depending on the country, these two products may fall under different lines of business and must then be managed separately. Both products are assumed to have the same technical interest rate δ.

For the whole-life insurance cover, the single premium Π_{wl} is also the amount of benefit paid in case of death in autonomy, in case of loss of autonomy, or in case of surrender in autonomy. The policyholder is allowed to surrender at any time (but this automatically cancels the loss of autonomy cover if the policyholder is in state a). The optimal behavior thus consists in surrendering the whole-life insurance contract at entry in state i so that the surrender value can be used as LTC benefits during the deferred period of length d_x.

According to the policy conditions, the insurer charges expenses proportional to the reserve at rate δ on the whole-life insurance product. These expenses serve as premiums for the LTC cover (the whole package thus requires a sufficiently high interest rate to be effective). This results in a zero interest rate for the whole-life insurance contract, as deduced from Thiele's equation: the calculations can be carried out for the life insurance component as if the technical interest rate was set to zero. For the whole-life insurance component, if the state s denotes surrender (so that we implicitly work here with a 4th state), the equivalence equation

$$\Pi_{wl} = \int_0^{\omega_x} {}_t p_x^{aa}(\mu_{x+t}^{ai} + \mu_{x+t}^{as} + \mu_{x+t}^{ad})\Pi_{wl}dt$$
$$= \Pi_{wl}$$

is obviously valid whatever Π_{wl}. Here, we take $\Pi_{wl} = md_x$, which is the full price of the package. The LTC component is then paid by the expenses charged on the whole-life insurance cover, that is,

$$\pi_a(t) = \delta md_x.$$

This is because the reserve V_{wl} of the whole-life insurance contract is just the expected present value of future benefits (because the contract stipulates a single premium, see the next section for the formal definition of the reserve), i.e.

$$V_{wl}(t) = \int_0^{\omega_x - t} {}_s p_{x+t}^{aa}(\mu_{x+t+s}^{ai} + \mu_{x+t+s}^{as} + \mu_{x+t+s}^{ad})\Pi_{wl}ds$$
$$= \Pi_{wl}$$
$$= md_x.$$

For the loss of autonomy component, as premiums are paid continuously in state a at rate δmd_x, the length of the deferred period is the unique solution of

$$\int_0^{\omega_x} {}_t p_x^{aa} \exp(-\delta t)\delta md\, dt$$

$$= 12m \int_0^{\omega_x} {}_t p_x^{aa} \mu_{x+td/12}^{ai} p_{x+t;0}^{ii} \exp\left(-\delta(t+d/12)\right) \overline{a}_{x+t+d/12;d/12}^{ii} dt. \quad (5.1)$$

The uniqueness of the solution of (5.1) results from the following argument. The left-hand side of the equivalence relation (5.1) increases linearly in d, starting from the origin. The right-hand side decreases in d, starting from the strictly positive value $12m\overline{a}_x^{ai}$. Therefore, by continuity, there must be a unique value fulfilling the

Fig. 5.10 Left-hand side of the equivalence relation (5.1) in the top panel, right-hand side of the equivalence relation (5.1) in the bottom panel, age $x = 65$ at policy issue for a yearly interest rate of 3% or 5%

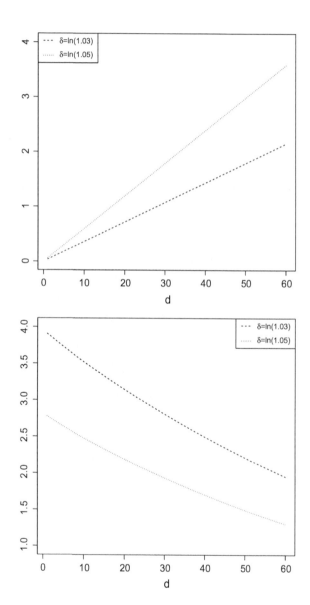

equivalence constraint (5.1). Notice that m cancels on both sides of (5.1) so that there is a unique value of d for each age x at policy issue.

Assume that the policyholder aged 65 has selected the desired value of $m = 1,000$. We then have to find the unique value of d solving the equivalence relation (5.1). The left-hand side of the equivalence relation (5.1) increases linearly in d, starting from the origin, as shown in the top panel of Fig. 5.10. The right-hand side decreases in d, starting from a strictly positive value, as shown in the bottom panel of Fig. 5.10. The unique value of d fulfilling the equivalence constraint (5.1) is shown graphically in Fig. 5.11. This results in $d_{65} = 58$ months for a yearly interest rate of 3% and $d_{65} = 33$ months for a yearly interest rate of 5%. In order to be conservative, d_{65} is chosen as the first integer for which the function is negative. As expected, d_{65} decreases with the interest rate, because a higher interest rate means a higher premium and a higher discounting for the LTC component.

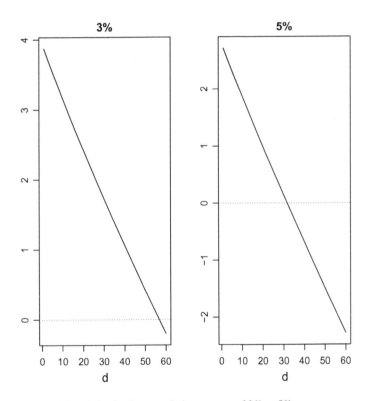

Fig. 5.11 Graphical search for d_{65} for a yearly interest rate of 3% or 5%

5.10   Reserves

5.10.1   Principle

LTC contracts are generally lifelong with a level premium fixed at contract initiation
so that the annual paid amount does not vary during the contract. This constant
premium or level premium depends on the underwriting age x.

As can be seen in Fig. 5.12, the annual risk premium (i.e. annual expected claim
amount for an active individual) is an increasing function of age except at very
advanced ages. Therefore surpluses are constituted in the first part of the contract as
level premiums exceed annual risk premiums. This surplus is called reserve and is
kept aside to meet future needs.

In the case when a single premium is paid at policy issue, a reserve must immediately be kept aside, and then "used" throughout the whole policy duration to meet
the insurer's expected costs.

By status the insurer should have available a reserve at any time. It is defined
prospectively as the actuarial value of future benefits less the actuarial value of
future premiums (and, in the case of a single premium, is simply defined as the
actuarial value of future benefits). Therefore, it depends on the state occupied by the
policyholder at the date of calculation. In LTC insurance, we distinguish a reserve
in state a at time t, henceforth denoted as V_t^a, and a reserve in state i at time t, with
autonomy lost at time $t - z$, henceforth denoted as $V_{t:z}^i$. Of course, there is no need
to define a reserve in state d as the policyholder's death automatically terminates the
contract.

Fig. 5.12 Annual risk
premium by age, stand-alone
LTC cover, and level
premium π_a for a
policyholder aged 65 at
policy issue

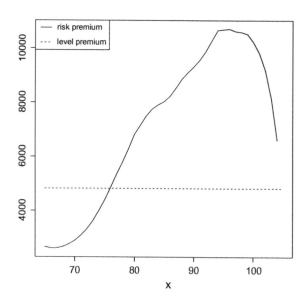

The equivalence principle states that, at policy issue, the expected present value Π of the premiums paid by the policyholder matches the expected present value B of the benefits included in the contract, i.e. $V_0^a = 0$. This equivalence no longer holds in the course of the contract. The reserve is the amount needed to restore financial equilibrium at any time $t > 0$. Specifically, let

$$
\begin{aligned}
\Pi_a(t) = &\int_0^{\omega_x - t} {}_s p_{x+t}^{aa} \pi_a(t+s) v(t, t+s) ds \\
&+ \int_0^{\omega_x - t} {}_s p_{x+t}^{aa} \mu_{x+t+s}^{ai} \\
&\left(\int_0^{\omega_x - t - s} {}_z p_{x+t+s;0}^{ii} \pi_i(t+s+z, z) v(t, t+s+z) dz \right) ds
\end{aligned}
$$

be the expected present value of the future premiums paid by an individual in state a at time t. Similarly, let

$$
\begin{aligned}
B_a(t) = &\int_0^{\omega_x - t} {}_s p_{x+t}^{aa} b_a(t+s) v(t, t+s) ds \\
&+ \int_0^{\omega_x - t} {}_s p_{x+t}^{aa} \mu_{x+t+s}^{ai} \\
&\left(\int_0^{\omega_x - t - s} {}_z p_{x+t+s;0}^{ii} b_i(t+s+z, z) v(t, t+s+z) dz \right) ds \\
&+ \int_0^{\omega_x - t} v(t, t+s) {}_s p_{x+t}^{aa} \mu_{x+t+s}^{ai} c_{ai}(t+s) ds \\
&+ \int_0^{\omega_x - t} v(t, t+s) {}_s p_{x+t}^{aa} \mu_{x+t+s}^{ad} c_{ad}(t+s) ds \\
&+ \int_0^{\omega_x - t} {}_s p_{x+t}^{aa} \mu_{x+t+s}^{ai} \\
&\left(\int_0^{\omega_x - t - s} {}_z p_{x+t+s;0}^{ii} \mu_{x+t+s+z;z}^{id} c_{id}(t+s+z; z) v(t, t+s+z) dz \right) ds
\end{aligned}
$$

be the expected present value of the future benefits for an individual in state a at time t. Clearly, the quantities Π and B entering the equivalence principle are

$$
\Pi = \Pi_a(0) \text{ and } B = B_a(0).
$$

In case the waiting period w for LTC claims is not exhausted yet at reserve calculation, the second integral in $B_a(t)$ does not start from 0 but from $w - t$ (the remaining part of the initial waiting period w specified in policy conditions). The deferred period is accounted for by appropriately defining the rate of benefit $b_i(t+s+z, z)$, setting it to 0 as long as $z \le d$.

The reserve in state a is then defined as

$$V_t^a = B_a(t) - \Pi_a(t)$$

and represents the share of future benefits not covered by future premiums (so that the insurer must have accumulated this amount from past premiums). Adding the reserve to the expected present value of future benefits, the product is in financial equilibrium as

$$\Pi_a(t) + V_t^a = B_a(t)$$

holds for all $t \geq 0$.

Let us now consider a policyholder in state i at time t, who entered the LTC state at time $t - z$. We assume that $t - z > w$ so that the LTC claim is not excluded because of the waiting period. If $t - z \leq w$ then the contract usually terminates and the insurer sometimes pays the total premiums paid so far, i.e. c_{ai} accumulates all the premiums paid in state a, until the transition to state i during the waiting period. For $t - z > w$, the reserve is given by

$$V_{t;z}^i = B_i(t; z) - \Pi_i(t; z),$$

where

$$\Pi_i(t; z) = \int_0^{\omega_x - t} {}_s p_{x+t;z}^{ii} \pi_i(t + s, z + s) v(t, t + s) ds$$

and

$$B_i(t; z) = \int_0^{\omega_x - t} {}_s p_{x+t;z}^{ii} b_i(t + s, z + s) v(t, t + s) ds$$
$$+ \int_0^{\omega_x - t} v(t, t + s) {}_s p_{x+t;z}^{ii} \mu_{x+t+s;z+s}^{id} c_{id}(t + s, z + s) ds.$$

In case policy conditions specify a deferred period d, the latter formula for $B_i(t; z)$ is valid as long as $z > d$. For $z < d$, the policyholder must first spend an extra time $d - z$ in the LTC state before benefits start to be paid. The first integral appearing in $B_i(t; z)$ then becomes

$${}_{d-z} p_{x+t;z}^{ii} \int_0^{\omega_x - d} {}_s p_{x+t+d-z;d}^{ii} b_i(t + d - z + s, d + s) v(t, t + d - z + s) ds.$$

5.10.2 Reserve Formulas for Some LTC Insurance Products

In this section, we assume that the premium is paid continuously, at constant rate π_a, as long as the policyholder stays in state a.

5.10.2.1 Stand-Alone LTC Cover

The reserve at time t for an autonomous individual is equal to

$$V_t^a = b_i \bar{a}_{x+t}^{ai} - \pi_a \bar{a}_{x+t}^{aa}$$

whereas the reserve for an individual in the LTC state at that time, who lost autonomy at time $t - z$, is equal to

$$V_{t;z}^i = b_i \bar{a}_{x+t;z}^{ii}.$$

The reserve for an autonomous individual V_t^a is represented in Fig. 5.13 as a function of t, for an initial age $x = 65$. The amount of reserve increases until the age of 100, before falling to 0 due to the high mortality risk.

Let us now examine the reserve $V_{t;z}^i$ in the LTC state, as a function of z. Figure 5.14 displays the curve $z \mapsto V_{15;z}^i$ for a policyholder aged 65 at policy issue (so that the age at reserve calculation is 80). We see that $z \mapsto V_{15;z}^i$ first increases until the end of the first year spent in LTC (i.e. for $z \leq 1$) and then decreases. This results from the high mortality during the year following the entry into the LTC state.

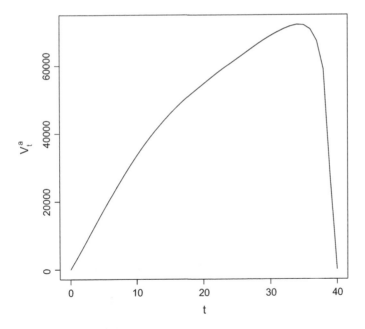

Fig. 5.13 Evolution of the reserve V_t^a for the stand-alone LTC cover, as a function of time t for a policyholder aged 65 at policy issue with $b_i = 12{,}000$

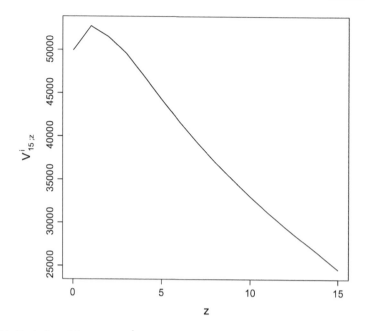

Fig. 5.14 Evolution of the reserve $V^i_{15;z}$ for the stand-alone LTC cover, as a function of the time z spent in the LTC state for a policyholder aged 65 at policy issue

5.10.2.2 Enhanced Pension

The reserve at time t for an autonomous individual is equal to

$$V^a_t = b_a \overline{a}^{aa}_{x+t} + b_i \overline{a}^{ai}_{x+t} - \pi_a \overline{a}^{aa}_{x+t}$$

Considering an individual who is in the LTC state at time t, who entered that state at time $t - z$, the reserve is given by

$$V^i_{t;z} = b_i \overline{a}^{ii}_{x+t;z}.$$

5.10.2.3 Packages of LTC and Lifetime-Related Benefits

When $t < n$, the reserve at time t for an autonomous individual is equal to

$$V^a_t = b_a v(t,n)_{n-t} p^{aa}_{x+t} \overline{a}^{aa}_{x+n} + b_i \overline{a}^{ai}_{x+t} + c_d \left(\overline{A}^{a;a \to d}_{x+t} + \overline{A}^{a;i \to d}_{x+t} \right) - \pi_a \overline{a}^{aa}_{x+t}.$$

The reserve for an individual in the LTC state is equal to

$$V^i_{t;z} = b_i \bar{a}^{ii}_{x+t;z} + c_d \overline{A}^{i;i \to d}_{x+t;z}.$$

5.11   Conclusion

In this chapter, we have explained how premiums and reserves for LTC insurance contracts can be computed. The equivalence principle inherited from life insurance remains at the heart of LTC insurance pricing. It has been applied here in a three-state, Semi-Markov framework. Analytical expressions have been obtained for the premiums and reserves of different LTC products, including combined products. The impact of specific contract conditions on premiums and reserves has been quantified. For more details, we refer the interested reader e.g. to [6, 7] or [9]; French-speaking readers may find it convenient to refer to [4].

Only annuity-type payouts have been considered here (a predetermined benefit level is assumed to be payable periodically to eligible individuals). This kind of product is typically sold in the EU. In some other countries, insurers sometimes reimburse the cost of assistance required because of the loss of autonomy rather than paying a predetermined monthly benefit. The three-state model worked out in this chapter remains useful to forecast the likely start of payments and their duration but an additional model for claim severities is needed for pricing and reserving. In that respect, future trends in claim costs must be taken into account. Because medical inflation is typically hard to forecast, the management of such LTC insurance products becomes even more difficult.

In this chapter, we have assumed that actual interest rates, morbidity and mortality rates remain equal to their assumed values entering actuarial formulas. In practice, these assumptions may be violated, sometimes to a large extent, and should be revised periodically in a dynamic perspective. We refer the reader to [3, 5] for recent contributions on this topic. Very few studies have investigated time trends in transition rates for multistate actuarial models. Renshaw and Haberman [11] identified time trends using separate Poisson GLM regression models for each transition. [2, 8] allowed for possible correlations by means of multivariate versions of mortality projection models (Lee–Carter and Cairns–Blake–Dowd models). The underlying risks (biometric, interest rate, inflation) and risk mitigation strategies are detailed in Chap. 9.

References

1. Brown, J., Warshawsky, M.: The life care annuity: a new empirical examination of an insurance innovation that addresses problems in the markets for life annuities and long-term care insurance. J. Risk Insur. **80**, 677–704 (2013)

2. Christiansen, M., Denuit, M., Lazar, D.: The Solvency II square-root formula for systematic biometric risk. Insur. Math. Econ. **50**, 257–265 (2012)
3. Denuit, M., Dhaene, J., Hanbali, H., Lucas, N., Trufin, J.: Updating mechanism for lifelong insurance contracts subject to medical inflation. Eur. Actuar. J. **7**, 133–163 (2017)
4. Denuit, M., Robert, C.: Actuariat des Assurances de Personnes: Modélisation. Tarification et Provisionnement, Collection Audit-Actuariat-Assurance, Economica, Paris (2007)
5. Dhaene, J., Godecharle, E., Antonio, K., Denuit, M., Hanbali, H.: Lifelong health insurance covers with surrender value: updating mechanisms in the presence of medical inflation. ASTIN Bull. **47**, 803–836 (2017)
6. Dickson, D.C., Hardy, M.R., Waters, H.R.: Actuarial Mathematics for Life Contingent Risks. Cambridge University Press (2013)
7. Haberman, S., Pitacco, E.: Actuarial Models for Disability Insurance. CRC Press (1998)
8. Levantesi, S., Menzietti, M.: Managing longevity and disability risks in life annuities with long term care. Insur. Math. Econ. **50**, 391–401 (2012)
9. Pitacco, E.: Health Insurance: Basic Actuarial Models. Springer International Publishing (2014)
10. Pitacco, E.: Premiums for long-term care insurance packages: sensitivity with respect to biometric assumptions. Risks **4**, 1–22 (2016)
11. Renshaw, A., Haberman, S.: Modelling the recent time trends in UK permanent health insurance recovery, mortality and claim inception transition intensities. Ins. Math. Econ. **27**, 365–396 (2000)
12. Waters, H.R.: The recursive calculation of the moments of the profit on a sickness insurance policy. Insur. Math. Econ. **9**, 101–113 (1990)

Part III
Determination of the Solvency Capital

Bob Yee

Introduction

The formulaic approach to solvency standards is in the process of being replaced by a principle-based method, where risks pertaining to a specific insurance offering from an insuring entity and differences among entities are accounted for. In this regard, France is at a more advanced stage than the United States.

Risks in long term care insurance for the insuring entities include product risks (claims, persistency, investment returns, and expenses) and non-product risks. The latter categories include regulatory risk and management risk. Long term care insurance is highly visible to the regulators due to protectionism for the elderly. Retrospective regulations can adversely impact the insurance entities. The long-tailed nature product risks mean that the long term care insurance portfolio will outlast generations of management. Often, valuable knowledge is lost when a new management team replaces the old.

A number of these risks are controllable or bounded. Certain product risks can be reduced through the use of reinsurance, hedging, and swapping techniques. Management risk can be mitigated by organized knowledge transfer and succession planning. Voluntary lapse rate are low enough that it is no longer a major concern in assumption setting. Future prediction on claims and mortality, however, remain more of an art than a science. In addition to the undefined relationship between claim and mortality as discussed in a prior subchapter, technology and artificial intelligence will have a profound impact on the future delivery of care. Moreover, pandemic lurks in the background despite medical advances.

With the great uncertainty of future insured events, principle-based approach is especially appropriate for long term care insurance. The stochastic methodology described in Chap. 7 would naturally be expanded to evaluate solvency.

Chapter 8 examines current solvency capital requirements for long term care insurance in the context of Solvency 2 framework in France and risk-based capital in the United States. As a risk management tool, the use of reinsurance to diminish product risks is also described. This chapter sets the stage for more extensive discussions on the development of future risk management and solvency regulatory standards.

Chapter 6
Construction of an Economic Balance Sheet and Solvency Capital Requirement Calculation in Solvency 2

Anani Olympio and Camille Gutknecht

6.1 Introduction

The Solvency 2 Directive, adopted in 2009 by the Council of the European Union and the European Parliament, officially became effective on January 1, 2016. Developed to improve the assessment and control of risks, it modifies in-depth the prudential guidelines applicable to insurance and reinsurance companies present in the European Union.

Long Term Care insurance is a particular branch in terms of insurance risk. The borderline between dependence and health problems is porous, since limitations of autonomy often result from current or past health conditions. Like health insurance, private Long Term Care insurance in Europe supplements the benefits or guarantees offered by public plans. In France, "Long Term Care" guarantees marketed by insurers have their own risk assessment system for dependence. Benefit amounts and their payment are independent of benefits paid by social insurance benefits. Benefits offered by the various plans that exist on the market may be of a different type. Some insurance products of group pension plans provide yearly Long Term Care coverage in their contract. During the transition to retirement the contract can continue optionally. The majority of individual or group contracts with optional participation offer lifetime benefits. At the end of a period defined in the contract, premium payment lapse no longer causes termination of the contract, but its reduction. Contracts offer cash benefits, the amount of which is defined and does not provide for the reimbursement of the cost of care, an amount which may evolve over time.

Contract pricing and reserving depend naturally on the type of coverage (lifetime or yearly) and the type of data available when premium rates are calculated. If yearly

A. Olympio (✉) · C. Gutknecht
CNP Assurances, Paris, France
e-mail: aaolympio@yahoo.fr

A. Olympio
University Claude Bernard Lyon 1, Villeurbanne, France

© Springer Nature Switzerland AG 2019

E. Dupourqué et al. (eds.), *Actuarial Aspects of Long Term Care*,
Springer Actuarial, https://doi.org/10.1007/978-3-030-05660-5_6

coverage can be priced and accrued from tracking of the ratio of incurred claims to premium issued, the rate of lifetime benefits will require a significant number of assumptions. Where available data permit, insurers rely on past claim experience to rate their death or disability benefits.

In the case of Long Term Care, the very long-term nature of the proposed guarantees makes the collection of these data spread over time and causes the need to take into account other factors. Indeed, the Long Term Care risk is linked to many social parameters:

- The evolution of family unit, lifestyle and intergenerational bounds;
- The evolution of medical science, daily living, access to care.

Assumptions made in the establishment of premium rates involve the mortality of the insured population, "active state", the likelihood of becoming dependent depending on different levels of dependence called "dependent states", continuance in the dependent states, and transition probabilities between states. The complexity and number of these assumptions mean that insurers cannot rely on the validity of their premium rates. The time dimension takes a very fundamental role in this type of coverage.

Monitoring and control of the coverage is indispensable in a world in perpetual evolution. So, if the life span is lengthened in France for men and women since 2006, the latest studies of DRESS[1] indicate that healthy life expectancy (without disability) has stagnated for 10 years (2006–2016).

The most common diagnostics encountered in retirement homes are neuropsychiatric disorders according to the DREES report 22,[1] in 2007, 36% of EHPAD[2] residents suffered from Alzheimer's.

Medical advances in the treatment of this condition could therefore have the effect of a rapid change in the likelihood of dependence or on the contrary certain lifestyles could result in an increase in the occurrence of certain ailments and would have an impact many years later. For all these reasons, regular monitoring and adjustments of the guarantees and/or benefits of Long Term Care contracts are indispensable throughout the duration of the coverage.

The calculation of the capital requirement as envisaged in Solvency 2 provides for the measurement of the impact on the sufficiency of a company's capital from a set of adverse events over a one-year horizon to a probability corresponding to the 99.5% quantile.

The long-term nature of the coverage offered by a Long Term Care Insurance contract and the consequent significant delay between the collection of the first premium and the payment of claims naturally leads to a high potential cost of the mere fact that the projection framework foreseen in the case of Solvency 2 provides for the maintenance of the unfavorable situation until the policies are matured. Long-term guarantees in the case of Long Term Care insurance thus see their unfolding experience impacted for a long time by adverse actual to expected claims.

In addition, the long-term character leads to a greater dependence of profitability on the interest rate environment. The financial soundness becomes an issue when several years or even a few decades elapse between collection of premiums and benefit payment.

For example, we can highlight the intrinsic leverage effect on lifetime products.
Suppose the portfolio's Active Life Reserve (ALR) is €100 m. It decomposes into:

- *a premium Expected Present Value (EPV) of €800 M*
  and
- *claim EPV of €900 M.*

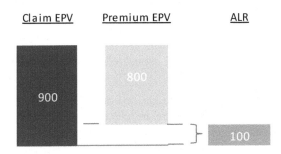

Assuming that the Claim EPV can be written as

$$ClaimEPV \sim \alpha \cdot Benefit^{Annual} \cdot duration^{dependent\ state}$$

If the average duration of the benefit period is 4 years, then a 20% decrease in the mortality of the dependents, all things being equal otherwise, would increase the duration by 0.8 years.

This would represent an estimated cost of around €180 M [(0.8/4) × 900 M], or 1.6 times the central ALR [180/(900 − 800)], using a zero reserve interest rate.

As we have just seen, in the case of Long Term Care, an event occurring today can have very important effects much later. Thus, a 15% longevity increase per year that would be maintained over time has a major impact on the soundness of a Long Term Care insurance product. However, insurers monitor their risk. Thus, once the risk has been identified the insurer has a set of possible measures, for example:

- Some group contracts may be terminated by the insurer.
- If contractual clauses permit, premium can be revised upward to compensate for the deterioration of the risk.
- The value of the benefits paid for reduced contracts or in the event of a premium lapse for current contracts, in accordance with the contractual clauses, may be revised downward.
- Pricing of new contracts may include a prudential margin to ensure a return to the underwriting soundness under the principle of mutualization.

The framework provided by the Solvency 2 standard does not allow the latter mechanism to be taken into account since new cases are not part of the projection framework. Other measures can be modeled and taken into account when calculating projected cash flows if there is a strategy described and validated by the board of directors to account for this type of situation. If this strategy is implemented in calculating the solvency of the company, a regular report of its impact on the results and its occurrence in the projections must be performed (in accordance with article 236 of the Commission Delegated Regulation (EU) 2015/35).

Long Term Care is a newly recognized risk, increasing with age and poorly known to insurers; they generally reinsure using coinsurance for several reasons:

- Risk is not well understood due to lack of market data (relatively new risk) and lack of regulatory tables.
- The insurer's commitment can be very high due to the length of the coverage (long-term) and requires a significant amount of solvency capital.
- The impact of coinsurance on risk capital reduction are relatively easy to take into account.

In terms of reinsurance-induced reduction of solvency capital, there is no ceiling under Solvency 2. So, reinsurance can be fully taken into account to reduce the risk.

However, contractual clauses must not be omitted. Indeed, the insurer must take into account cessation clauses, profit-sharing clauses, and asset transfer provisions in the best estimate reserve valuation.

6.2 Main Principles for the Economic Balance Sheet Calculations

6.2.1 Summary of Solvency 2 Provisions

Solvency 2 is based on three pillars:

- Pillar 1 contains quantitative financial requirements, for both balance sheet and solvency.
- Pillar 2 deals with more qualitative aspects, such as governance and risk management in the broad sense, but also the role of the regulator and the regulatory process.
- Pillar 3 describes the regulatory reports and information to be published.

The calculation of capital requirements, reserves in the sense of Solvency 2 and the contribution of Long Term Care insurance contracts to eligible equity for the purposes of capital requirement will be detailed in the following section.

6.2.1.1 Elements of an Economic Balance Sheet Standard

The economic balance sheet as defined in the Solvency 2 Directive is based on the concept of the best estimate of the value of balance sheet entries. The best estimate

Table 6.1 Statutory balance sheet and prudential balance sheet

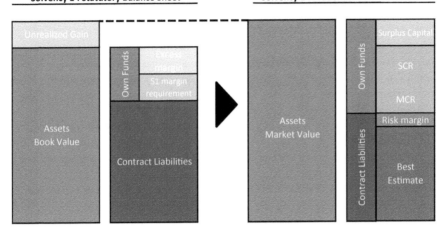

is obtained when the **different entries in the balance sheet are valued at their market value**. When such a market value is not directly available or observable in a reasonably liquid market, the valuation must be based on a model that uses the information available in the markets in an optimal and coherent manner, or on an approach based on the valuation of the income generated. The market value is thus considered as the most relevant indicator of a realistic economic value at any time.

The above diagram shows the main differences between the Solvency 1 and Solvency 2 balance sheets for a typical Long Term Care insurer (Table 6.1).

In French statutory accounting, assets are recorded at book value, i.e. on a value based on the depreciated acquisition price. Inversely, under Solvency 2, assets are estimated on the basis of their market value at the valuation date.

Under Solvency 1, reserves consisting mainly of Active Life Reserves (ALR) and Claim Reserves (CR) shall be determined from:

- Characteristics of current policyholders and their contracts
- Generally prudent experience tables due to the lack of regulatory tables and the limited data available
- A conservative non-life valuation interest rate.

The accounting provisions correspond to the value of the insurer's guaranties toward the insured. Under Solvency 2, the reserves shall consist of:

- The Best Estimate (BE), which represents the economic value of future flows relating to contracts in stock as of the valuation date,
- The Risk Margin (RM), which creates a supplement added to BE.

Reserves incorporate all the liabilities of the insurer engendered by the contracts: these include, in particular, beyond policyholder guarantees, liabilities to third parties (commission paid to the sales forces, various costs paid to Third Party Administrators, wages and taxes arising from the contracts...).

The calculation of the capital eligible to be applied to the capital requirement under Solvency 2 derives from the economic balance sheet. In fact, the total market value of assets decreased by reserves, potential deferred tax liabilities and the value of any other low-balance liabilities makes it possible to obtain the elements allowed to be included in the capital before application of any limits arising from the classification rules related to the Solvency Capital Requirement (SCR) or Minimum Capital Requirement (MCR).

The balance sheet based on market values is a "snapshot", a representation at a precise time of the financial situation of the organization.

This is subject to a range of risks, such as the underwriting risk (or insurance technical risk) or the market risk (or investment). To ensure that the organization will be able to fulfill its commitments with policyholders, the Solvency 2 framework sets out to **identify these risks and measure their effects on the balance sheet** and on the eligible capital.

Under Solvency 2, the effects of significant quantifiable risks are determined and aggregated within SCR to estimate the cost of the maximum loss arriving once every 200 years. MCR corresponds to the amount of Solvency 2 capital under which the organization is not authorized to go below in order to continue its activities [withdrawal of approval by the French insurance regulatory authority, *Autorité de Contrôle Prudentiel et de Résolution*, (ACPR)].

The Solvency 2 framework offers the possibility of estimating these capital requirements either by the so-called standard formula or by setting up an Internal Model (IM) by the insurer.

In the following, we will assume that the standard formula option has been retained. The use of an internal model could indeed be interesting on the perimeter of Long Term Care risks. However, this approach requires a large amount of data and a history sufficient enough to be able to observe the experience of insureds at extreme old age. As these conditions are difficult to meet for this type of risk, this point is not dealt with in this chapter.

The standard formula provided for in Solvency 2 requires calculating the impact on the company's capital of a set of risks corresponding to the financial, technical and operational risks to which an insurance undertaking is subject. Aggregation of the estimated cost of these risks using diversification factors allows the estimation of the capital cost of the 99.5% quantile for the SCR calculation.

6.3 Description of Solvency 2 Balance Sheet Basics

6.3.1 Calculation Main Elements

6.3.1.1 Assets

For the balance sheet assets, or at least some of them, directly observable market prices are available: this is the case for sovereign bonds or quoted stocks. For other investments, valuation models may be necessary. For real estate, the value derived from recent assessments can be used. Reinsurer reimbursements are shown as assets and reinsurance is not deducted from liability reserves, which improves the transparency of the process and of the *Reporting*.

6.3.1.2 Reserves

Under Solvency 2, the concept of coherence with the market extends to reserves in principle, for consistency in the construction of the balance sheet. However, there is no liquid market that would provide directly observable market prices for insurance portfolio transfers. Under Solvency 2, the value of reserves is conceived as the sum of two elements:

$$S2\,Reserves = Best\,Estimate + Risk\,Margin$$

The Best Estimate (of the current portfolio of liabilities) is mathematically the expected value of future payments flow less future income flow. These are estimated using realistic probabilistic assumptions about the risk factors that may affect these future flows and updating them with the risk-free rate curve relevant at the valuation date. More simply, the Best Estimate represents the expectation of the cost to the organization to pay its contractual commitments as expected on the expected due date, while taking into account premiums yet to be received. In calculating the best estimate, the underlying assumptions used correspond to the best estimate of future flows and should not contain additional risk margins. Another difference with the statutory reserves in French regulation is that the discount rate applies to all reserves, using the market rate curve without regard to the particular risk. For Non-Life organizations, a separate Best Estimate must be calculated for Active Life Reserve and Claim Reserve.

The contract limit defined in the standard formula (Commission Delegated Regulation (EU) 2015/35 article 18) defines the projection horizon of premiums for inforce contracts (boundary). In the event that the insurer has the unilateral right to refuse the premium, premiums subsequent to the date on which that right may be exercised are not taken into account in the projection. This situation can occur in the case of yearly cancellable group insurance contracts. In this case the insurer is freed from its commitments relating to contracts which are not in claim status.

For lifetime benefit contracts, the insurer's ability to revise the premium rate can lead to limiting the projection of premiums. However, the relevant acts provide that the projection of premiums stops when[3]:

> (c) the future date where the insurance or reinsurance undertaking has a unilateral right to amend the premiums or the benefits payable under the contract in such a way that the premiums fully reflect the risks. Point (c) shall be deemed to apply where an insurance or reinsurance undertaking has a unilateral right to amend at a future date the premiums or benefits of a portfolio of insurance or reinsurance obligations in such a way that the premiums of the portfolio fully reflect the risks covered by the portfolio. However, in the case of life insurance obligations where an individual risk assessment of the obligations relating to the insured person of the contract is carried out at the inception of the contract and that assessment cannot be repeated before amending the premiums or benefits, insurance and reinsurance undertakings shall assess at the level of the contract whether the premiums fully reflect the risk for the purposes of point (c).

Depending on the contract, internal limits to the company in terms of premium revision policy or the inability to know precisely the fair cost of individual risks could lead to the view that premiums must be projected on a lifetime basis.

This approach, depending on the contracts, may be prudent and probably leads to a price closer to that which could constitute a cash value of the contract (or surrender value). In fact, a premium lapse may result, depending on the contracts and their duration, in releasing the insurer from its contract or in a reduction in benefits. These situations may create margins for the insurer without connection to the likely future experience of the contracts.

Therefore, the limit of Long Term Care contracts depends on the nature of the contract and the interpretation of the regulation. This is a major issue that has a significant impact on the value of reserves and the risk profile within the meaning of Solvency 2.

A risk margin is added to the Best Estimate to obtain a value of the reserves consistent with the market. It is calculated using the "Cost of Capital" method (CoC). This method is based on the idea that the organization (or those who finance it) hopes to be paid for the risks taken. The method can be summarized as follows: Suppose the organization wants to transfer or acquire an existing portfolio, the assignee will not accept the transfer if the value of the assets transferred with the portfolio is equal to the Best Estimate. Actual outgoing future payments can significantly defer, and even exceed, future flows associated with the Best Estimate, and the assignee will not accept the risk without an expected positive yield. The assignee will therefore request an additional amount (in addition to the Best Estimate), the risk margin.

The amount of the risk margin may be determined as follows: The organization must not only hold reserves; it must also have a risk capital to absorb unforeseen losses and comply with the regulations. The organization (or those who finance it) wants to be paid for the capital asset. The risk margin is simply equal to the product of the capital required and the cost of capital. The required capital used in calculating the risk margin contains only the SCR inherent in the current insurance portfolio; it also includes the operational risk and the default risk for the transferred reserves. The important thing is that the "avoidable" market risk, because it depends on the investment policy of the insurer and not on its insurance business, is not to be included. The amount of the risk margin therefore depends on the characteristics of the insurance portfolio. The risks that can be covered in the markets are not taken into account in the risk margin.

The cost of capital is set at 6% (Commission Delegated Regulation (EU) 2015/35, Section 3, Article 39). The Best Estimate will suffice, on average, to exactly ensure the run-off of the portfolio. The risk margin will then be released gradually and will be sufficient to ensure a 6% return (beyond the risk-free rate) for those who fund the activity. Only the excess yield beyond the risk-free rate is included in the risk margin. The risk capital made available to the organization will be able to be invested in products which earn the riskless interest rate at the time.

6.3.1.3 Risk Segmentation

For reserve calculation, insurance and reinsurance liabilities must be segmented at a minimum for each activity line. This segmentation applies to both the Best Estimate and the Risk Margin. Finer segmentation in homogeneous risk groups can be used if it improves the accuracy of the reserve and capital cost valuation. Guarantees are allocated to the segment of activity that best represents the nature of the underlying risk. It should therefore be noted that this segmentation may differ from the branches of activity as defined to reach agreements or other accounting elements. For example, guarantees that are technically treated as life contracts (reserving made with assumptions of underlying distributions as opposed to a claim run-off triangle approach or Loss Ratio) should be considered Life coverage, even if they arise from non-life contracts. Conversely, certain life insurance contracts could give rise to non-life liabilities if the risk selection rules are similar to those of non-life risks. In particular, annuities paid under non-life contracts are considered life insurance. Depending on the technical nature of the risks, health (in the meaning of Solvency 2 i.e. all coverage of lifetime care, except for death) is split into two types of coverage: similar to life health (so-called Similar Life Techniques, or SLT, Health), based on techniques similar to that used in life insurance; and not similar to life health (Non-Similar Life Techniques, or NSLT, Health) in accordance with article 55 of the Commission Delegated Regulation (EU) 2015/35.

6.3.1.4 Discounting

As part of the Best estimate evaluation, cash flow discounting is carried out at a risk-free interest rate: the adjusted swap rate curve adjusts to the credit risk (provided by EIOPA), with the possibility of:

- Extrapolating the rate curve to evaluate long-term coverage;
- Taking into account a Volatility Adjustment to the risk-free rate curve or a Matching Premium on this rate curve (it is not possible to combine the two). These

adjustments help to confront the volatility of the equity induced by the liquidity of assets backing the insurance coverage.[4]

6.3.1.5   Contract Maturity

As part of the determination of the incoming and outgoing flows constituting the BE, the strong hypothesis adopted by the Directive is to value the contracts to maturity while assuming the continuation of the insurer's activities in determining projection assumptions. This economic vision is a very different vision of the insurer's future liabilities, depending on its ability to cancel its contracts.

6.3.1.6   Capital Requirements: SCR, MCR

Solvency 2 provides for two different capital requirements, the SCR (Solvency Capital Requirement) and the MCR (Minimum Capital Requirement). Although it is not possible to compare them exactly with Solvency 1 requirements, the role of SCR can be roughly compared to the solvency margin requirement and MCR to the Guarantee Fund (1/3 of the margin requirement). As in the Solvency 1 guidelines, MCR will also be subject to an absolute floor (Absolute Minimum Capital Requirement, or AMCR).

- SCR: The SCR is the regulatory capital to be held by an insurance organization to limit the insolvency of the underlying risk to a maximum of one in 200 over a one-year horizon. Insurers have the possibility to calculate the SCR either by using the standard formula (the use of a methodology and calibration parameters are described in the technical specifications) or by using an internal model.
- MCR: This is the minimum level of capital that the organization must hold permanently, under penalty of immediate action by the regulatory authority, which may lead to withdrawal of approval.

6.3.2   Block Structure for Calculating Economic Capital in the Standard Formula

The SCR of an insurance organization corresponds to the aggregation of all the following standard risks (Table 6.2).

The SCR calculation in the standard formula is expressed as follows:

$$SCR = BSCR + Adj + Op.$$

The basic SCR (BSCR) corresponds to the gross SCR of the absorption capacity of losses by profit sharing and deferred taxes. It is obtained by aggregating the SCR

Table 6.2 Analysis grid of Solvency 2 risks

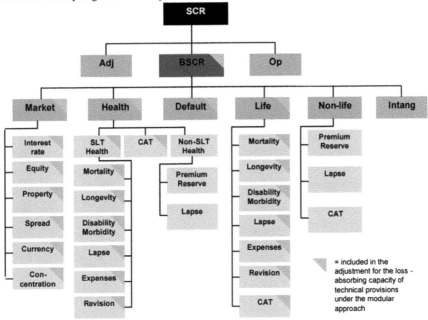

associated with the six standard modules (Market, Health, Counterparty Default, Life, Non-Life, Intangible):

$$BSCR = \sqrt{\sum_{i,j} Corr[i,j] \cdot SCR_i \cdot SCR_j} + SCR_{intangible}.$$

This BSCR takes into account the correlation coefficients between risks (i.e. between SCRs), defined by the correlation matrix Corr (annex IV, point 1 of Directive 2009/138/EC) (Table 6.3).

Table 6.3 Correlation structure

SCR global	Market	Default	Life	Health	Non-life
Market	100%				
Default	25%	100%			
Life	25%	25%	100%		
Health	25%	25%	25%	100%	
Non-Life	25%	50%	0%	0%	100%

6.4   Application to Dependence Risk

6.4.1   Classification of Long Term Care Insurance in Solvency 2 Standard

Long Term Care is a perfect illustration of the difference in classification between the French standard and the standard Solvency 2.

In the French standard, Long Term Care can be considered as a non-life risk because the covered hazard is neither life nor a life risk.

In Solvency 2, Long Term Care is classified as health because it is a contingency risk different from death. Within this module, Long Term Care can be classified into two sub-modules depending on the nature of the coverage and the technical basis:

- Sub-module "Health Similar To Life (SLT)": In the event that the Long Term Care coverage is permanent, this requires modeling by techniques similar to life insurance.
- Sub-module "Health Non-Similar To Life (non-SLT)": In the event that the Long Term Care coverage is yearly renewable.

Hence, the risks of the corresponding standard formula are as follows (Table 6.4).

Table 6.4 Risk identification

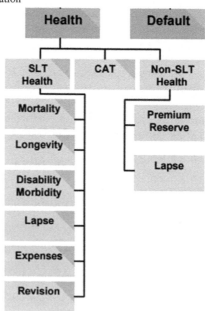

6.4.2 Characteristics of a Long Term Care Insurance and Assumptions Used

In the case of Long Term Care insurance under the "Non-SLT Health" line, the SCR calculation would be based on a fixed cash benefit and depends on Best Estimate (BE) premium and reserve. These contracts fall under the income protection category. The calculation of the SCR in this case results from the simple application of coefficients provided for in the standard formula, the case considered in the following is that of a Life-like product.

For a given Long Term Care insurance product, the calculation of each SCR in the "Health SLT" module requires calculating a "shocked" BE, i.e. a BE with adverse projection assumptions. This methodology requires modeling the Long Term Care risk: we will therefore define a Long Term Care product with its modeling framework.

The choice is for a lifetime product which, therefore, does not result in a Health Non-SLT SCR. The modelled product provides for the payment of a lump sum and a monthly annuity in the case of Partial Dependence[5] as recognized by a medical consultant, according to the Iso-Resource Grid.[5] The recognition of a state of Total Dependence[5] would double the amount of the annuity provided in the case of Partial Dependence. The premium is monthly and depends on the issue age.

In the modeled product: each year, an active policyholder can remain active, become dependent, or die. A dependent is supposed to remain dependent or dies.

In this scheme, entry into a state of Partial[5] or Total[5] Dependence is considered definitive without possible recovery; the condition of an individual can only deteriorate. Indeed, the reversibility of the insured's dependent status is a real modeling challenge, linked to the lack of data. It differs from the temporary inability to work and permanent disability with an improvement of the insured's health. The simplifying assumption in this product may be questioned in accordance with the definition given to the Partial Dependence[4] state.

The diagram illustrates the transitions between the different states of the product (Fig. 6.1).

Finally, we note that the SCR calculations presented below do not take into account the specific treatments related to reinsurance. We do not model treaties in this chapter, and the interested reader is referred to the reinsurance chapter.

6.4.3 Actuarial Modeling of a Long Term Care Insurance Contract

In the context of Solvency 2, it is necessary to have a deterministic model of cash flow projections allowing us to dynamically evaluate the solvency of the contract, incorporating the different assumptions. These different streams make it possible to project the balance sheet over several years, to estimate the *Best Estimate* reserves and calculate the economic capital of the contract.

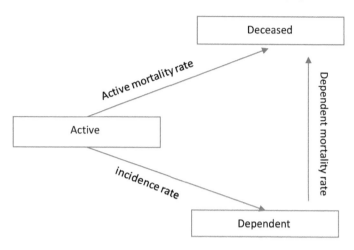

Fig. 6.1 Description of the states

6.4.3.1 Simplified Projection Model

Generally speaking, for a given insured with age X at the valuation date, the cash flow projection is described by a system describing successively:

- The number of active premium-paying actives:

$$N_{x,t}^a = N_{x,t-1}^a \times \left(1 - q_{x+t-1}^a\right) \times (1 - i_{x+t-1}).$$

- Premium income:

$$P_{x,t} = \pi_{x,t} \times N_{x,t}^a.$$

- Incidence:

$$C_{x,t} = i_{x+t-1} \times N_{x,t-1}^a.$$

- Claims:

$$N_{x,t}^d = \sum_{k=0}^{t-1} \left[C_{x,k} \times \prod_{s=0}^{t-k-1} 1 - q_{x+k,s}^d \right].$$

- Finally, benefits (annuity payments can vary depending on a parameter or can remain constant):

$$B_{x,t} = r_\theta \times N_{x,t}^d.$$

The preceding equations allow one to immediately calculate the expected cash flows:

$$\rho(t) = \sum_x \left(P_{x,t} - B_{x,t} \right).$$

Thus, on any date T, a projection of the Best Estimate Liabilities (BEL) is given by the following formula:

$$E[BEL(t)] = \sum_{s \geq t} \frac{\delta(s)}{\delta(t)} \times \rho(s),$$

with $\delta(t) = (1 - r_t)^t$ the discount factor (deterministic). As:

$$BEL(t) = E_t \left(\sum_{s \geq t} \frac{\delta(s)}{\delta(t)} \times \tilde{\rho}(s) \right) = \sum_{s \geq t} \frac{\delta(s)}{\delta(t)} \times E_t(\tilde{\rho}(s))$$

we reach the equality:

$$E[BEL(t)] = E \left[E_t \left(\sum_{s \geq t} \frac{\delta(s)}{\delta(t)} \times \tilde{\rho}(s) \right) \right] = \sum_{s \geq t} \frac{\delta(s)}{\delta(t)} \times E(\tilde{\rho}(s)) = \sum_{s \geq t} \frac{\delta(s)}{\delta(t)} \times \rho(s).$$

In the context of Solvency 2, to dispose of these cash flows allows us to calculate the margin requirement on each date, applying to the calculation assumptions the shocks at the different dates to deduct, by measuring the Best Estimate variation, the required capital as part of the standard formula. It should be noted that when the Best Estimate is negative, which corresponds to anticipation of future gains, the insurer cannot use them as shock reduction. It will be appropriate to add the different loads, commissions and expenses to calculate the BEL value.

The annual approach presented here for simplification purposes should be taken into account in the drafting of distributions if it is retained. Indeed, given the nature of the risks, the order in the application of the probabilities has a significant impact on reserve values.

Finally, the need for unit capital in the event of a shock having no impact on the value of the assets is therefore of the form:

$$SCR = [+BEL_{shock} - BEL_{central}]$$

The calculation of a Long Term Care insurance product SCR requires three calculations for underwriting, counterparty, and market risks. The counterparty risk arises when liabilities are transferred to a reinsurer or from certain types of assets held. Market risk would require modeling the company's general assets and, at the start, the cash flows from other marketed products. In the absence of these elements, the market risk was calculated for a single product, more for illustration than to estimate minimum capital. Future premiums are taken into account, which may eventually be revised as contractually planned. If the absence of a projection of premiums can be entertained for an existing portfolio, and in view of a transfer, in a pricing approach and to measure the required capital, the shareholder will wish to know its future remuneration from the perspective of retention to maturity.

Underwriting gain is not modeled in the basic scenario (insureds behave as expected in the premium rate calculation, except in the context of shocks), or administrative gain (premium loads are equal to expenses, except in the expense risk shock, and in the sensitivity scenarios).

In the Solvency 2 standard formula, the prescribed shocks considered are as follows:

- Longevity risk (20% reduction in active and dependent mortality).
- Morbidity risk (35% incidence increase in the first year and 25% for subsequent years). The 20% shock on either the recovery or the persistency rates should be taken into account only if, in the valuation of the best estimate, another option than death can lead to a change of state in case of dependency. This would be the case in multi-state models, for example.
- Expense Risk (10% expense increase).

This simplistic model would lead, in some stressed scenarios, to the collection of premiums that generate underwriting losses without ever revising premium rates, which is clearly unrealistic. As a result, it is important to propose improvements to the model to make it more realistic by taking into account certain risk reduction factors.

To illustrate the impact, we can look at the very simple example shown in the "Introduction" Sect. 6.1:

The cost of a 20% decrease in the mortality of the dependent people has a cost of €180 M. Assuming that the cost is transferred to the premiums, they should increase by 22.5%. Practically, the increase can be done over several years.

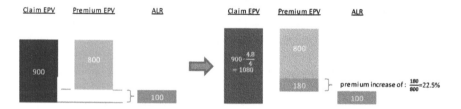

The main sources of improvement of the proposed model are:

- taking into account revalorization and financial assets;
- premium rate revision; and
- taking reinsurance into account.

6.4.3.2   Revalorization and Financial Assets

French Long Term Care Insurance contracts generally provide for the creation of a "revalorization fund", fed by possible gains (underwriting/financial), in order to revalorize the guarantees over time. It may be envisaged that the fund thus constituted should be resumed following a deterioration of underwriting results (linked to an "external shock"). The recovery of the established financial reserve has the effect of mitigating the impact shocks on the *Best Estimate*, and therefore reduces the required SCR.

The modeling of this measure for the SCR calculation implies projection, operational and management assumptions, which must be justified. It also passes through the explanation of the results on the projection horizon.

- Projection assumptions:

Increasing the revalorization fund by projecting underwriting gains is hardly justifiable. On the other hand, it is possible to add to this fund through the projection of financial gains and in this case be able to reach over time higher financial return rates than the reserve interest rate.

- Operational assumptions:

This assumes that the insurer has the means to detect an Actual to Expected (A/E) deterioration of underwriting results. This assumption is reasonable once the insurer implements a recurring system of risk monitoring better/worse than expected analysis to detect deteriorations in the risk.

- Management assumptions:

The assumption of revalorization funds recapture is justifiable if the insurer is contractually allowed to do so.

Financial gains will be projected in the framework of this model. Formally, this amounts to setting up the following formula:

$$FR_t = \beta_t - IR_t \times (P_t + B_t) \times \left(ror_{fin} - i_{val}\right),$$

$$FDR_t = \sum_{s=0}^{t} FR_t \times \left(1 + ror_{fin}\right)^{t-s},$$

with FR_t: financial result, β_t: financial asset income, IR_t: valuation interest income, P_t: premium, B_t: benefit, ror_{fin}: financial returns, i_val: valuation interest rate, FDR_t: revalorization fund.

The introduction of financial assets into the model logically decreases the SCR. Moreover, the establishment of a revalorization fund (increased by financial gains), used to mitigate underwriting losses in stressed scenarios, allows a steeper SCR decrease.

6.4.3.3 Modeling Future Decisions: Premium Rate Revisions

Premium rate revision is a possible action in case of reserve deficiency and is generally foreseen in the general provisions of [French] Long Term Care insurance contracts. In the application of the standard formula, the projections made involve collecting premiums that generate underwriting losses without ever revising premium rates again, which is obviously unrealistic. According to the procedures for setting up premium rate revisions, all or part of the additional margin linked to the stressed scenarios is transferred to policyholders, thereby reducing the required SCR.

Modeling this provision for the SCR calculation implies operational assumptions which must be justified.

• Detection of risk deterioration assumption:

This is justified if a risk monitoring process and a better/worse than expected analysis is set up to detect worsening risks.

• Claim follow-up assumption:

This consists in setting up ways to measure the magnitude of shocks on biometric assumptions (incidence, continuance and severity shocks) and management costs (claim expense shock). It is justified if the insurer defines a method of constructing/adjusting biometric assumptions in order to account for the changes in risks.

• Premium rate revision assumption:

This is justified if the insurer is contractually able to carry out premium rate revisions, taking into account any ceilings.

• Implementation assumption:

This deals with the time to identify and set up a risk monitoring and adjustment process. In addition to the strategies for identifying and measuring shocks, it is reasonable to make an assumption about the time (months/years) required for its implementation.

At least four categories of insured persons are to be taken into account:

• Premium paying actives,
• Future insured,
• Paid Up (limited premium) and Reduced Paid Up (lapsed premium) actives,
• Claimants.

The transfer of risk through premium rate revisions only occurs through the first two categories, whereas the shocks apply to all categories. The restoration of underwriting balance is limited to the additional premium that can reasonably be transferred to premium paying participants (current or future) and is sensitive to the maturity of the portfolio and assumptions about new issues.

If these conditions are fulfilled, taking into account future decisions in the projection model may reduce the magnitude of continuance and incidence SCRs. For this it is necessary to model:

- the time at which the insurer will be able to measure the risk and implement a premium rate revision,
- the decision on the magnitude and duration of premium rate revisions,
- the impact over time on the outcome of the implementation of the strategy.

The model will need to be able to handle two claim assumptions. One, called "experience", that allows to model the cash flow of pricing benefits (those that make up the Best Estimate); and so-called "reserve", which allows the calculation of statutory reserves and corresponds to a distribution known to the insurer.

The time required for the insurer to be able to measure the deterioration of premium rates can thus be estimated in the projection model by tracking the numbers of deaths or the incidence count. The time limit can thus be calculated from the following elements:

- The period from which the difference between the expected numbers (calculated with the so-called valuation distribution) and the actual numbers ("Best estimate" distribution) exceeds an action threshold in relation to thresholds defined in the current management action policy.
- The time required for the insurer to analyze the data.
- The time limit to effectively implement the premium rate revision, the latter includes policyholder communication.

To determine the duration and value of the premium rate revision, an indicator that can be retained in the model is a deficient statutory reserve. The latter corresponds to the difference between the statutory reserve calculated under the valuation distribution (or before shock in the case of the SCR calculation) and the reserve calculated with the experience distribution (or shocked distribution in the case of an SCR calculation). It will then be necessary to determine the duration and the annual value of successive revisions. These elements are based on the maximum threshold of revisions practiced by the company or contractually practical. To estimate the amount of the premium rate revision to be used, the rate of increase to be applied to the probable present value insured in the ALR calculation allows us to measure the revision to be applied to the net premium to restore balance at the statutory reserve level. The implementation of this increase will also have an impact on the future administrative margins. It will be appropriate to determine whether the insurer's strategy in the case of a claim deterioration aims to recover its margins or to cancel losses.

Once these elements are determined, the model must implement the revision by modifying the premium flows after the start of management action and recalculating the ALR taking into account these changes in order to establish the successive results and updating that result.

The difficulty in implementing such a strategy often does not lie in modeling. The question is to clarify precisely the strategy that would be applied in a deterioration situation and to ensure that the company governance provides:

- The effective monitoring of risk indicators, that these indicators are published, analyzed and communicated to management to make an informed decision.

- A clear policy of premium rate review practices incorporating both marketing and actuarial considerations.
- The possibility for management to make decisions about premium rate revisions over several years.

These conditions fulfilled, the premium rate revision policy implemented in the model will constitute a "management action" which must meet criteria laid down in the delegated activities (in accordance with article 23 of the Commission Delegated Regulation (EU) 2015/35: management validation, proof of "use test" in the event of a real situation and consistency with internal policies, regular communication of the use of the strategy and its effect on the SCR).

6.4.3.4   Reinsurance

The Best Estimate must be calculated gross of reinsurance, without deducting liabilities arising from reinsurance contracts and securitization tools and without taking into account the amounts covered by reinsurance. In return, a reinsurance asset equal to the difference between the Best Estimate gross and net of reinsurance is recognized. This asset generates a capital charge for the counterparty risk. Moreover, in the case of the standard formula, the risk margin must be calculated net of reinsurance, which is consistent with Article 101-5 of the Directive stipulating that the SCR must be calculated net of reinsurance.

The reinsurance asset must be calculated using the same principles as the Best Estimate calculation by incorporating the probability of the reinsurer default. This consideration results in an adjustment of the probable present value of future cash flows which is calculated on the basis of the probability of default of the counterparties and the resulting average amount of losses.

When the counterparty benefits from high ratings, the matching default adjustment should be fairly low compared to reinsurance recovery. In this case, the adjustment for the counterparty default can be evaluated in a simplified manner by the following formula:

$$Adj_{CD} = -\max\left((1 - RR) \times BE_{\text{Rec}} \times \text{Dur} \times \frac{PD}{1 - PD}; 0\right)$$

With RR the recovery rate, PD the one year probability of default, Dur the duration and BE_{Rec} the best estimate of recoverable amounts.

- Parameter determination

The Probability of Default (PD) and Recovery Rate (RR) are set by the regulator (Table 6.5).

The recovery rate represents the portion of the debt that will be recovered in the event of default of the reinsurer and is set at 50% for reinsurance. All risks assumed by the reinsurer from the insurer (collateral, etc.) result in deduction of the exposure.

Table 6.5 Probability of default by credit quality

Credit quality step	0	1	2	3	4	5	6
Probability of default P_iPD_i (%)	0.002	0.01	0.05	0.24	1.20	4.175	4.175

- The risk of failure of reinsurance counterparties

The counterparty risk is the risk of losses resulting from unforeseen failure or downgrade of the credit rating of counterparties or debtors in risk reduction contracts, such as reinsurance treaties. It must cover risk reduction contracts, such as:

- Reinsurance treaties;
- Derivative and securitization products;
- Claims from intermediaries;
- Any other credit exposure not covered in the spread risk sub-module.

In the case where the Loss-Given-Default (LGD) for counterparty i (non-random) is denoted as y_i, the expectation and variance of the overall loss are calculated as:

$$M = \sum_{i \in I} p_i y_i \quad V = \sum_{i,j \in I} \omega_{ij} y_i y_j$$

with the following notations:

$$\omega_{ij} = \frac{\theta(1 - b(\theta, p_i))(1 - b(\theta, p_j))}{\theta + \frac{1}{b(\theta,p_i)} + \frac{1}{b(\theta,p_j)}} - (p_i - b(\theta, p_i))(p_j - b(\theta, p_j)) \quad b(\theta, p_i) = \frac{p}{1 + \theta(1 - p)}$$

By assuming that the distribution of global losses follows a Lognormal distribution and is defined by:

$$\sigma = \sqrt{\ln\left(1 + \frac{V}{M^2}\right)} \quad \mu = \ln(M) - \frac{\sigma^2}{2}$$

we derive the target capital as follows:

$$SCR = \min\left(\sum_{i=1}^{N} Y_i, q \times \sqrt{V}\right)$$

with q = 3.

In the case of a single counterparty, the calculation is simplified to:

$$SCR = \min\left(1, q \times \sqrt{p \times (1 - p)}\right) \times Y$$

For an AAA rated counterparty it is therefore found that the SCR amounts to 2.9% of the present value of the ceded cash flows (compared to the 58% measured on average for the underwriting risk).

Table 6.6 Balance sheets with and without reinsurance

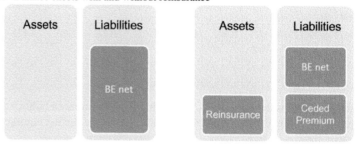

It is possible to synthesize the logic behind incorporating the reinsurance counterparty risk as follows: without reinsurance, we have the best estimate on liabilities and after implementation of a coinsurance treaty, there are the receivables from the reinsurer as an asset and the payables to the reinsurer as a liability (Table 6.6).

- SCR determination implications
  With respect to the calculations to be carried out to determine the SCR level:

  - In the absence of reinsurance, the underwriting SCR is calculated by measuring the variation of the best estimate as a function of applied shocks;
  - In the presence of reinsurance, the underwriting SCR is calculated by measuring the change in the value of the best estimate net of the reinsurance asset (i.e. one explicitly manipulates here a variation of Net Asset Value, or NAV) and this calculation is completed by the determination of a counterparty SCR.

6.4.3.5 SCR Calculations for Each Sub-module

The calculation of the capital required for the components of *Health SLT* underwriting risk for Long Term Care is not based on the nature of the risk but on the application of shocks to the underlying risks: the required capital corresponds to the impact on the level of the eligible surplus (*BOF, Basic Own Funds*) in a shocked scenario. The impact on BOF for the underwriting risk is measured as the impact on the *Best Estimate* reserves:

$$\Delta BOF = \max\left(0; BOF^{central} - BOF^{shock}\right) = \max\left(0; Reserves_{BE}^{shock} - Reserves_{BE}^{Central}\right).$$

The calibration of the shocks to be applied reflects the SCR definition, namely $VaR_{99,5\%}$ of the BOFs on a one year horizon. With the many shocks to be applied in the standard formula and their interactions on the active and dependent funds, a detailed and prospective model is required. It must be able to calculate a robust SCR for each sub-module, a real challenge for the insurer. Moreover, special care must be taken on the assumptions for premium and benefit revalorization, if any, which could lead to particularly high capital requirements.

The standard formula alone does not take into account all Long Term Care underlying risks.

We detail below the shocks according to the standard formula for a cash benefit "Long Term Care" product.

Mortality Shock—EIOPA [1] 01/2015—Article 152:

> The capital requirement for health mortality risk shall be equal to the loss in basic own funds of insurance and reinsurance undertakings that would result from an instantaneous permanent increase of 15% in the mortality rates used for the calculation of technical provisions.

Given the nature of the Long Term Care contract, the mortality shock has two opposite effects:

- Reduction of the insurer's liabilities, which would result in a BE decrease, in particular due to the increase in the claimant mortality.
- Reduction in premiums received by the insurer, with the decrease in the active population, thus increasing the BE amount.

The SCR will therefore be non-zero, if the second effect is the most important. Two product characteristics will be decisive for the weight of this second effect:

- The first element is the duration of the product. The more recent the product, the greater the amount of expected premium.
- The second element is the reserve margin of the product. The more important this margin is, the more premature the lapse, the more significant the loss of profits.

Health Longevity Shock—EIOPA [1] 01/2015—Article 153:

> The capital requirement for health longevity risk shall be equal to the loss in basic own funds of insurance and reinsurance undertakings that would result from an instantaneous permanent decrease of 20% in the mortality rates used for the calculation of technical provisions.

In order to determine the amount of capital required to cope with a decrease in mortality rates, the impact applied is similar to the longevity shock of the module "life" or a permanent 20% decrease of mortality rates for all ages.

In the case of Long Term Care, this decrease affects both actives and dependents.

As with the mortality shock, this shock has two opposite effects for the BE calculation: the increase of the liabilities of the insurer, which increases the BE, and the increase of expected premium, which decreases the BE.

Note:
The longevity shock has a very strong weight with an increase of BE which is related to:

- The premium paying active population living longer.
- More active lives reach ages where the dependence incidence is the highest.
- More benefits following the heavier shift from Partial Dependents to Total Dependents.

Health disability-morbidity shock—EIOPA [1] 01/2015—Article 154:

> The capital requirement for health disability-morbidity risk shall be equal to the sum of the following: (a) the capital requirement for medical expense disability-morbidity risk; (b) the capital requirement for income protection disability-morbidity risk.

Long Term Care insurance products with expense reimbursement benefits would be subject to requirement (a). Long Term Care insurance products with cash benefits are income protection insurance and are subject to requirement (b). This chapter concentrates on cash benefits.

EIOPA [1] 01/2015—Article 156:

> The capital requirement for income protection disability-morbidity risk shall be equal to the loss in basic own funds of insurance and reinsurance undertakings that would result from the following combination of instantaneous permanent changes:
>
> (a) an increase of 35% in the disability and morbidity rates which are used in the calculation of technical provisions to reflect the disability and morbidity in the following 12 months;
>
> (b) an increase of 25% in the disability and morbidity rates which are used in the calculation of technical provisions to reflect the disability and morbidity in the years after the following 12 months;
>
> (c) where the disability and morbidity recovery rates used in the calculation of technical provisions are lower than 50%, a decrease of 20% in those rates;
>
> (d) where the disability and morbidity persistency rates used in the calculation of technical provisions are equal or lower than 50%, an increase of 20% in those rates.

The morbidity/disability shock consists of two combined sub-shocks: an increase in incidence, which means a deterioration in health status and a decrease in the recovery rate. This recovery rate in the case of dependence can be interpreted as the exit from a dependence state to a dependence-free state or to a lower dependence level. But a dependent state termination due to mortality must not be subject to the application of this shock since it is already shocked in the submodule mortality and longevity. If other cases where changes between states are considered, the transition to an active (healthy) state is often not supported by models. Indeed, a return to the active state after a stay in a dependence state, given the levels of dependence covered by the coverage, is rarely accounted for, if at all.

For the case of change between states of dependence, the application of the recovery shock differs significantly according to the approach taken in terms of modeling. Thus, in the case of contracts offering different guarantees depending on the level of dependence, if the model provides for passage distributions between the different states, a shock will have to be applied to these transition probabilities. If the model does not foresee a change of state, the shock cannot not be applied.

Note:
This shock increases the liabilities of the insurer and decreases the insured's.

Lapse shock—EIOPA [1] 01/2015—Article 159:

> The capital requirement for SLT health lapse risk referred to in Article 151(1)(f) shall be equal to the largest of the following capital requirements:
>
> - capital requirement for the risk of a permanent increase in SLT Health lapse rates;
> - capital requirement for the risk of a permanent decrease in SLT Health lapse rates;
> - capital requirement for SLT health mass lapse risk.

The lapse shock is used to assess the amount of capital necessary to cover the risk of unfavorable changes in the insured's behavior in terms of voluntary termination.

The SCR amount of SCR retained for this shock is the maximum of the 3 SCR calculated under the following scenarios:

- The permanent increase scenario is applied by increasing lapse rates by 50%,
- The permanent decrease scenario is applied by lowering lapse rates by 50%,
- The massive termination scenario corresponds to a total lapse rate of 40%, which replaces the total first year lapse rate in the first year of the projection.

Remarks:
From the modeling point of view, the construction of a biometric distribution on the insureds of the reduced portfolio can be a real challenge given the lack of data and the sometimes unpredictable behavior of the insured. On the other hand, taking into account future claim reductions requires additional projections to take into account a smaller portfolio.

This shock leads to a change in the liabilities of the insurer and those of the insured. In the case of lifetime coverage, the voluntary exit of the insured can result in a reduction in benefits or even an end of coverage. It should be noted that the level of underwriting margin included in the gross premium will have a strong influence on the costliest scenario. In particular, the massive lapse scenario deprives a large portion of future premiums and therefore leads to a very high SCR for products with high underwriting margin.

Warning: the choice of the scenario chosen for the SCR calculation does not depend solely on the results of the Long Term Care product. The scenario chosen is the one with the highest SCR for all contracts and risks covered by the insurance company. This means that the lapse SCR of this product will depend, among other things, on the profile of the insurer's portfolio.

Expense shock—EIOPA [1] 01/2015—Article 157:

> The capital requirement for health expense risk shall be equal to the loss in basic own funds of insurance and reinsurance undertakings that would result from the following combination of instantaneous permanent changes:
>
> - an increase of 10% in the amount of expenses taken into account in the calculation of technical provisions;
> - (b) an increase by 1% point to the expense inflation rate (expressed as a percentage) used for the calculation of technical provisions.

The expense shock is intended to estimate the impact on the insurer's liabilities of an inadequate valuation of the expenses associated with the insurance contract. The shock calibration is similar to the expense shock of the "Life" module.

Note:

The methodology for calculating this shock mechanically increases the insurer's liabilities, so the SCR is strictly positive.

Revision Shock—EIOPA [1] 01/2015—Article 158:

> The capital requirement for health revision risk shall be equal to the loss in basic own funds of insurance and reinsurance undertakings that would result from an instantaneous permanent increase of 4% in the amount of annuity benefits, only on annuity insurance and reinsurance obligations where the benefits payable under the underlying insurance policies could increase as a result of changes in inflation, the legal environment or the state of health of the person insured.

The revision shock aims to assess the impact on the insurer's commitments of an unanticipated change in the amount of benefit payments due to an evolution of:

- Inflation,
- Regulation: regulatory changes are only taken into account when the regulation is already passed on the valuation date,
- The deterioration of the claimants' health status.

In the case of a pure cash benefit Long Term Care product, annuity payments cannot vary. This shock could therefore be considered as not applicable. However, benefits that would increase with worsening in the dependent status could be part of the application of this shock.

Catastrophe (CAT) Shock—EIOPA [1] 01/2015—Article 160:

> The capital requirement for the health catastrophe risk sub-module shall be equal to the following:

$$SCR_{HealthCAT} = \sqrt{SCR_{ma}^2 + SCR_{ac}^2 + SCR_p^2}$$

- SCR_{ma} denotes the capital requirement of the mass accident risk sub-module;
- SCR_{ac} denotes the capital requirement of the accident concentration risk sub-module;
- SCR_p denotes the capital requirement of the pandemic risk sub-module.

For a risk classified in the "SLT Health" module, the dependence is subject to CAT shock.

The case of dependence is not explicitly dealt with in the texts. Thus, the transition from active state to dependent state, following the occurrence of one of the three scenarios constituting the SCR CAT, could be considered not to be part of the assumptions contained in the Solvency 2 regulation. Hence, the SCR CAT would be zero.

Another approach would be to consider the transition to dependence as a form of disability that could correspond to a state reached due to an accident. In this case the dependence risk should be included in the concentration and mass accident SCR calculations.

6.4.4 Global SCR Calculation

From the SCR calculation of each of the sub-modules described in the previous section, we will determine the steps to calculate the SCR.

The following two points are essential to fully understand the aggregation mechanism under the Solvency 2 standard:

- The final SCR that is sought to be determined is the SCR of all the risks born by the insurance company which has a Long Term Care insurance portfolio. Therefore, at each stage of aggregation in the "Octopus" (see Table 6.2), one must sum the SCR calculated for the Long Term Care portfolio with the SCR of the insurer's other risks.
- At each stage of aggregation in the octopus, i.e. at each stage where one wishes to calculate an SCR using lower level SCR in the octopus, a correlation matrix must be used. This octopus should not be seen as a simple sum of successive calculations.

Step 1:
For each of the sub-modules which are the subject of an SCR calculation (mortality, longevity, incidence, lapse, expenses), the SCR obtained on the Long Term Care portfolio with the SCR of the other risks insured by the company are calculated. The SCR of each of the 5 SCR Health-SLT sub-modules is obtained.

Step 2:
By adding the SCR of the Revision sub-module, and applying the following formula and matrix, we obtain the SLT-Health SCR (Table 6.7):

$$Health\ SLT\ SCR = \sqrt{\sum_{i,j} Corr[i,j] \cdot SCR_i \cdot SCR_j}.$$

Step 3:
We still go up in the octopus chart and we determine the Health SCR (Table 6.8).

To do this, we add the CAT SCR and the Non-SLT Health SCR, calculated elsewhere on the company's portfolio, using the formula and the following matrix (Table 6.9):

Table 6.7 Aggregation matrix

SCR health SLT	Mortality	Longevity	Disability/Morbidity	Lapse	Expenses	Revision
Mortality	100%					
Longevity	−25%	100%				
Disability/Morbidity	25%	0%	100%			
Lapse	0%	25%	0%	100%		
Expenses	25%	25%	50%	50%	100%	
Revision	0%	25%	0%	0%	50%	100%

Table 6.8 Structure of the "health" SCR

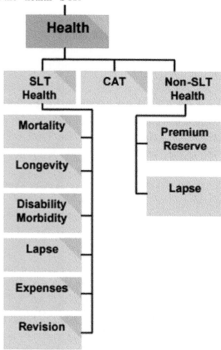

Table 6.9 Sub-module aggregation

SCR Health	SLT	Non-SLT	CAT
SLT	100%		
Non-SLT	50%	100%	
CAT	25%	25%	100%

$$Health\,SCR = \sqrt{\sum_{i,j} Corr[i,j] \cdot SCR_i \cdot SCR_j}.$$

Step 4:

The formula and correlation matrix of part I.B. 4 is applied by adding all SCRs except the Health SCR determined in the previous step. This gives the Basic Solvency Capital Requirement (BSCR).

Note:

Before applying the formula and the matrix, we must add the impact of our portfolio in the Market risk module.

The "Market" module reflects the risk associated with the level or volatility of the market value of the financial instruments impacting the value of the assets and liabilities of the company. It adequately reflects any structural mismatch between assets and liabilities, particularly in terms of their duration. In fact, with an average dependent age of 83^6 years on average (France), the insurer may be required to compensate the insured many years after the asset duration. However, it is sometimes difficult for a Long Term Care insurer to achieve its financial and liability cash flows over a long period, which can cause a high liquidity risk. Even if sovereign bonds are of long maturity, they rarely exceed 20 years.

Hence, we have two contributions to the calculation of this SCR market module:

- Assets and liabilities of the portfolio,
- Impact of a change in discount rates on the BE.

The description of the method of calculating this Market SCR is quite long and is only partially part of this work because the bulk of this module depends on the financial securities in relation to the Long Term Care portfolio. However, the nature of these securities depends essentially on the investment strategy and not on the Long Term Care portfolio.

Step 5:
In this last step, the following formula is applied:

$$SCR = BSCR + Adj + Op.$$

Operational (Op):
The OP SCR is a calculation based on the company's premium volumes and statutory reserves. Premium income and reserves of a Long Term Care insurance portfolio are thus added to those of the company's other portfolios to calculate the operational SCR.

Adjustment (Adj):
The ADJ component corresponds to the loss mitigation effect.

Insurers have the opportunity to reduce the basic SCR for a loss-absorption capacity, linked to a profit-sharing mechanism and deferred taxes.

Indeed, if the risks underlying the SCR calculation turn out as expected, variations may still result due to:

- The fiscal profile of the organization and therefore on balance sheet deferred taxes,
- The value of future optional benefits.

The deferred tax represents future tax gain or charge on the result reached by the variation brought to the balance sheet of the company. In the case of a gain, we add the amount of deferred tax to the asset, then it is referred to as Deferred Tax Asset (DTA). If it does not, it is assigned to liabilities and is called Deferred Tax Liability (DTL).

These variations are likely to absorb some of the capital losses.

6.5 Discussions on the Difficulties Associated with Solvency 2 Implementation

The technical Solvency 2 specifications have not been specifically defined for the Long Term Care risk. As a result, insurers must in some cases make their own interpretations to measure the risk associated with this type of portfolio.

In order to better measure the Long Term Care risk, the Solvency 2 Directive authorizes insurers to use a Partial Internal Model (PIM) which allows them to opt for the standard formula by derogating from it for certain risk modules. However, lack of statistics and history on the Long Term Care risk makes calibrating such a model fastidious.

6.5.1 Premium Revision

The interpretation of premium revisions under Solvency 2 is subject to question. In the event that future premiums are reviewable "without restrictions" and that the criteria of Section 6.3 of Article 18 of the Commission Delegated Regulation (EU) 2015/35 are fulfilled, they should not be taken into account in the projections beyond one year for the calculation of the standard formula. However, such a revision allows the correction of worse than expected experience, which seems necessary for the long-term risks.

If one considers that the ability to revise premiums leads to the retention of a contract maturity of one year, then the contracts become reduced paid up. This situation is often favorable to the insurer, as the premiums can no longer amortize the SCR. On the other hand (see contract maturity), it should be noted that the ability to revise premiums is not the only criterion to be retained in the case of a Long Term Care insurance contract.

6.5.2 Indexing or Revalorization

Under Solvency 2, a revalorization of premiums or benefits, if it exists, is taken into account in the Best Estimate. Often conceived as a profit sharing mechanism, it could require stochastic modeling, which can become costly in calculation time and complexity in the interpretation of the results obtained.

However, contract revalorizations should be dissociated from indexing. In fact, the latter increases premium and benefits in the same proportions and therefore can lead to worsening actual to expected results. Benefit revalorization, on the other hand, can generate a cost for the insurer, especially if they are guaranteed in the contract.

6.6 Conclusion

In this chapter, we have mainly detailed the methodology for assessing the capital requirement for the risk of underwriting the Long Term Care risk. Even if this chapter does not deal in depth with Market SCR and operational SCR, one should supplement the issue risk consideration by an assessment of financial and operational risks. Thus, the insurer will have a holistic assessment of the required capital to deal with the underlying risks of the Long Term Care coverage.

As mentioned in this chapter, the rules for calculating capital required under Solvency 2, the prudential standard in force since January 1, 2016, have not been specifically designed for the Long Term Care risk. They seem to be poorly adapted to this risk, which poses a number of problems of coherence and measurement of the actual level of capital needed for this risk. Nevertheless, regulators offer the insurers the opportunity to deviate from these rules by using a Partial Internal Model. As mentioned above, designing PIM presents its own challenges. As a result, the majority of insurers subject to this standard are not satisfied with the treatment of the Long Term Care risk under Solvency 2 but are forced to apply the standard formula that generates a very high capital requirement.

Given the long duration of liabilities and the significant level of the associated SCR, it is indispensable for an insurer to hold a projection model taking into account management actions. These choices allow one to take into account monitoring tools to be used in order to calculate the appropriate capital requirement. Thus, the integration of management actions is a means for the insurer to meet the prudential requirements by ensuring the continuity of a product with lifetime liabilities.

Finally, on February 28, 2018, the European Insurance and Occupational Pensions Authority (EIOPA) published new recommendations to amend the rules used to calculate the standard SCR formula. Unfortunately, these recommendations do not make any changes to the treatment of lifetime Long Term Care insurance benefits. However, it should be noted that a new standard formula revision is planned for 2020.

End Notes

1. DREES—Study 1046—The French live longer, but their healthy life expectancy remains stagnant (Jan. 2018, French).
   Solidarity and health report N ° 22—elderly people in institutions (2011, French)
   For DREES, see notes 11 in 'The Long Term Care Risk' chapter.
2. EHPAD
   Établissement d'hébergement pour personnes âgées dépendantes, or
   Establishments to accommodate dependent elderly.
   Since 2002, a national law defines and regulates these types of nursing homes.
3. Official Journal of the European Union, Regulation 10/10/2014, Article 18 3 (c).
4. Taking into account an adjustment to the volatility of the risk-free rate curve impacts the Best Estimate valuation, the impact is neutral on the SCR. For the calculation of the risk margin, the rate curve is taken without adjusting for volatility.

5. Total and Partial Dependence
   Total Dependence: 3 ADL out of 5 or Level 1 or 2 for IRG.
   Partial Dependence: 2 ADL out of 5 or Level 3 or 4 for IRG.
   IRG (Iso Resource Grid) see Chap. 3, Sect. 3.1.
6. OCIRP (2017) Baromètre OCIRP Autonomie 2017
   OCIRP: Organisme Commun Institutions Rente Prévoyance.

References

1. EIOPA: Règlement délégué n°2015–35 complétant la directive 2009/138/CE du Parlement européen et du Conseil sur l'accès aux activités de l'assurance et de la réassurance et leur exercice (solvabilité II) (2015)
2. Courbage, C., Roudaut, N.: Empirical evidence on long-term care insurance purchase in France. In: The Geneva Papers on Risk and Insurance—Issues and Practice, October 2008, vol. 33, Issue 4, pp. 645–658 (2008)
3. DREES—Étude 1046—Les Français vivent plus longtemps, mais leur espérance de vie en bonne santé reste stable (janv. 2018)
4. CEIOPS: Consultation Paper no. 51-Draft Level 2 Advice on SCR Standard Formula-Counterparty Default Risk (2009)
5. OCIRP: Baromètre OCIRP Autonomie 2017 (2017)
6. Croix, J.C., Planchet, F., Thérond, P.E.: Mortality: a statistical approach to detect model misspecification. Bull. Français d'Actuariat 15(29) (2015)
7. Guibert, Q., Juillard, M., Planchet, F.: Un cadre de référence pour un modèle interne partiel en assurance de personnes. Bull. Français d'Actuariat 10(20) (2010)
8. Lusson, F.: L'équilibre actuariel de long terme en assurance dépendance en France. Revue d'Analyse Financière (47) (2013)
9. Planchet, F., Tomas, J.: Multidimensional smoothing by adaptive local kernel-weighted log-likelihood with application to long-term care insurance. Insur. Math. Econ. 52, 573–589 (2013). http://dx.doi.org/10.1016/j.insmatheco.2013.03.009
10. Planchet, F., Tomas, J.: Uncertainty on survival probabilities and solvency capital requirement: application to LTC insurance. Scand. Actuar. J. (2014). https://doi.org/10.1080/03461238.2014.925496
11. Sator, N., Sother, G.: Approche solvabilité 2 et ERM du risque dépendance. In: Proceedings of the AFIR Colloquium (2013)

Chapter 7
Solvency Capital for Long Term Care Insurance in the United States

James C. Berger

7.1 Introduction

An insurance company holds a responsibility to its consumers and to the general public to maintain its solvency so that the insurance needs of its customers will be met. Insurance is inherently an uncertain business. What level of financial buffer should be held to be sure the business remains solvent and able to pay future claims? This buffer is referred to as the capital and surplus of the company.

Capital is based on some metric. It may refer to regulatory capital or some internal capital level of a company. Anything beyond this capital is referred to as surplus.

This chapter delves into the various issues in setting the level of capital for an insurer. Capital and surplus do have a cost to a company—the difference between the opportunity cost in holding the capital and surplus and the marketplace return that can be obtained on these funds. Minimizing the capital and surplus is a reasonable goal for an insurer. This starts with solid risk management.

7.2 Risk Management

Risk management considers the risks to a company and the potential mitigation of those risks. As discussed elsewhere in this book, enterprise risk management looks more broadly at the risks and how they impact the whole of the enterprise, not just the segment where the risk initially develops.

When thinking about solvency capital, the process begins with the identification of the risks to a company. Solvency capital discussions do not deal with risk mitigation, per se, but they certainly surface these risks and bring mitigation discussions to

J. C. Berger (✉)
North American Life and Health, GE , 7010 College Blvd,
Overland Park, KS 66210, USA
e-mail: jimberger321@gmail.com

© Springer Nature Switzerland AG 2019
E. Dupourqué et al. (eds.), *Actuarial Aspects of Long Term Care*,
Springer Actuarial, https://doi.org/10.1007/978-3-030-05660-5_7

the front of everyone's minds. Thus, solvency capital levels contemplate more than just the risk of claims being higher than anticipated, or investment income being lower than previously projected. Questions about, for example, economic cycles and immigration patterns come to bear. The many possibilities of the world in which we live should be examined for their impact on not just long term care insurance (LTCI) but upon the company as a whole.

A risk management department has become a standard part of insurance entities. This department may lead the solvency capital conversation, directing thought toward risk mitigation. Then the discussion moves to solvency capital levels for risks that cannot be fully mitigated and must be risk-accepted without full mitigation. Consider some examples.

- During a recession, will the policyholders choose to seek LTCI benefits they otherwise might have waited to pursue? This might increase incidence, and benefit utilization may decrease if the less chronically-ill choose to gain benefit rather than continuing to pay premiums.
- How will investment yields and default rates change during a recession? Is the investment portfolio taking the appropriate level of risk given the company risk appetite?
- What if the government (federal or state) chooses to change LTC benefits? What if qualifying for Medicare LTC benefits become easier? Or harder?

Mitigation tactics might include interest rate swaps, reinsurance, counterparty risk considerations by setting boundaries on acceptable contract terms, premium rate increase actions and regulatory outreach.

Solvency capital is considered in several common ways. First, U.S. risk-based capital is a formulaic construct derived from values that support a life insurance company's blue book,[1] the annual statutory accounting document filed each year with the state regulators.

Next, stress testing capital uses a model that will consider shocks to assumptions and their impacts on the company. These are sensitivity tests done at a severe level. Typically, stress tests start with a change to a single assumption, then assumptions may be combined as insights into the singular assumptions are absorbed. A related test may be called "scenario testing" whereby the shocks are related to specific scenarios addressing possible real-world events, e.g., oil shocks, wars, financial crash. Or what would the Present Value of Ending Surplus (PVES) be if there happened to be mortality improvement of 20% over five years at all ages caused by a pharmaceutical breakthrough with an associated morbidity improvement of 10%. Consider what would happen to the PVES in this scenario if there was no associated morbidity improvement, or what if mortality improvement did not impact morbidity improvement except that those with cognitive claims simply survived longer on claim. A scenario might combine multiple items, and, in fact, usually will. For example, a scenario might consider if immigration in the U.S. should slow considerably result-

---

[1] In the US, there are other colored books for other insurers: yellow for property and casualty (general insurance) companies, orange for health insurers, brown for fraternal societies, etc.

ing in fewer workers to fill nursing home jobs, thus increasing costs, while at the same time the investment return of the company for new money dropped considerably due to Federal Reserve actions to stimulate the economy that simply needs more workers.

Calibrating such tests is not a straightforward endeavor. The goal is to ask what sorts of things are happening in the world and how those events impact the company. Broad thinking is needed, and potentially outside-the-box scenarios should be described. While we cannot think of the Black Swan events by definition, some refer to Grey Swans as those events we could conceive of but which we may too quickly discount, thinking them to be too unreasonable to address. Perhaps the scenario creators are being too dismissive of real possibilities no matter how unlikely those scenarios might be.

Another type of solvency capital is economic capital, which is the amount a company determines it needs to be able to weather a series of negative events of a period (e.g., one year or lifetime) at a confidence level chosen by management (say, 99.9%). At the 99.9% confidence level, the company is saying it would have enough capital for any set of events coming its way in all but one instance in one thousand. This calculation is the most complex of those discussed here. Economic capital is detailed later in this chapter.

7.3 Positioning the Risk

The risk appetite of a company describes how much and what type of risk a company is willing to accept in order to meet its strategic objectives. Will it invest in assets below investment grade? Will it pursue lines of business that are more difficult such as LTCI? Will it invest in a captive sales force, meaning that the agents will only sell its products but will have certain other benefits?

The risk tolerance is the measurable limit of the risk appetite, e.g., what portion of assets can be from a single obligor, or what portion of a company's LTCI policies can be written by any one of the company's general agents. **Jill Douglas, head of Risk for Charterhouse Risk Management** stated "*The risk appetite statement is generally considered the hardest part of any enterprise risk management implementation. However, without clearly defined, measurable tolerances the whole risk cycle and any risk framework is arguably at a halt*". With LTCI, tolerance levels can include elements of pricing such as maximum sales levels or the challenges that may exist from the state regulatory environment and the need to limit policy issuance in any given state to control risk. A company may decide to write no more business in a state that is not forthcoming with rate increases the company believes appropriate. As there are not many LTCI writers these days, losing one may significantly decrease the volume of business sold in a state. (This may be the harbinger of higher participation in Medicaid LTC benefits in later years, something states should wish to avoid.)

When pricing an LTCI policy, assumptions must be made as to the future of claim frequency and severity, the persistency of policies both regarding voluntary lapsation and policyholder deaths, and the level of investment return a company can expect. Early LTCI experience in the United States showed a significant overestimation of policy termination assumptions. In the mid-1990s, ultimate lapse rates were commonly projected to be 5% per year but, in fact, have been generally less than 1%. The impact of missing this annual assumption has a similar impact to missing the investment return assumption by the same percentage. In the numbers above this would translate to a 4% lower annual investment return than expected, a large miss. This adds to the pressure of investment returns estimated in the late 1990s to be in the 6–7% range for the next fifty years but turning out perhaps 3% lower than anticipated. While some companies employed interest rate swaps to hedge against this gap, many did not since they expected rates to return to the higher historical levels.

When sensitivity testing the pricing of LTCI, companies would have been well served to note how badly things could go and to express a risk appetite that limited sales to a manageable level. It was common to think of the need to grow the block to drive down administrative expenses. Were risk tolerances set and adhered to?

The pricing risk of the claim costs can include slopes on claim incidence rates that are too flat by attained age and severity assumptions that were too low. It is not common today to issue policies after age 80, so there will be a lag between when new business moves from the low issue ages for new business into the higher ages of 80 and older, where a large portion of claims occur. Actual levels of incidence and severity can also change over time, thus adding to the uncertainty. If assumptions were sound when the block was written with an average age of 60, will they still be true twenty or thirty years later? If not, a risk margin would be warranted. One positive note can be found in the experience of the utilization rate of the Maximum Daily Benefit (MDB). With lower interest rate environments, the thought is the claim inflation rate is also lowered. This causes reimbursement benefit claims to be less expensive, especially for policies with a compound inflation rider. This does not make up for the loss of investment income but does somewhat dampen its impact.

Strategically, a writer of LTCI must ask what can go wrong and how any problems will be handled. The insurer may wish to limit its exposure by setting risk tolerances for sales until it is able to understand the business. The newness of the LTCI risk in the U.S. marketplace would suggest this could take some time. Consultants may have access to better data than the insurer, but they still may be limited. One consultancy has regularly updated its claim cost curve, steepening it at each iteration. But consultants do not have crystal balls, just a bit more knowledge of what others are seeing as of today. With inherent knowledge limitations, how much exposure should be accepted? While the marketing department may want to write as much as possible, commissions can be adjusted to get the right level of strategic exposure—enough to meet fixed costs of information systems and marketing, but not so much it would cause significant harm if things turn out worse than projected, a very real risk for LTCI. On the other hand, limiting commissions may discourage the very agents who bring in the best risks. These agents may write for another company. One way around the administrative issue is to hire a Third-Party Administrator (TPA). A TPA will do underwriting per the direct writer's specifications, administer the policies, and manage the claims. The direct writer will need to audit this effort but typically

TPAs are knowledgeable and continue to adjust to new information as it arrives in the marketplace.

One classical solution to the risk exposure issue is reinsurance. In the current U.S. market, reinsurance is quite limited, if available at all. As of the beginning of 2018, there is only one reinsurer with any interest in taking new business. Occasional boutique reinsurers come along with interest in acquiring older blocks that direct writers wish to dispose of, but these boutique entities have generally not been successful in matching their perceived investment prowess with the risks on the liability side.

The writer of LTCI will want to consider exposure related to ages, product types, markets, and any product innovations. LTCI pricing is not always done with the goal of equal return on all ages or all product design; this statement is from a deterministic pricing perspective. If stochastic pricing is used and there is an assumption that one knows the proper distribution of the risk, this may cause the writer to want to focus on, for example, ages 60–70 and specific types of inflation protection, such as guaranteed purchase options where a policyholder can increase coverage every three years by predetermined amounts with an accompanying premium increase. There may be no interest in Home Care only policies that may have unfavorable risk distributions. Group products may carry too much risk as they tend to take a long time for results to develop. Does the company want to be a market leader in policy design or would it rather be a follower in hopes of avoiding being on the leading edge, often called the "bleeding edge"? These are all part of setting a risk appetite with specified risk tolerances.

A company should determine if it has a unique expertise. Perhaps it has well-developed marketing channels to seniors or to the worksite. These do not necessarily mean that the expertise will transfer well to LTCI. Many Medicare Supplement writers thought it would be natural to add LTCI to their portfolio of products, only to find that the skill sets are not as similar as initially thought. The sale of LTCI looks more like a wealth-protection opportunity than a medical care solution, while its administration is more like disability income insurance. It has been said that in life insurance, claims are paid; in medical insurance, claims are processed; but in disability insurance (and LTCI), claims are managed. With the sales process of LTCI different from that of Medicare Supplements, the administration of claims is different, the underwriting is different, and it is much more complex, it is easy to look back and see that the fit was not as anticipated.

Assets are a place where a well-crafted tactical approach can greatly assist. Some assets are not a good fit for LTCI, such as short-term maturities. When a LTCI policy is written, its liability may start with a duration of roughly 100 years. This creates a huge exposure to interest rate risk. This duration quickly falls to a more manageable level, but a level still in need of long-duration assets. LTCI assets may gain from private placements of good investment quality. Due to the long risk exposure, equity may be of value, except that it is penalized in the U.S. risk-based capital formula. To manage this asset portfolio, the manager must consider how long this risk is. A policy sold to a 50-year-old doesn't reach the age of the highest claim costs for around 30–45 years. The longest maturity bonds are typically 30 years, and these are in demand from many sectors such as life insurance and pension funds. But investing in shorter terms may not get the level of return needed, and may leave the company with the risk of dropping interest rates just when reinvestment of the final bond payout occurs. In addition, the asset default risk requires constant attention. Thus, it may be desirable to increase the rating of the asset portfolio.

Risk appetites and tolerances need to be well considered and then adhered to so that the portfolio of LTCI risks will not cause major damage to the company in subsequent years.

7.4   Components of Risk Management

Risk management considers the many perspectives of risk, ranging from strategic to legal to insurance risk. Strategic risk was discussed in the previous section. A company should ask "Is LTCI is a good fit for this company?". Considerations include company expertise, markets, and risk tolerance metrics.

Consider the market. LTCI covers mostly seniors. Older citizens are given special regard in society, as they should be. If something goes wrong in how a company is perceived to treat seniors, there could be a strong reaction. This did occur with one U.S. LTCI carrier which ultimately decided it was better to let the state of domicile, Pennsylvania in this case, take over the portfolio than to continue to attempt to manage a situation that was hurting its reputation. The company received negative press, and it appeared that reputational damage could only get worse if the company was unable to fully meet its obligations in that block of LTCI. With the addition of some surplus, the company negotiated the state takeover of the business so that claims could be managed per the state's policies, and rate increases and benefit revaluation could be requested by fellow regulators rather than by the for-profit company.

In another situation, LTCI led the company Penn Treaty/Network America into receivership. As a mono-line company (PNTA substantially only offered insurance in one product line), it found premium was insufficient and premium rate increases were not at the level required for long-term survival. LTCI is a product line with significant leverage, so that a company without deep pockets may find itself in trouble.

Another risk to manage resides in the legal realm. This can be particularly true when dealing with claims administration. As indicated above, when challenged by a senior, a company is already on the defensive. Claims adjudication is complex, and there are many subjective elements to be judged in the process. If a company is sued, its options may be limited. Good processes in the claims area are paramount in avoiding these issues. These processes would involve solid communication protocols, clear claim adjudication guidelines, and experienced leadership to deal with the inevitable ambiguities.

Regulatory risk is found in LTCI policy filing, and later in any requests of rate increases. In the 1990s, some states would require specific policy provisions, or would demand lower premium rates than initially submitted. This increased the risk to writers of business in these states. Rate increase actions have been a sore point for states, companies, and policyholders. Companies feel they are justified in asking for increases, but states appear to feel that the policies should have been better priced so rate increases of a smaller amount, if any, should have been needed. LTCI policies are almost always filed as guaranteed renewable, meaning that rate increases are allowed for a class of policyholders within a state given the approval of the state insurance commissioner. States responded to rate action requests in quite diverse ways, with some states granting most actuarially justified increases, while other states appear to do all they can to not give an increase. Of course, those states that have been more accommodating to the companies feel their policyholders are paying for the rate increases other states have denied—an unlevel playing field is being created. Even the rate stability regulations of the early 2000s, while helping to move rates to a more sustainable level, did not fully correct the situation. In part, this reflected the developing knowledge of LTCI, both in the company and regulatory communities, and the inability to predict investment returns many years in the future. For policyholders who had paid a level premium for many years, the prospect of their premium perhaps doubling or more led to the policyholder dropping their policy. This created a difficult situation as many of them had reached retirement with fixed income, with a more critical need for their policy coverage as their health may have deteriorated.

Business risks involve underwriting, policy issuance, policy administration (including rate increase implementations), and claims administration. Also, the risk of knowledge management must be well-thought-out. Knowledge of LTCI is not common in the industry, so maintaining qualified staff and transferring the knowledge to the next generation will be important for risk mitigation. The challenges of administrative systems being able to unify the policy data, any rate increase adjustments, regulatory requirements, and the administration of claims require a major investment. Is it worth it to a company? As discussed, one possibility is to outsource the policy administration to a TPA. This outsourcing can extend to the use of turnkey products, which are designed, priced, and administered by the TPA, perhaps reinsuring part or all of the risk. If a company starts into the business and then decides that it no longer wishes to participate in this market, what recourse do they have? It is unsure they would ever be able to sell the business to another entity at a later date. In one case even with the sale by reinsurance to a boutique reinsurer, a deal fell through two years later, so the business and all its risks returned to the original insurer.

The design of the stand-alone LTCI product has typically included the company's acceptance of the long-term investment risk, as well as the risk that the long term care delivery system will change in an unpredictable way over the life of a policy. Several approaches to move some of the investment risk to the policyholders have been considered, and in some limited cases implemented. These include a product that featured increases to benefits when the company investment portfolio did well. Another hedge is found in so-called combo products which are a combination of a life insurance plan or an annuity with an LTCI rider. In theory, combining these risks reduces the overall volatility of outcomes by using offsetting risks of mortality and LTC. Early experience has emerged favorably for the combo products when compared with stand-alone LTCI products.

Product design is a risk management tool. Insurance of any kind transfers some or all of a risk from the policyholder to the insurer. With stand-alone LTCI, the company takes on all the policyholder's risk with certain boundaries (e.g., daily and policy benefit limits, claim adjudication structures). Transferring some of the investment risk is more difficult but can be somewhat accomplished. An example of how the industry became smarter over time is with the restoration of benefits (ROB) provision. This provision attempts to give a recovered claimant a fresh start at benefits. If the claimant had a policy with a 3-year benefit period and had used one year's benefits before recovering, a six-month period of wellness may provide a restoration of all three years of benefits. Early policies stated that if no benefits were claimed for a six-month period, benefits were restored. Savvy claimants simply stopped a day before exhausting benefits and waited six months to restart claims. This met the requirements of weak ROB policy language but certainly not the spirit. Later policies required evidence of wellness, not just lack of claims. This illustrates that both policy design and policy language are risk management tools.

The broad community of LTCI including companies and regulators have been seeking design alternatives. There are few companies currently marketing LTCI which will mean more pressure to the state Medicaid roles, so it is in all parties' best interests to create new avenues to cover the risk of LTC.

Insurance risks—morbidity, lapse, and mortality—are inherent to the insurance world. Morbidity is a complex assumption. Many companies deem it best to use a first principles model which parses a claim cost into its claim incidence rates, its claim termination rates, and its utilization rates of the maximum daily benefit. In the most detailed first principles models, an assumption is required for the transitions between care settings (home care vs. assisted living facilities vs. skilled nursing facilities). The first principles method is more precise than using a claim cost that aggregates these claim elements together along with a discount rate, but it has a liability in its inability to tie the assumptions to financial statements, the ultimate source of "reality" for any company, as can be done with a claim cost approach. In addition, there are considerably more first principle assumptions to be made than with claim costs, which could lead to inadvertent bias. This said, both the first principles and claim cost methods are valid. A company's management must decide what makes the most sense in their situation. Many actuarial software packages can do both, and perhaps using both can give greater confidence in a model.

It is emerging that LTCI has a mortality rate unlike any standard industry mortality table. This may be due to its unique market and underwriting, but also may be an artifact of the mortality data which lacks clarity on the reason for a policy termination. Did the policyholder simply opt to stop paying premiums (especially in an environment of rate increases) or did the policyholder die? The company may have no straightforward way to gain that answer with certainty and may never know its true mortality experience. Even expensive-to-use public databases may not be complete. Using a first principles model requires mortality to be separated into healthy lives and disabled lives, further challenging a company's understanding of its mortality experience.

All the insurance risks point toward a risk that the model assumptions are in some way flawed. Is the modeler thinking about the LTCI risk correctly with all of its intricacies? Is the modeler over-thinking or under-thinking the model? How can she know of what she doesn't know?

7.5   Management Tools

Company management will need tools with which to understand the risks of the company. These tools come in various forms. Since the early 1990s, the results of Asset Adequacy Testing (AAT) in the U.S. have been a primary tool in setting regulatory capital levels. The test asks if the assets of a company are adequate to mature its liabilities. This testing is commonly done using Cash Flow Testing (CFT), where liabilities and assets are projected along interest rate scenarios. Then some metric such as the Present Value of Ending Surplus is used to examine the resulting cash flows. The interest rate scenarios may be deterministic or stochastic. Deterministic scenarios often include the New York 7 (NY7) which were put forward by the state of New York in the early 1990s. The NY7 includes seven scenarios:

- a level scenario where there is no change to interest rates,
- pop-up and pop-down scenarios where the interest rates suddenly increases (decreases), say, by 3%, to a regulatorily-specified level and stay increased (decreased),
- up-down and down-up scenarios where the interest rates increase (decrease) for five years at, say, 1% per year, and then return to the original level over the next five years, and
- up scenario and down scenario where there is a gradual increase (decrease) in the interest rates of, say, 0.5% per year for ten years followed by level rates at the increased (decreased) level after the ten years.

These deterministic scenarios are often augmented by scenarios that are half as severe, or by other rate patterns of interest to management. Down scenarios may need adjusting to not produce negative interest rates. The key is to show how the company performs in adverse scenarios, so that the adverse elements may be mitigated in some way. For LTCI writers, a large risk is that investment returns stay low over the life of the business. Scenarios where rates increase do not tend to be of concern, though with the interplay of interest rates, investment returns, and the rate of increase in the cost of care which may be correlated with interest rates, the "up" scenarios should not be completely ignored.

Asset-Liability Management (ALM) is a part of the management effort for which CFT gives insight. LTCI tends to favor long-maturity high-grade assets since the liabilities of LTCI are long-maturity. Matching the cash flows is not completely possible since most blocks of LTCI have policies lasting longer than the typically longest-maturity fixed income assets of 30 years. Investment in equities would provide a potential solution but these are penalized in the U.S. Risk Based Capital (RBC) formula to the point that they are typically not worth holding. If cash flows cannot be matched, companies tend to work to match duration (the sensitivity to changes in interest rates) with an eye on asset convexity (a second-order sensitivity to interest rates) as well. Best intentions in this regard may give way to the realities of what investments can be found. The universe of investment-grade 30-year securities is currently under significant demand, thus, depressing yields and quality.

Other management tools include stress testing and scenario testing as discussed in Sect. 7.2.

The Own Risk Solvency Assessment (ORSA) is addressed in a separate chapter of this book. The U.S. implementation of ORSA may vary slightly from the European version; at the time of this writing U.S. ORSA is still a work in progress. Conceptually, ORSA is still about a company's view of its risks rather than a regulator's view or a rating agency's assessment.

For any management tool, it must be considered whether the tool is founded on solid assumptions. Economic scenario generators, such as interest rate generators, are based on many assumptions that are difficult for all but the expert to fully appreciate. The interest rate scenarios that are generated are then simply one possible set of rates based on some past interest rate experience, and it is hoped they represent possible future realities in a robust way, but this cannot be completely certain. Then these interest rate scenarios go into a model of the company using a myriad of assumptions about mortality, voluntary lapse, benefit exhaustion, morbidity, expenses and commissions structures. Many of these are based on past experience which is hoped to be progressing in a stable way into the future, but will it? Through additional sensitivity testing an insurer learns where to focus its risk mitigation efforts. Do we know of parameters that may be subject to significant change in the future? For LTCI, this may be particularly challenging due to the long timeframe of the product and assumptions that have significant elements of consumer choice embedded, e.g., how care is used or what is lapse behavior when a rate increase is implemented. Can we adequately predict changes? Or can we at least attempt to model several possible future paths? This moves into the realm of expert judgment. The question should be asked if we can truly measure the things which we are attempting to calibrate. With limited data points on dynamically changing parameters, how much is knowable? Still, modeling must be done, and ranges of assumptions can give meaningful guidance in setting capital and surplus levels.

Experience has taught that there are no easy judgments. Approaching a question from multiple directions and with multiple parties tends to bring the best results in the long haul. For example, LTCI modeling can be done with a first principles approach or with claim costs, as previously mentioned. By approaching the problem of benefits with both the first principles and the claim cost methods, we gain a clearer understanding of the risk and more confidence that we have a reasonable model; it is reasonable with respect to the experience developing on each component of a claim, and at the same time is reasonable with respect to the financial statements of the company. This same approach of using multiple views bears fruit when used in the development of solvency capital, whether it be for stress testing, economic capital, the ORSA, or elsewhere.

7.6   Economic Capital

In the world of solvency capital, economic capital ("e-cap") is both the most complex and the most difficult to execute. Its sophisticated view of solvency cuts across all the risks of a company and considers all the correlations of these risks. Conceptually, it is answering the question of how much capital is needed to keep a company solvent at some confidence level, i.e., probability level. Typically, this level will be a very high probability such as 99.5% and be expressed over a given time period, typically one year or over the life of the business. While these seem to be clear guidelines, they become less so when the very pertinent question is asked as to what management thinks e-cap is and how they will be using it. If the modeler is modeling something different from what the company management is expecting, poor decisions may result.

Value-at-risk (VaR) is a common measure of tail risk used in e-cap. The European regulatory capital standard is Solvency II which uses economic capital set at a 99.5% VaR, this being the value at which 99.5% of simulated outcomes fall below the value and only 0.5% above. In other words, only 0.5% of modeled losses exceed the 99.5% VaR value; only events of the 1-in-200 nature give a solvency issue. Some companies have chosen a level of 99.97%, a very high bar indeed. These confidence levels are determined by management or even the company board of directors and are influenced by the rating agencies that give higher ratings to companies with high confidence levels. One rating agency sets 99.97% as the bar for an AAA rating, while 99.95% is needed for an AA rating, and so on.

Another measure used for economic capital is the Tail VaR (TVaR)[2] which is the *average* of the exceptional events in the company risk distribution in excess of its threshold, say, 99.5%. Where a 99.5% VaR would set the economic capital level at the 99.5% point on the company risk distribution, a 99.5% TVaR looks at all points in excess of 99.5%, averages them, and sets the e-cap value at the level of the average. Clearly, TVaR is a larger number than VaR at the same confidence level. Therefore, TVaR may be set at 99.5% to give roughly the same e-cap value as VaR at a higher confidence level, say, 99.9%. While VaR is more intuitive, it is not as mathematically robust, a matter outside the scope of this discussion.

A typical economic capital calculation would involve the creation of a correlation matrix among the identified risks of the company. Such a matrix is difficult to derive from company data due to the limited data inherent in the process. This is less an issue with market risks such as interest rates and credit defaults which do have broad industry data, but for insurance risks where data might be thought to lose its relevancy after ten or twenty years and be partially dependent on the company

---

[2]TVaR is also known as the Conditional Tail Expectation (CTE).

itself, then there may be only ten or twenty data points representing the ten or twenty years of data.[3] A situation like this calls for significant judgment and will require the input of people across the organization to determine with some level of comfort what these correlations may be. There are alternatives. Solvency II provides just such a set of correlations though they may need adjustment for a company's specific risks. Solvency II correlations are also set at the 99.5% confidence level rather than 50%; in the aggregation process discussed below, this must be considered. Perhaps some of the Solvency II correlations should be reduced if the matrix is to be calibrated to a 50% confidence level. Other sources of correlations will be discussed below.

The accounting basis of economic capital must be defined. Typically, it is said to be calculated on an economic basis meaning a best estimate basis. An economic balance sheet is to be used in this formulation. While this is appealing on the surface, one must ask how decisions in the company are actually made. Are they based on economic values, on statutory or GAAP financials, or some other internal measure? As stated above, economic capital should ideally align with how decisions are made.

Are decisions made based on actual experience or on how the financials change? In other words, if a reserve change is required, is this part of the economic capital calculation or is it not considered since the economic capital is only based on economic and not financial events? One can easily argue that reserve changes are an actual event which depicts a change in the expectation of the present value of future cash flows. Some may think that changes in a one-year timeframe to the actual experience are the target but with a long-tailed product such as LTCI, this simply kicks the can down the road without addressing the issue at hand. This is not how an insurance company should be managed.

The time horizon is of importance. Will the calculation be about changes over a one-year period or over a longer or shorter period? One-year is common though by no means universal. Many companies look at events that may occur over the lifetime of the business, not just one year. The answer to these questions is not usually the domain of the economic capital modeler, rather it is a management decision based on how the economic capital results will be used.

Another modeling decision will be whether to use a market-consistent approach or a real-world approach. Market-consistency refers to the distributions being calibrated to reproduce actual market prices. A real-world approach ignores this requirement and choses distributions that fit the company's view of the world. It tends to be that market-consistent approaches go with one-year time horizons while real-world approaches go with lifetime horizons.

Various ways of doing e-cap exist. It can be done on a deterministic basis through stress testing at various levels of uncertainty. More often, e-cap is found through simulation using the individual risk distributions. For example, with LTCI, the probability distribution around voluntary lapse, benefit exhaustion, mortality, and morbidity may be determined. For each of these risk elements, a first pass may use a lognormal dis-

---

[3]Time periods of less than one year would be inconsistent with a one-year timeframe. While, say, quarterly data can be combined into annual effects, it contains more limited data in each period so the volatility is anticipated to increase.

tribution, a distribution which is fully determined by only two points. The question being answered is how the present value of ending surplus is impacted by changes to, for example, the lapse rate. These values may be found in sensitivity tests run during the CFT work. If a factor of 0.9 applied to lapse rates is determined to be one standard deviation and no change is the distributional mean, then these two points have fully determined the lognormal distribution. A company may decide to start with total policy terminations rather than breaking down the terminations into lapse, death, and benefit exhaustion. This may be expedient as the determination of distributions for all three components may be too much, at least initially and correlations among the three components would also be needed. It is better to have a working baseline model which is understood than one that is too complex to know if it is really doing what it is thought to be doing.

LTC Morbidity Modeling

Similar to total terminations, morbidity can be considered in multiple ways. There are three components: incidence, continuance, and utilization. A distribution can be formed by a Monte Carlo simulation around overall morbidity that changes the values of each of the three components randomly with each trial. But are there correlations among the three components? If so, a correlation matrix must be introduced. Parameterizing the correlation matrix is an exercise in judgment that may be of less value than stopping with the aggregated claim cost estimates.

A risk can have multiple dimensions: random fluctuation, parameter misestimation (discussed below), and trend risk such as morbidity improvement or cost of care inflation. These must all be considered. A possible distribution for the first two risks is the beta-binomial distribution. This distribution can be readily researched, and other probability distribution options explored. These other options depend on the modeler's degree of uncertainty about the variability that exists in each component. Judgment is a critical part of this exercise. Another way to tackle random fluctuation is to use standard actual-to-expected (a/e) studies. Since it would be common to use these studies in, say, five-year age bands to set the estimates of the rate of claim cost, these studies would be a natural place to look for variance. Segment the policies into, say, four random segments of, ideally, equal size and explore the standard deviation in the resulting a/e between the four segments. Multiple divisions of the data into four segments, or even more segments, will give several views of the variability.

Let V = Variance [a/e(1), a/e(2), a/e(3), a/e(4)]. Then the Standard Deviation SD is sqrt(V). But this SD is among the four segments when what is wanted is the standard deviation of the whole which is four-times the size. So the desired standard deviation is SD/sqrt(4) or SD/2.

At this point, the modeler will have standard deviations by age band and perhaps gender. These standard deviations can be used in a simulation of new claim cost curves based upon the base curve. By weighting the ages by their current exposures times their present value of future claims, a ranking of the simulated curves can be done and the variance among these rankings used to determine a distribution of percentage variances from the base claim cost curve. Thus, random fluctuation is quantified through this distribution. It may be convenient to find the best fit of the simulated data using another distribution such as a lognormal or normal, and using the subsequent parametric distribution for the economic capital modeling. Purists may prefer to use the actual simulated values as the distribution rather than the parametric representation.

Parameter misestimation examines the parameter used in, say, a binomial distribution. The probability of not going on claim will be denoted as p. Suppose there are 25 claims out of an exposure of 1000. This would suggest $p = 975/1000 = 97.5\%$ and $q = 1 - p = 2.5\%$. But the 25 claims represent only one realization of what may happen. Due to randomness, the claim count could have been 28 or 24 instead of 25. $q = 2.5\%$ is picked as we have no other information with which to calibrate but the "true" underlying rate may be 2.4% or some other value. To consider this misestimation problem and to see what distribution of possible values for the rate q may be, a common device is the Beta distribution, which can be easily researched.

Trend is more difficult yet to model. Modeling starts with the delineation of the two major trend elements, Morbidity Improvement (MI) and Cost of Care inflation (CoC). The author's experience suggests that these two risks dominate the LTC morbidity risk making random fluctuation and parameter misestimation secondary issues. While MI and CoC do impact the level of claims and thus the parameters used in modeling claims, they are distinguished from parameter misestimation; MI and CoC are long-term changes to claim levels while parameter misestimation examines the volatility of experience used in setting rates, not the general, over-time drift (i.e., trend) of these rates. Unfortunately, these trend risks have more judgment involved. Data by which to set a MI assumption is quite imprecise and some actuaries feel there is not enough data to determine anything. It is the author's opinion that MI decreases with attained age, but beyond this, the data holds little in the way of clear answers. And the less data a company has, the less clear the answers. Judgment will require the input of other knowledgeable parties both inside and outside the company, and few will have solid opinions.

Once some basic idea of MI is arrived upon, alternative scenarios of how MI may play out over the life of the business are required. The calibration of these alternative views is not obvious but is necessary. Deterministic scenarios are one alternative while stochastic models based on judgment is another. It is sobering to repeat that this assumption is quite significant.

Cost of care inflation refers to the increase in the cost of nursing homes, assisted living facilities, home health care, or any other item covered as a benefit under a company's policies. This inflation is almost always relative to a Maximum Daily Benefit (MDB) cap. An exception is an indemnity policy which pays the full daily benefit no matter the amount of service being used. For policies that only reimburse actual charges, the utilization rate is the rate at which the daily benefits are being used. If a policyholder has a MDB of $200 but is only utilizing $120 each day for home care services, then the utilization rate for this policy would be 60%. If the overall rate of cost of care inflation (before caps) is 3%, then the next year this $120 would inflate to $123.60. If the policy has an Inflation Protection Option rider (IPO) that inflates the MDB at a compounding rate of 5% each year, the $200 would inflated to $210 so the utilization rate would fall from 60% to $123.60/$210 = 58.86%. On the other hand, if there were no IPO, the utilization rate would rise from 60% to $123.60/$200 = 61.8%.

Claimants may already be receiving their MDB. In the example above, the claimant might be in a nursing home which costs more than the $200 per day benefit, say, $202. The utilization rate would be 100%. The next year at 3% inflation the policy with no IPO still has a benefit of $200 even though the nursing home expense has increased to $202 × 1.03 = $208.06. If the policy has an IPO, the $200 MDB would increase to $210 and the utilization rate would fall from 100 to 99.08%.

Not mentioned so far are the various other IPOs that exist. Above, the assumption is made of what is known as a compound IPO where benefits increase at, say, 5% compounded each year, a common level due to regulatory requirements. Also available from most companies are simple IPOs where the increase is, say, 5% of the *original* MDB, not the *current* MDB as with the compound IPO. Yet another alternative is a guaranteed purchase option (GPO) which could take many forms. A common structure would allow for an increase in benefits every three years of 5% compounded annually over those three years, or 15.76%, if the policyholder pays for the increased benefits using premium rates for the policyholder's attained age, not the issue age. Neither the simple IPO or the GPO will be covered further here but each adds complexity to the problem.

Modeling the CoC necessitates a way to inflate benefit payments while keeping in mind the MDB caps on the policies. Thus, the overall block utilization rate will not vary simply with the type of IPO but also with the policy duration. Can a company's data show how this might work for various levels of current utilization? Further, the impact of the CoC inflation rates on the calendar year may be modeled through correlation with Treasury yields or company investment returns; these should be contemplated in some manner. As a modeling challenge, this may be the largest one for LTC morbidity. CoC may have the largest variance of any LTC component, so the choices made are paramount in modeling LTC morbidity.

There are elements of model parameterization that are uncertain. Uncertainty is different from random variability in that the variability has a distribution of possible values while uncertainty describes a situation where there is no clear distribution. For example, if model risk is incorporated at the morbidity level or at a higher level, how is this parameterized? There may be a few data points to consider, but their relevance and their associated confidence levels are uncertain elements.

This author's experience has shown that the development of the individual or marginal risk distributions is a time-consuming and challenging endeavor. A satisfying result may not come in any reasonable amount of time. Seeing e-cap as a process with much learning about company risks along the way is very much a part of the experience.

Correlations
Once the marginal risk distributions are determined, the aggregation process begins. A simple aggregation would add up each risk determined at the desired confidence level, say, 99.9%. This would ignore the diversification benefit among the risks.

The first step in taking diversification of risks into consideration would be the use of a correlation matrix. Such a matrix contains the correlations between each risk so that, for example, ten risks would have 45 correlations. The diagonal entries of the matrix are all 1.00. The matrix is symmetric so that $x_{i,j} = x_{j,i}$. Developing such a matrix is a potentially daunting challenge but there are suggested matrices from sources such as Solvency II, the Australian Prudential Standards, and the CRO Forum. These sources give roughly the same results but do have differences in the structure of the risk correlations and in the values of some of the correlations. Note the confidence level these sets of correlations assume. Overall, this is more art than science. Nonetheless, once a correlation matrix is established, sensitivity tests on the matrix can find the most important assumptions, worthy of the most thought and care. A good test would be the sensitivity to a broad shift to the matrix of, say, adding 25% to all correlations with a maximum of 100%. Generally, these matrices are developed with values which are integer multiples of 25%. More precision than this may be spurious. This author has used a rough thought process which says 0% means there is no correlation; 25% means there is some but little correlation; 50% means there is certainly correlation but many other influences, too; 75% means there is strong correlation, but not complete correlation as would be the case with 100% correlation. Negative correlations are similarly devised. Once again, this is quite judgmental.

The above suggested sources for correlations ignore LTCI. It would be easy to look to disability income for the needed relationships. French actuaries have correctly identified this approach as improper, noting that the Solvency II values would make nonsense of LTCI risks. Disability correlations are too severe for LTCI and should not be used. The LTCI actuary will need to do some thinking to plug this hole in publicly available matrices.

Correlations must be thought of in the context in which they will be used. For example, the calibration of the interest rate/credit default correlation might start with the assumption that interest rates drop in challenging investment times which have higher defaults. Thus, these are negatively correlated. But interest rates dropping is adversarial for LTCI, in addition to higher default rates. Now this becomes a positive correlation. Yet in some situations, rising interest rates can cause problems for defaults as companies can't turn over their loans at the same low rate. This particular correlation deserves much attention.

Once correlations are determined, they should have regular periodic reviews, say, biannually. This is necessary because correlations may change over time. For example, the correlations between hedge funds, banks, brokers, and insurers have increased over the last 20 years. They may continue this trend in subsequent years. This may also happen with the other LTCI risks.

Simply aggregating the risks with the correlation matrix may reduce the e-cap value by roughly half. While the correlation matrix is a large step forward in correctly understanding the level of overall risk, one final step can give a superior picture of the performance of the risks at the tails, where economic capital resides. Copulas are mathematical functions which combine marginal risks with their associated correlation matrix while potentially giving mathematical credence to the statement "in a crisis all correlations go to one". Details about copulas are beyond the scope of this book but may be readily found, though their intuition is not trivial. The most basic copula is the Gaussian copula, which is based on the normal distribution. It provides no change to the correlations in the tails and is often eliminated as an option due to this property. When one thinks carefully about the nature of insurance risks (though not considering market risks) this elimination may be a hasty decision. Insurance strives to avoid moral hazard, thus decorrelating risks from one another.

Another common choice is the t-copula with its degrees of freedom used to calibrate tail correlations. It has the small drawback in that it impacts both the left and right tail, but only one tail is the "crisis" tail. The other tail is where things are going quite well and no tail correlation would be expected. The t-copula can be readily implemented, even in a spreadsheet. It combines all risks using the same tail convergence, i.e., degrees of freedom. An alternative is the individuated t-copula. It allows that some risks are more tail-correlated than others, i.e., that the degrees of freedom among risks may vary.

A third choice is the Gumbal copula which has heightened tail dependence at only one tail. Its implementation is more complex and is not discussed here.

There are other purposes for which e-cap can be used besides setting capital levels. It does give a clearer view of the nature of the risks to a company through its rigorous conceptualizing. It is also quite useful in determining the relative risks between product lines so that riskier products can be loaded with greater capital requirements. This process allows for risk allocation which keeps a company from unwittingly selling too much of a product without having the appropriate capital to back it. Another use would be in compensating product line managers. If there is more risk brought on by the sale of a product, should a manager be compensated as much for good years as the manager who brings on low-risk exposure? Probably not. While perhaps difficult to explain to managers, once it is set, managers know the parameters by which they will be judged.

7.7 Risk-Based Capital

Risk-Based Capital (RBC) is a formulaic risk mitigation calculation used by insurance companies and banks in the U.S. and elsewhere. It establishes minimum capital requirements meant to protect these firms and their stakeholders by ensuring sufficient capital on hand in case of operating losses, while protecting the market as a whole.

RBC was implemented for US insurance companies by the National Association of Insurance Commissioners (NAIC) in 1992. The development of RBC for insurers was in large part a response to the junk bond and real estate concentrations of the 1980s. In the U.S., insurance is regulated to the largest extent by each state. This is unlike banks, which are largely regulated nationally by the Federal Reserve though there is significant state regulation of banks when only operating in a single state. The NAIC coordinates insurance regulation among the states with a solvency focus. While regulations are not uniform, they are broadly the same among states with model regulations being created by the NAIC and then given to states to put into their regulations or to modify before implementing.

RBC attempts to give one structure for determining regulatory capital levels. This structure will vary depending on the type of insurer: health, property and casualty, life, mutual, fraternal. LTCI may be found under any of these insurer types. The formula is based on factors applied to various statutory (vs. GAAP) values found in an insurance company's annual statement. The four major categories of life insurance company risk in RBC are

1. Asset risk (known as C-1 risk), which encompasses market and credit risks, and is based on asset amounts shown on the insurers annual statutory financial statement,
2. Underwriting, insurance pricing, or claim risk (C-2), which uses the net amount at risk, premium, and reserves,
3. Asset/liability risk (C-3), which considers the risk of changes to interest rates, and
4. Miscellaneous business risk (C-4), based on premium, expenses, and other measures.

These four risk categories are then combined through a formula that contemplates correlations among these risks, what is often referred to as a diversification benefit. In other words, if one risk goes negative it does not necessarily follow that all the other risks go negative, especially going negative at the same rate as the first risk. As noted earlier, it is said that "in a crisis, all correlations go to one." This appears to have been true in the 2008–09 financial crisis and would give pause that correlations should be granted in the calculation of RBC. But RBC covers risks at a lower level than might be needed for a crisis. Crisis-level capital is more the realm of economic capital which can be set with a much higher threshold.

RBC is an early warning instrument for regulators. A company first calculates its Total Adjusted Capital (TAC) which is made up of its statutory capital and surplus plus the asset valuation reserve, any reserves it is voluntarily holding, and one-half of any policyholder dividend liability. Then the RBC components for the four above-mentioned risks (C-1 through C-4) are calculated to give the Authorized Control Level (ACL). If the TAC is greater than twice the ACL, no action is required by the company with regard to its capital level. If the TAC is 150–200% of the ACL, a company action plan is required in which the company lays out its plan to address the low capital level. If the ratio of the TAC to the ACL is 100–150%, then not only is the plan required, but a company examination is required, and the regulator may issue Corrective Action orders. If the ratio is below 100%, the regulator may place the company under regulatory control. Finally, if the ratio does not exceed 70%, the regulator must place the company under regulatory control.

For LTCI, the pre-tax before covariance-adjustment life insurance ACL RBC is based as follows.

- C-1 (credit): 0.90% of assets
- C-2 (insurance):

- 15.4% of the first $50 million of premium and 4.6% of additional premium, plus
- 7.7% of claim reserves, plus
- 38.5% of the first $35 million of incurred claims and 12.3% of additional incurred claims

- C-4 (business): 0.524% of premiums.

This formula is under review by the NAIC at the time of this writing. Updates can be found at the NAIC website, www.naic.org.

Typically, a company will target a level of the ACL such as 400% or 600%. This gives them a cushion if there is a negative event. The company may also be interested in what rating agencies like Moody's and Standard & Poor's might say about their capital strength in setting their ratings. Rating agencies may have their own RBC-like formula which is somewhat meant to mimic economic capital on a formulaic basis, not a stochastic basis.

The RBC calculated by the formula is an approximation of the "true" capital that should be held by an insurer. Thus, the levels above have some flexibility. A more ideal method would be for the insurer to determine its capital requirements based on its intimate knowledge of its own risks. This is the reason for the regulatory evolution to the ORSA that is described in another chapter. The ORSA may use either stress testing or an economic capital calculation in its risk assessment, these being more sophisticated tools than the RBC calculation. Still, RBC plays a significant role as a more objective measure of risk, which when used in conjunction with the ORSA is expected to provide even more regulatory insight and thus insurer and marketplace stability.

RBC does not consider liquidity risk, operational risk (which some see as the largest risk of a company), fraud risk, the strength of the management, customer loyalty and the company's competitive advantages. All the non-RBC risks are significant; they should be captured in the ORSA.

7.8 Summary

If all risks were well understood, the reserves of an insurance company would be sufficient to cover all future claims. Yet the world changes over time as do the risks undertaken by insurers, sometimes to a company's benefit, other times to its detriment. Capital and surplus are there to help when the actuary's best estimate adversely misses the mark.

Capital levels in the insurance marketplace have maintained its stability. Even as banks were rocked in the 2008 financial crisis, insurers weathered the event with damaged balance sheets but doors still open.

LTCI remains a difficult line of business with new revelations of inadequate pricing appearing every so often, from lower-than-expected lapse rates to historically low investment returns to higher-than-expected claims at the upper ages. Having sufficient capital and surplus is a challenge both from a regulatory and a company perspective.

Should RBC for LTCI be increased? Can we truly understand what a 99.9% level of risk would look like? How is it to be calibrated with any certainty?

Still, these are the needs of the insurance industry and especially for LTCI. A perfect answer does not exist but understanding a company's risks and its subsequently developed risk mitigation strategies will help to maintain stability in the uncertainty of the insurance world.

Bibliography or Resources

The field of solvency capital is constantly evolving. Reference material is constantly being updated. I doubt this chapter will prove to be any different. It is written with general concepts in mind. These references the author has found useful in pushing deeper into the concepts and mathematics that are part of this field

1. International Actuarial Association: Stochastic Modeling: Theory and reality from an actuarial perspective. *A strong overall view of stochastic modeling. A good place to start. A lengthy set of references for company specific practice is given in Appendix B* (2010)
2. Klugman, S.A., Panjer, H.H., Willmot, G.E.: Loss models: from data to decisions, 4th edn, The detailed mathematics of many risk distributions is presented with clarity. Used for actuarial education. The art of picking a distribution for a specific risk is a matter for other sources. Wiley (2012)
3. https://www.wikipedia.org/. This invaluable online reference is a go-to for details about distributions or other mathematics. Its detail on regulatory items is not as strong indicating that going to specific regulatory websites would be preferred
4. https://en.wikipedia.org/wiki/Copula_(probability_theory). Copulas are commonly used in economic capital work. This reference may serve as an introduction to a topic that will likely require several attempts. The internet is filled with good material
5. https://eiopa.europa.eu/regulation-supervision/insurance/solvency-ii. Solvency II is a benchmark for all capital work. It is extensive and not always intuitive but when there is no clear path forward for the practitioner, Solvency II provides a "safe harbor". As stated in the chapter, long-term care insurance is not addressed in solvency II and the use of disability as a substitute is not appropriate
6. https://www.apra.gov.au/. Australian Prudential Standards are another view of how capital is determined
7. https://www.thecroforum.org/. The CRO Forum provides up-to-date thinking about various insurance risks. It provides alternatives to Solvency II benchmarks. Of particular interest would be papers on "Calibration recommendation for the correlations in the Solvency II standard formula", "Scenario Analysis Principles and Practices in the Insurance Industry", and "Establishing and Embedding Risk Appetite: Practitioners' View"

Chapter 8
Impact of Reinsurance: Qualitative Aspects

Guillaume Biessy and Ilan Cohen

8.1 The Role of Reinsurers

Since the first development of Long Term Care Insurance in the markets that are the most developed today, the reinsurance industry has always been a strong support to insurers. It is indeed via the knowledge of the reinsurers that the insurers are able to understand how to create a new product, how to price the product and how to manage this long-term risk through time.

Nowadays, many insurers often ask for support to the reinsurers, mainly for the following reasons:

- Make the best use of the technical and medical expertise of the reinsurer on the Long Term Care risk
- Make the best use of its expertise on conducting experience analysis on the Long Term Care risk
- Make the best use of the acquired expertise via some external (i.e. non-actuarial or medical) experiences and partnerships.

Whatever the chosen reinsurance structure is, reinsurers provide support to insurers through the life cycle of the Long Term Care Insurance product:

- Market research and product design
- Medical underwriting
- Statistical analysis
- Pricing and Market distribution
- Claim management
- Reserving and risk monitoring

G. Biessy · I. Cohen (✉)
SCOR, 5 avenue Kléber, 75795 Paris Cedex 16, France
e-mail: icohen@scor.com

G. Biessy
e-mail: GBIESSY@scor.com

© Springer Nature Switzerland AG 2019
E. Dupourqué et al. (eds.), *Actuarial Aspects of Long Term Care*,
Springer Actuarial, https://doi.org/10.1007/978-3-030-05660-5_8

This is the reason why reinsurance solutions are seen, for the Long Term Care risk, as a partnership on the whole development of the portfolio. There are three main reasons that push insurers to reinsure their Long Term Care Insurance contracts:

1. To be covered against the uncertainty around the Long Term Care risk via some risk transfer solutions
2. To avail themselves of actuarial support from reinsurers, including pricing and risk monitoring via experience analysis studies
3. To have access to their medical expertise, in order to develop medical question-naires and manage claims.

8.1.1 Access to Long Term Care Data

One of the key advantages to asking for the services of a reinsurer is the access to experience studies, on a market wide basis. In fact, reinsurers, as they reinsure multiple insurers through the market—or even on different markets through the world, have access to Long Term Care data from those different insurers. Reinsurers are then able to study the Long Term Care risk on a "market basis", in order to help insurers to correctly price or monitor their insurance portfolios.

Some insured lives portfolios provide enough credible information to establish actuarial assumptions. Others, although more modest in size and length of experience, can nevertheless provide a non-negligible amount of information. As in the handling of any data for statistical purposes, if several sources of information are aggregated, very close attention must be paid to various parameters influencing the assessment of the risk, as these parameters may vary from one source to another. Some of these parameters are:

- the definition used for the risk of Long Term Care, parameters such as the waiting period (whether there is one, whether it differs or not according to the cause of the loss of autonomy, relative or absolute waiting period, duration…),
- any deferment periods,
- risk selection at the start of coverage in the case of insurance data,
- claim management,
- characteristics of population used,
- the insurance policy itself in the case of insurance data (for example: known anti selection on this policy, distribution method).

Data can sometimes be aggregated to build up experience-based probabilities once parameters have been analyzed in depth and resulting statistical bias is evaluated.

Reinsurers, by reinsuring different insurers, are then able to achieve an under-standing and a knowledge of the risk that is much larger than one insurer can acquire by itself on its portfolio.

8.1.2 Medical Expertise

Reinsurers tend to work on biometrical risks from a technical point of view, but also, due to their global dimension, from a medical point of view. This is one of the areas, developed by the reinsurers, which also bring insurers to seek reinsurance. To understand and know more about the risk, reinsurers will study the Long Term Care risk from a medical point of view, but also through therapeutic, epidemiological and social factors that can influence the risk. Medical and social reports and studies from around the world contribute information to help understand the Long Term Care risk, and its possible future evolvement.

The medical related areas on which reinsurers can bring support to the insurers are mainly:

• Medical underwriting/selection

Medical research by reinsurers includes how to adapt medical selection to the Long Term Care risk, which is still new to insurers. A deep dive into the Long Term Care risk is needed to understand this risk and its risk factors, in order to create the best medical underwriting, and adapt the main medical selection strategy to each insurer seeking support from reinsurers.

• Pricing of substandard lives

Underwriting teams of reinsurers help insurers to make the most effective risk decisions in their underwriting of applicants, with strong selection strategies in the medical field. They can provide support in helping insurers to identify high risk level profiles, to spot substandard risk and to determine the conditions under which to accept certain risks, and limit adverse selection.[1]

• Claim management

Reinsurers have an extensive claim management expertise, and vast portfolio analysis experience to help insurers analyze existing claims and develop personalized selection strategies suited to the need of each company. Reinsurers will provide support to insurers to:

– Analyze and clarify causes and circumstances of a claim
– Confirm that there was no undisclosed history of illness related to a claim
– Verify that a claim complies with applicable risk assessment policies.

8.1.3 The Influence of Reinsurers on Long Term Care Insurance Markets

Many reinsurers tend to have a global scope (one reason being the matter of diversification of their risks), and operate on all five continents. This global system/view can also be considered as having an impact on the transfer of knowledge and experience on the Long Term Care risk through the world.

- Reinsurers act as advisors for new developments to local insurers by sharing world-wide experience

In a world in constant movement, the role of the reinsurers is to help insurers to provide to their insureds the best insurance solutions to help them to face risks. One way to do so is to look at the experience in other insurance markets. In the case of Long Term Care risk, one can see the Asian continent as an ageing continent (China, for example, due to the impact of the "one child" policy). In order to find adequate solutions to face the cost related to the inevitable ageing of the population (i.e. Long Term Care costs), insurers can find all the international experience in one place, by turning to reinsurers. With their international experience and knowledge, the reinsurers can help the insurer to:

- Find the best insurance product that will fit the needs and expectations of the insureds
- Share the experience (both actuarial and medical) that reinsurers have from other markets, in order to assist the insurer in the pricing of the insurance product, and the implementation of medical underwriting and claims management policy
- Share best practices based on experience, in order to support insurers on how to monitor and manage a Long Term Care Insurance portfolio through time, due to its long-term nature.

- Reinsurers help regulators and governments understand public solutions that exist through the world

The phenomenon of population ageing in most of the countries of the "western world" and of some countries in Asia, and the costs related to this issue (Long Term Care costs) is fairly new. In that context, governments and regulators are always keen to understand how this topic is handled by governments in other countries that are facing this shared issue.

- Some governments are looking for the best (and difficult to find!) solution to face the issue of long term care. Some of them may be interested in discussions with reinsurers, who act as advisors. This shows the impact that reinsurers can have in both the establishment of public initiative and in understanding how they can promote the role of private insurance.
- Reinsurers tend to educate the insurance industry as a whole (including regulators), in order help the industry to build a framework in which solutions to Long Term Care can be best implemented. This education can be done via industry seminars, as well as publications, in such a way that reinsurers can share with the insurance industry their experience and knowledge.

8.1.4 Reinsurance Framework: Special Conditions and Technical Aspects

The long-term nature of the Long Term Care risk forces reinsurers to adapt some of their reinsurance conditions to insure a perfect alignment of underwritten risks, both by insurers and by reinsurers.

Specifics are of two kinds: special conditions of the reinsurance treaty, and technical aspects that link the insurer and the reinsurer.

8.1.4.1 Special Conditions of the Reinsurance Treaty

A reinsurance treaty is a convention between an insurer and a reinsurer that defines the conditions under which the insurer is going to cede part of its risk.[2] For the Long Term Care insurance contracts, some specific modalities have to be defined, in order to be sure that the specificities of this risk (among which the long-term nature, as well as the uncertainty around the risk) are well taken into account, and do not impact negatively the insurer or the reinsurer.

– Gross reinsurance premium or net reinsurance premium

In most cases, the reinsurance premium, for a coinsurance, will be:

- Gross level premium based—the reinsurer then reimburses the insurer for the commissions that correspond to its marketing and administrative loads.
- Pure level premium based—the reinsurer will not be linked to the insurer for its marketing and administrative loads.
- Risk premium level—the reinsurer will receive the premium for the risk of the year, this for each year of the contract. The correlation with the insurer will then be slightly different.

Reinsurance premium payment modality must be clearly specified, and the premium table (usually per issue age) added to the treaty.

– Reserves calculation

Due to its long-term nature, on a level premium basis, actuarial modeling requires computing mathematical reserves. How these reserves are calculated must be detailed in the reinsurance treaty.

- The calculation of each of the liabilities (of the insurer and of the insured) are detailed, in order to make sure both parts (insurer and reinsurer) agree on how to compute it.
- The reinsurer will propose reserve methodologies to be included in the treaty, so that both parts follow the same calculation.

Modalities concerning the management of the reinsurer share of the mathematical reserves must also be specified in the reinsurance treaty. These reserves can remain in the insurer's books (modco). If such is the case, conditions of financial remuneration conditions must be stated in the reinsurance treaty.

– Claim investigation

Long Term Care claims management is not obvious to all insurers, and reinsurers may benefit from studying all insurer medical records, to build their global experience. In that context, reinsurance treaties should specify the right of the reinsurer to carry out such an analysis, at least on a sample of claims received by the insurer (for approved claims, as well as for rejected claims).

Usually, it is stated that during the first years, the reinsurer will provide support to the insurer in their claims management.

– Technical and financial mechanisms

All technical and financial mechanisms that link the insurer and the reinsurer (stability reserve, profit sharing, revalorization fund etc., detailed below) must be detailed and clearly explained. Each part of the mechanism should detail its calculation, with both insurer and reinsurer results.

One important point is the level of those funds once insurance contracts end. If the funds are not zero, the allocation of the money should be well detailed.

– Modalities on reinsurance rates adjustment

Some reinsurance does not provide full rates guarantee to insurers. In that context, it should be stated under which terms the reinsurer can or cannot increase reinsurance rates, how the reinsurer is linked to the insurer, and if this concerns new business only, or also in-force portfolios.

If a specific mechanism is put in place to detail a rates increase trigger, it should be specifically detailed.

• Cancellation/cut-off

Reinsurance cancellation has to be detailed in the treaty. Two different cases should be stated:

• Cancellation of new business: in case reinsurer or insurer does not want to pursue the reinsurance of the risk for new business starting from a specific time, the conditions under which a cancellation is possible have to be specified.
• Cancellation of in-force portfolio: in the case of cancellation of a run-off portfolio, whether it is from insurer or reinsurer initiative, liability transfer conditions must be clearly stated and explained. Transfer methodology should be detailed, as well as actuarial and financial assumptions.

8.1.4.2   Technical Aspects of the Reinsurance Treaty

– Loss carry-forward

The loss carry-forward is the transfer of debit or loss from one accounting period to another under a reinsurance treaty. This is an instrument that is often used in the reinsurance treaty to ensure adequate cost sharing between insurer and reinsurer.

– Equalization fund

The equalization fund is designed to face any potential variations in claims through time. It is usually agreed upon and shared between insurer and reinsurer. In such case, the insurer can "pick" into the equalization fund of the Long Term Care contract when the Profit & Loss (P&L) account is unprofitable, on the specific period.

In other words, the equalization fund is like a reserve that is built using a positive P&L balance and can be used as a buffer if the experience deteriorates. In such case, the equalization fund can be released partially or totally to offset negative results.

The equalization fund is set according to the following specificities:

- A certain percentage of the P&L account is set to the equalization fund in case of positive P&L results.
- The equalization fund can be capped by a percentage of the mathematical reserves.

When this ceiling is reached, the surplus goes back as profit (split between insurer and reinsurer). This surplus then represents additional profits.

The figure below shows a simplified example of the mechanism of an equalization fund:

Debit	Credit
Year N claims paid (including revalorizations)	Year N premium
12/31/N premium reserves	12/31/N − 1 premium reserves
12/31/N claim reserves	12/31/N − 1 claim reserves
Expenses (acquisition costs, management fees…) (% of the premium)	Interest earned
	Calculated at the technical rate and applied to the premium and claim reserves at 12/31/N − 1
Possible loss carry forward	
TOTAL Debit	TOTAL Credit
Debit balance	Credit balance

Scor Global Life source for the table

If the balance of that technical account is positive, then:

- X% of this amount goes to the insurer equalization fund
- Y% of this amount goes to the reinsurer
- 1 − X% −Y% of this amount goes to the equalization fund.

If the technical account balance is negative, then:

- 100% of the negative result is imputed to the equalization fund
- In case the equalization fund is not large enough to cover all the losses, a loss carry-forward can be created and added to the equalization fund debit side for year N + 1.

A reinsurance treaty often states that the equalization fund is capped to a percentage of the mathematical reserves. If so, surplus goes, as profit, under a profit sharing mechanism that must be defined in the treaty (defining which percentage goes to insurer and reinsurer).

The equalization fund can then be written as:

Debit	Credit
12/31/N positive balance of the technical account	12/31/N − 1 equalization fund
	12/31/N x% of technical account positive balance
Balance: 12/31/N *equalization fund*	

It is important to note that the equalization fund belongs to the insurer, but a reinsurance treaty should detail how a non-zero equalization fund is divided at the end of the contract.

– Revalorization fund

It is often usual to define the revalorization fund in a reinsurance treaty. The revalorization fund will be used to finance the possible indexation (or revalorization) of premiums and benefits (annuity if the benefit is an annuity, for example).

The revalorization fund mechanism is the same as the one for the equalization fund.

8.2 The Impact of Different Reinsurance Structures for Insurers

The impact of reinsurance for insurers consists in two effects:

- Protecting itself from any potential changes in the risk, or extreme claims pattern, that can negatively impact its underwriting results
- Reducing Risk-Based Capital.

Both effects are among the most important ones that lead insurers to seek reinsurance for their Long Term Care insurance portfolios.

All reinsurance schemes propose a real partnership between the insurer and reinsurer. Whether it is a proportional treaty (coinsurance for example) or not proportional (stop loss), the future of both entities is linked.

The structure of the reinsurance premium can be of two kinds:

- Level premium (in the case of coinsurance only)
- Risk premium (in all cases, including coinsurance).

8.2.1 Reinsurance Structures for the Long Term Care Risk

8.2.1.1 Coinsurance

A proportional treaty is an agreement between a reinsurer and a ceding company (the reinsured) in which the reinsurer assumes a given percent of losses and premium. The simplest example of a proportional treaty is called "coinsurance". In a coinsurance treaty, the reinsurer receives a flat percent, say 50%, of the premium for the reinsured policies. In exchange, the reinsurer pays 50% of losses, including allocated loss adjustment expenses, on the book. The reinsurer also pays the ceding company a ceding commission which is designed to reflect the differences in issue and underwriting expenses.[3]

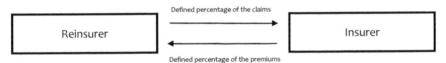

Defined percentage of the claims

Reinsurer

Insurer

Defined percentage of the premiums

Coinsurance treaties are straightforward and therefore easier to price than other treaties. Given the complexity of the Long Term Care risk, this makes them ubiquitous on the market. In its simplest form, a coinsurance treaty reduces the risk proportionally to the percentage of the coinsurance. Therefore, it can be used to take a bigger market share while maintaining the same level of liability. However, by doing so the insurer renounces an important part of the profit. Coinsurance treaties usually include a ceding commission and a profit-sharing clause which may be used for the insurer to retain part of the profits.

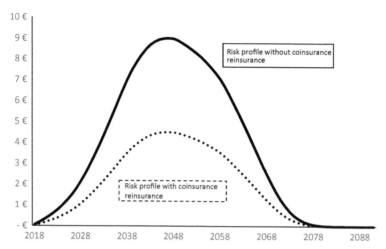

Coinsurance treaties are nevertheless not really designed to reduce the solvency capital requirements as in the best case they have no impact on the shape of the loss distribution. In a coinsurance treaty as in other reinsurance treaties, the financial rating of the reinsurer really matters as the default risk from the reinsurer may also impact the tail of the loss distribution for the insurer. However, let us not forget that reinsurer experience may allow for better monitoring of the Long Term Care risk, allowing more reactive actions such as increases in premium rates if the level of the risk was not in line with the initial assumptions or if it shifted over time, something which, given the high duration for this risk, will occur at some point in time. Hence, most insurers on the French market have so far made the choice to be reinsured by at least one or even several reinsurers.

From a reinsurer point of view, it is paramount that the insurer keeps a reasonable proportion of the premium after all treaties have been applied or the insurer may lose the incentive to monitor the risk or increase premium if need be. Indeed, the insurer may decide to incur some loss on a specific Long Term Care product rather than increase premiums to preserve its relationship with the insured and its reputation, especially if the Long Term Care product only represents a small share of the insurer business.

8.2.1.2 Long Term Care Swap

A Long Term Care swap is designed for the reinsurance of a closed block of Long Term Care policies. It swaps the estimated cash flows payment stream for an actual cash flows payment stream.

The insurer proportionately reinsures a block of in-force policies. It pays predetermined cash flows on a quarterly or annual basis to the reinsurer. These cash-flows are calculated on an agreed set of actuarial assumptions that define the Long Term Care risk (incidence rates, mortality rates…), applied to the insurer's Long Term

Care portfolio. They include expected cash flow payments and the reinsurer's loads for potential expenses and capital margin. Conversely, the insurer receives from the reinsurer actual cash flow payments. As a result, the insurer keeps the financial risk (managing reserves) while reinsuring the Long Term Care risk. This scheme is simple and transparent.

The Long Term Care risk is ceded to the reinsurer as shown on the following graph (under the assumption of a risk shift):

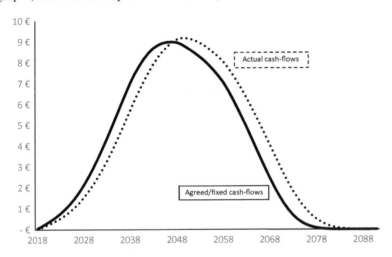

8.2.1.3   Long Term Care Stop Loss Reinsurance

Long Term Care Stop Loss reinsurance will protect an insurer, on a specific portfolio, against unusual or unexpected incidence or mortality risk. A Long Term Care Stop Loss aims to act as a cost-effective protection against an extra load in claims payment, due to a drift or an exceptional year on one of the risks that explains the Long Term Care risk:

– A change in the incidence risk, that will lead to more claims to pay
– A change in the mortality of active insureds, that will lead to more people exposed to the incidence rates
– A change in the mortality of disabled insureds, that will lead to disabled insureds staying longer than expected in the state of loss of autonomy.

   The insurer and the reinsurer will agree on a set of actuarial tables that will estimate a certain number of open claims (i.e. paid claims) that the insurer will have to pay. The Stop Loss reinsurance agreement will allow both parts to agree on a deviation of the risk, based on the number of open claims, say an extra number of open claims of 20%. For claims over this limit (i.e. 120% of the expected claims), the reinsurer will then pay for the higher than expected part of claim cost (i.e. cost above 120%).

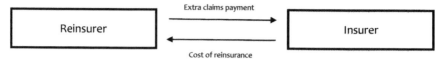

The insurer will then be covered against any shift of the Long Term Care risk or against any exceptional claims year during the development of the insurance portfolio. The reinsurance program will then allow the insurer to stop its losses following that specific event or drift in risk.

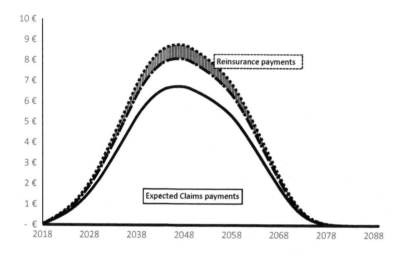

8.2.1.4 Tail Risk Reinsurance

Even though as of today all possible reinsurance structures do not exist for the Long Term Care risk, it is possible to imagine some reinsurance schemes that could be well adapted to this risk.

One of the main unknowns about the risk, which represents a large component of Risk-Based Capital required by the insurers, is the longevity in the state of loss of autonomy (continuance.) In most insurance markets on which we have seen Long Term Care insurance products develop, it appears that the behavior of claims within the first years is pretty well measured. What is less known is the ability of the insurers to answer the question: "how long can a claim last?"

By reinsuring the tail of the claim duration, the insurer can find a benefit on two points:

– Being covered against claims that last too long in their portfolio

- Reduce the Risk-Based Capital by minimizing possible maximum loss on a portfolio.

The insurer and the reinsurer would then define a claim duration from which
reinsurance would intervene. From that time, and until the natural expiration of the
claim, the reinsurer would cover the insurer. The participation of the insurer can be
designed up to 100% of the benefit amount per claim.

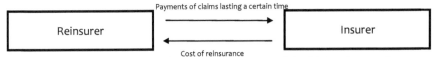

The insurer would then get a high coverage in case claim(s) last longer than an
expected time:

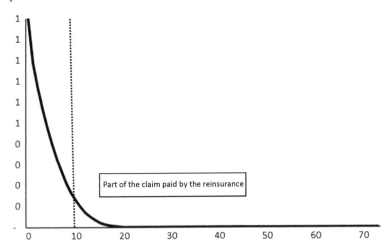

8.2.2   Impact of Reinsurance Structures on Risk-Based Capital

Each of the above structures have a different impact on Risk-Based Capital. In this
part, we propose to study loss distributions, depending on the reinsurance scheme,
in order to have a better understanding of the impact on the RBC from each of the
structures.

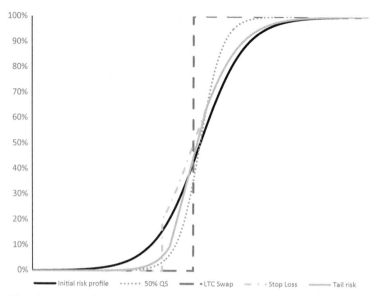

The above mentioned figure provides the distribution of losses for the insurer before and after each reinsurance treaty. The swap is the best option for reducing Solvency Capital Requirement (SCR) but it also negates any opportunity for profit. Stop loss is second and it preserves profitability best. Coinsurance drastically reduces both capital requirement and profitability. The tail risk solution provides the lowest reduction in capital requirement but on the other hand preserves profitability.

Coinsurance
Coinsurance is the more straightforward structure when it comes to impact on capital. Indeed, at first sight one may consider that capital requirement is reduced in proportion to the coinsurance. However, depending on the reinsurer financial rating, default risk may also have a significant impact on capital requirement. The profit sharing clause in coinsurances treaty increases profitability but has no impact on tail distribution. Overall, coinsurance treaties reduce exposure to long term care risk but do little in terms of capital requirement optimization.

Long Term Care Swap
As a Long Term Care swap makes the future cash flows deterministic, it nearly removes capital requirement for the book that has been swapped. The only remaining capital requirement is linked to the probability of default from the reinsurer, which may be minimized by the inclusion of collateral in the product.

Stop Loss
Stop loss makes the reinsurer carry any loss which goes beyond a given level, which basically removes the tail of the loss distribution. Capital requirement will therefore never be greater than the maximum loss beyond which the stop loss kicks in, default risk put aside. It makes this treaty a very efficient solution to reduce capital requirement while keeping a reasonable profit margin.

8.2.3 Tail Risk

Tail risk reinsurance reduces the exposure of the insurer to longevity risk in Long Term Care. However, it does not have any effect on the risk associated with a lower active lives mortality than anticipated or a higher frequency of claims. Furthermore, it does not completely remove longevity risk as a decrease in mortality in the dependent state would also affect even the first years of claims retained by the insurer.

End Notes
1. (Scor Global Life—www.scor.com).
2. (Apref—le petit glossaire de la réassurance—www.Apref.org).
3. (Basics of reinsurance pricing—actuarial study note—David R. Clark, FCAS—first version 1996—revised 2014).

Chapter 9
Impact of Reinsurance: Quantitative Aspects

Frédéric Planchet

9.1 Introduction

Non-mutualized risks associated with Long Term Care insurance contracts should be considered in a specific way and covered, whether through additional capital or the introduction of risk transfer solutions.

The risks involved are associated with capital outflows on the one hand and annuities on the other, born of a poor assessment of the distribution of incidence and/or continuance. The longevity risk is a significant risk in this assessment.

In order to determine optimum solutions in this area, approaches to measuring and covering these risks must be developed.

This document considers non-mutualized risks associated with continuance assumptions in a Long Term Care insurance portfolio and illustrates how to quantify continuance risks on the basis of simple models, to assess gains associated with a risk transfer solution (capital and solvency requirement).

9.2 Uncertainty Measurement on a Continuance Assumption

A risk associated with an error on the continuance distribution is considered, assuming the sampling risk is perfectly mutualized. The only hazard considered here is therefore the one associated with the underlying continuance table and the main random variable is the conditional expectation (from the underlying table) of the present value of future cash flows.

F. Planchet (✉)
Laboratory SAF, University Claude Bernard Lyon 1, Villeurbanne, France
e-mail: frederic@planchet.net

PRIM'ACT, 42, Avenue de La Grande Armée, 75017 Paris, France

© Springer Nature Switzerland AG 2019
E. Dupourqué et al. (eds.), *Actuarial Aspects of Long Term Care*,
Springer Actuarial, https://doi.org/10.1007/978-3-030-05660-5_9

9.2.1 General Framework

Formally, the risks analyzed here can be described by the fact that the distribution of a continuance variable (survival, dependence) is itself random. A durational model is in practice conditional on a state variable that describes the uncertainty about the distribution considered.

The description of this state variable is specific to the type of risk considered:

- an illustration in the context of Long Term Care insurance is proposed in Planchet and Tomas [4];
- references for longevity risk are numerous.[1]

In a more precise way, the models studied are based on the fact that a variable of duration T is replaced by T_θ, θ being a random variable for which a distribution must be specified.

For example, in Haas [2], the density function describing life expectancy is taken from the Log-Poisson model (*See* Brouhns et al. [1]), $\mu(x, t) = \alpha_x + \beta_x \times \kappa_t$ is made random by assuming that the vector $\theta = ((\alpha_x), (\beta_x), (\kappa_t))$ is a normal random variable.

In a Long Term Care insurance context, Planchet and Tomas [4] use

$$\ln\left(\frac{q_x^\varepsilon}{1 - q_x^\varepsilon}\right) = \ln\left(\frac{q_x}{1 - q_x}\right) + \varepsilon.$$

This model makes it possible to simply construct an opposing scenario at a given confidence level. Indeed, once ε is fixed, one gets a table whose directional impact on the variable of interest is known: for a risk in case of continuance, a negative ε degrades the situation by decreasing termination rates; for a risk in case of termination by considering a positive ε one builds an opposing scenario. So, it is sufficient to select ε according to normal distribution quantiles.

The practical difficulty is the determination of the distribution of the variable of interest (for contract reserves, *Best Estimate*, etc.), which is rarely possible analytically and must be carried out using simulation methods.

We can note that the construction of models is not necessary to set up a hedging solution, the price of which is proposed by the reinsurer. Such construction is, however, important to assess the equilibrium price of the scheme in order, where appropriate, to establish a partial internal model for the underwriting risk (or, in a simpler way, under Own Risk and Solvency Assessment).

---

[1] See http://www.ressources-actuarielles.net/C1256F13006585B2/0/398CCED53711615AC12578 06003375CB.

9.2.2 Illustrative Model

In order to measure the impact of the introduction of a non-proportional reinsurance scheme (or, more generally, risk transfer) on a Long Term Care insurance portfolio, we generalize the framework of Planchet and Tomas [4] by adding a priority and a scope to the present value of future cash flows.

9.2.2.1 Structure of the Present Value of Future Cash Flows

We consider a set of individuals in a dependent state who receive a benefit as long as they remain in that state. The cash flow occurring at time t is of the form:

$$F_t = \sum_{j\in J} r_j \mathbf{1}_{(t;\infty)}(T_j),$$

where T_j is the continuance distribution for policyholder j and r_j the cash benefit amount for j, which leads to the sum of the updated future cash flows:

$$\Lambda = \sum_{t=1}^{\infty} F_t(1+i)^{-t} = \sum_{t=1}^{\infty} \frac{1}{(1+i)^t} \sum_{j\in J} r_j \mathbf{1}_{(t;\infty)}(T_j) = \sum_{j\in J} r_j \sum_{t=1}^{\infty} \frac{\mathbf{1}_{(t;\infty)}(T_j)}{(1+i)^t}$$

$$= \sum_{j\in J} r_j X_j,$$

where $X_j = \sum_{t=1}^{\infty} \frac{\mathbf{1}_{(t;\infty)}(T_j)}{(1+i)^t}$. When the number of individuals is large enough, Λ is approximately Gaussian because of the independence of the variables X_j. Knowledge of the distribution of Λ is thus reduced to calculating the expectation and variance and since $E(\Lambda) = \sum_{j\in J} r_j E(X_j)$ and $V(\Lambda) = \sum_{j\in J} r_j^2 V(X_j)$, the expectation and variance of

$$X = \sum_{t=1}^{\infty} \frac{\mathbf{1}_{(t;\infty)}(T)}{(1+i)^t}.$$

It is easy to find $E(X) = \sum_{t=1}^{\infty} \frac{S(t)}{(1+i)^t}$ and $E(X^2) = \sum_{t=1}^{\infty} \frac{S(t)}{(1+i)^{2t}} + 2 \times \sum_{t=2}^{+\infty} \frac{1}{i}\left[\frac{1}{(1+i)^t} - \frac{1}{(1+i)^{2t+1}}\right]S(t)$.

A non-proportional reinsurance treaty of priority F and order P is such that the insurer retains $R = \Lambda - [(\Lambda - F) \wedge P]^+$ (see Sect. 9.3.3) and hence transfers to the reinsurer the amount $C = [(\Lambda - F) \wedge P]^+$. The distribution function of R is given by:

$$F_R(x) = \begin{cases} F_\Lambda(x) & x < F \\ F_\Lambda(x+P) & x \geq F \end{cases}$$

By using the Gaussian approximation of the distribution of Λ, we have $F_\Lambda(x) = N\left(\frac{x - E(\Lambda)}{\sigma(\Lambda)}\right)$, so the distribution of R is known.

It can be noted that the above approach is easily adapted to a full surplus contract (see III.a.1.1) and that, in this case, the distribution limit remains Gaussian, only the moments of this variable are affected by the terms of the treaty.

9.2.2.2   Introduction of a Random Event on Continuance Distributions

The model now introduces a random event to the underlying continuance distributions. The easiest way to introduce uncertainty about conditional termination rates is to introduce noise to the associated *Logits* (See Planchet and Therond [3]):

$$\ln\left(\frac{q_x^a}{1 - q_x^a}\right) = \ln\left(\frac{q_x}{1 - q_x}\right) + \varepsilon$$

with ε a randomly centered variable, which is assumed to have Gaussian characteristics. Equivalently, if we have:

$$q_x^a = \frac{a \times \exp(\lg(q_x))}{1 + a \times \exp(\lg(q_x))}$$

the shock is controlled by the volatility of ε, expressed as σ.

Under the framework of the Solvency 2 benchmark, the uncertainty on the reserve table is intended to be covered through the underwriting component of the Solvency Capital Requirement (and thus partially through the risk margin).

9.2.2.3   Solvency Capital Requirement Calculation

The elements introduced below provide information on the distribution of Λ and R, representing gross and net reinsurance premium. However, evaluating the quantile at 99.5% of this distribution provides a biased SCR evaluation, as it does not take into account:

- The one-year projection limitation (the model projection is carried to the ultimate duration);
- The risk margin.

We can integrate these constraints by using the approximation:

$$SCR = \frac{\frac{VaR_{99,5\%}(\chi)}{BEL_0} - 1}{1 - \alpha \times D_0 \times \left(\frac{VaR_{99,5\%}(\chi)}{BEL_0} - 1\right)} BEL_0$$

with $\chi = \frac{F_1 + BEL_1}{1 + R_1}$.

It can then be observed that the χ distribution can reasonably be approached by the Λ distribution in the absence of reinsurance and the R distribution when we introduce reinsurance. We therefore use:

$$F_\Lambda(x) = P(\Lambda \leq x) = E[P(\Lambda \leq x|a)] \underset{|J| \to +\infty}{\to} \int N\left(\frac{x - \mu(a)}{\sigma(a)}\right) F_a(da)$$

in the absence of reinsurance and

$$F_R(x) = \underset{|J| \to +\infty}{\to} \int F_R(x|a) F_a(da)$$

with reinsurance.

In practice we approach these repartition functions by simulation on the basis of a sample of the variable a:

$$F_\Lambda(x) \approx F_\Lambda^{(K)}(x) = \frac{1}{K} \sum_{k=1}^{K} N\left(\frac{x - \mu(a_k)}{\sigma(a_k)}\right) \text{ and } F_R(x) \approx F_R^{(K)}(x)$$

$$= \frac{1}{K} \sum_{k=1}^{K} F_R(x, \mu(a_k), \sigma(a_k)).$$

The quantile of a given level is then calculated by resolving by dichotomy the equation $F^{(K)}(x_q) = q$.

To this underwriting SCR, the counterparty SCR should be added, the calculation of which is specified in Sect. 9.4. The SCR aggregation is carried out as follows

$$SCR_{tot}(F, P) = \sqrt{SCR^2(F, P) + SCR_{def}^2 + 0.25 \times SCR(F, P) \times SCR_{def}}.$$

The implementation of the treaty impacts the Net Asset Value (NAV) and the SCR. Assuming that the only two risks are the underwriting risk and the counterparty risk, and by designating $\pi = \frac{NAV}{SCR}$ to be the coverage rate before the treaty is set up, we have:

$$\pi(F, P) = \frac{\pi \times SCR + Adj_{CD}}{SCR_{tot}(F, P)}$$

with $SCR = SCR_{tot}(+\infty, 0)$ and $Adj_{CD} = -\mathbf{max}\left((1 - RR) \times E(C) \times Dur \times \frac{PD}{1-PD}; 0\right)$.

9.2.2.4 Numerical Example

The above model is implemented with an illustrative continuance distribution (representative of a risk of high dependence) and assuming a 20% shock. Calculations are done without taking into account discounting. The probability of counterparty default is set at 10^{-4} (Class 1 of the scale described in article 199) and the recovery rate at 50%.

The results presented do not take into account the risk transfer cost and only measure at this stage the effect on the SCR level (reported in the *Best Estimate* gross amount) and the coinsurance level specified in the reinsurance treaty (Figs. 9.1 and 9.2).

This type of simple model makes it possible to highlight the basic structures and to select an optimum strategy.

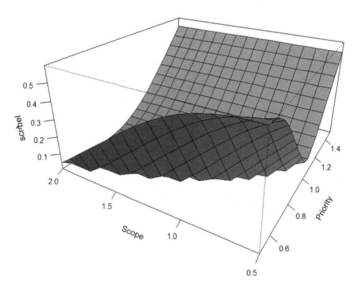

Fig. 9.1 Evolution of the SCR/BE ratio based on scope and priority

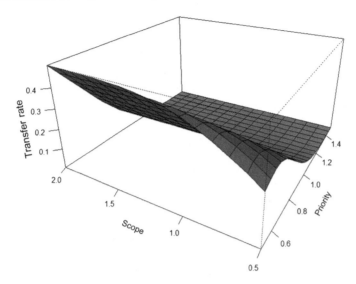

Fig. 9.2 Change in transfer rate based on scope and priority

9.3   Reinsurance

9.3.1   Traditional Reinsurance

Conventionally, we distinguish proportional and non-proportional reinsurance contracts.

In the proportional case, the reinsurer takes over a fixed ratio (the coinsurance rate) of the losses of the ceding company. This is the classical coinsurance contract.

Consider a particular case of proportional Long Term Care coverage, which is easily generalized to other risks considered.

9.3.2   Mortality Coverage by Risk

The type of premium paid to the reinsurer stems from differentiation between so-called proportional premium, or coinsurance contracts and Yearly Renewable Term (YRT), or risk premium, contracts. In the first case, the ceding company pays the reinsurer a fixed ratio of its premiums (the transfer rate), so the reinsurer follows the price of the cedant and reimburses a reinsurance commission. This is the classical proportional reinsurance. In the case of risk premium contracts, the reinsurer carries out its own pricing. This includes its management costs and its expected profit, so there is no reinsurance commission. The reinsurer will therefore set the premium rate that it deems necessary based on the age and gender of the insured. In practice, this can be a simple way to apply a fixed percentage to the standard claim rate table(s). This second option brings a considerable amount of administrative cost, since it involves per policy maintenance. In practice, insurer and reinsurer calculate a simplified aggregate premium (so-called level premium).

It may be noted that these coinsurance contracts engage the reinsurer for all claims, even the smallest. As a result, the portfolio profile of the ceding company is not altered, and is only a linear transformation away (in proportion to the coinsurance rate).

The second large class of proportional contracts is that of surplus share reinsurance. It is a proportional contract, in which the transfer rate varies with each policy, depending on the amount insured and a retention provided by the contract. Specifically, if we denote by S the claim amount of policy for which the insured amount is K, then as part of full coverage with deductible F, retention amount P and capacity C, the reinsurer will take over $\alpha \times S$, where

$$\alpha = \min([K - P]^+, C)/K.$$

It has already been noted that the rate of alpha transfer fully depends on the policy up to the insured capital K. On the other hand the transfer rate is nil for all policies whose insured capital is less than the full retention P, there is no risk transfer for these policies. Finally, C/K represents the maximum ratio for which the reinsurer wishes to commit.

Unlike coinsurance, surplus share modifies the company's portfolio profile, as it keeps the "small" risks in full. This allows it to balance its portfolio, thus ceding some of the larger risks, while maintaining a suitable premium level, since all the premium is kept on smaller risks. The premium mode for coinsurance and YRT applies in the same way to surplus share contracts.

These proportional contracts constitute the great majority of contracts in traditional reinsurance for Long Term Care risks.

Non-proportional reinsurance contracts are now briefly described.

9.3.3 Per Event Excess Contract

The conventional non-proportional reinsurance contract or Excess of Loss contract helps protect against a significant deviation from a continuance assumption with respect to the average (premium) assumption. The maximum offset that the insurer is willing to assume is directly related to the level of the deductible of the reinsurance treaty. The reinsurer takes over liabilities exceeding this threshold.

The exact form of quantities transferred to the reinsurer is described in Sect. 9.2.

These contracts are mostly per event contracts. This means that they allow the company, say in life insurance, to hedge against a risk of accumulation linked to the occurrence of a particular event, such as a natural disaster.

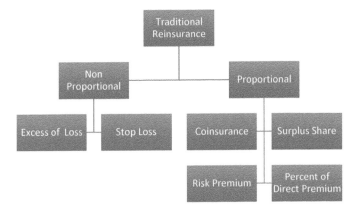

Fig. 9.3   Traditional reinsurance approaches

Finally, in order to protect its annual result, the insurer may also wish to purchase a *Stop Loss* contract. The operation is the same as for Excess of Loss contracts, but the non-proportional coverage relates directly to the Loss Ratio (Claim/Premium). The boundaries of the treaty, scope and priority, are therefore directly expressed in the form of Loss Ratios. In the particular case of continuance risks that interest us here, this type of contract is set up relatively exceptionally in practice.

The following diagram provides a summary mapping of possible risk transfers in the context of traditional reinsurance (Fig. 9.3).

A conventional reinsurance treaty, in proportional and non-proportional frameworks, is based on the overall result of the insurance operation without isolating the specific risk against which the insurer desires to protect itself. If one wishes to distinguish the continuance result, and focus only on the coverage of this risk, a first possibility is to consider the longevity swap.

9.3.4   Non-traditional Reinsurance: Longevity Swap

A Longevity *Swap* consists of a cash flow exchange of a fixed duration against a variable duration. The underlying risk is the future mortality of the portfolio. The Longevity *Swap* focuses therefore on the result of mortality which differentiates it from the traditional reinsurance.

However, the ceding company may also transfer the financial risk separately to another company (bank, financial institution). To secure the contract against the failure of one of the counterparties, a Special Purpose Vehicle (SPV) can be created.

A *Swap* pays the insurer an amount that is based on the potential loss generated by the actual lower mortality from the expected level derived from the tables used to calculate premiums. In exchange the insurer periodically pays a predetermined

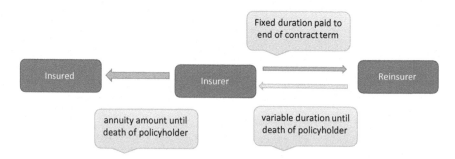

Fig. 9.4 Cash flow of a longevity swap

amount for a predetermined duration to the reinsurer. This mechanism is described below (Fig. 9.4).

In other words, the insurer pays the consideration on the basis of a contractually defined mortality table and receives the benefits actually paid to the annuitants. This mechanism has the advantage of being very easy to implement and does not require a complex model for the insurer.

One can cite the following operations using this instrument:

- August 2011: ITV with credit Suisse for £ 1.7 bn;
- November 2011: Rolls Royce with Deutsche Bank for £ 3 bn;
- February 2012: Aegon with Deutsche Bank for 12 MD €;
- February 2013: BAE Systems with Legal & General for £ 3.2 md
- July 2014: BT with Prudential Financial for £ 16 bn.

However, this device is relatively unused and the operations to date have been predominantly purchased by companies wishing to limit their own pension funds liabilities and that a priori the associated price levels can be considered high.

AXA also signed a Longevity *Swap* with the reinsurer Hannover Re for over 22,000 annuitants for nearly €700 million of reserve. In this contract, AXA pays Hannover Re a premium (fixed amount) fixed in advance on the basis of the TGH/F 05 mortality table including:

- Expected mortality of annuitants, including future improvements;
- Cost of capital;
- Management fees.

Hannover Re pays to AXA the individual annuities for a period depending on the life of the annuitant (variable). This Longevity *Swap* is contracted for a duration of 80 years and its duration has been in effect for over 10 years. This distant horizon suggests that AXA will be able to protect itself long enough against the risk of underestimating longevity due to the valuation tables, i.e. the reinsurer will take over potential additional reserve needed until the end of life of the annuitant. AXA therefore hopes that the forecasts of its actuaries, considering the actual existence of the longevity risk of the annuitant portfolio, should be confirmed. If this is not the case and the regulatory tables have sufficient margins, AXA would incur a reinsurance loss.

9.3.5 Index-Based Longevity Coverage

A major Long Term Care risk is continuance. A relatively new risk sharing tool to hedge the mortality risk is described below with potential applications to the continuance risk.

The development of a market for the transfer of the mortality longevity risk has been facilitated over the last few years by the development of tools to improve the level of transparency and standardization. This includes:

- *Longevity Toolkits*: For example, the software provided by the research group *Continuous Mortality Investigation* (CMI) in the United Kingdom or *LifeMetrics* Developed by JP Morgan.
- Longevity databases: for example, Club Vita Launched by the British consultant Hymans Robertson, data provided by CMI, or data from the German group Deutsche Borse.
- Longevity indexes: begun in the last decade, indexes evaluating longevity offers the possibility of marketing indexed bonds. Below are a few indexes on the market:

  - The longevity index developed by Credit Suisse which is based on the national population of the United States. It was created in 2005 and is broken down by age and sex.
  - The Longevity Index *LifeMetrics* Launched by JP Morgan in 2007 concerning American, English and Dutch populations. The methodology for constructing this index as well as the estimation of prospective mortality is publicly available.
  - QxX, a proprietary index, launched in December 2007 by Goldman Sachs on a sample of the population in the United States.
  - The Xpect Data, since 2008, on German and Dutch populations.

With these indexes, it is possible to create tradable bonds in the markets. The use of indexes, however, generates a basic risk for the insurer. Indeed, indexes reflecting the general population may not evolve as the mortality of the insured portfolio.

Compromise between basic risk, transparency for investors and simplicity

By definition, the type of exchange that contains the most basic risk is the transfer of cost risk, since the protection purchaser receives payments corresponding to the actual realization of a continuance distribution. The basic risk in this case is avoided. This is the case, for example, in the context of a longevity swap, as described above.

In contrast, index coverage provides for indexed payments on a given index, such as a national public index. These payments are not going to correspond *exactly* to the financial commitments of the purchaser of coverage, thus there is a basic inherent risk in this type of transaction. However, by going from cost to index, the basic risk has been increased, but transparency has been improved for financial investors and coverage sellers. In fact, the latter have easier access to the index on which the product is written than to the details of the portfolio and the financial commitments of the protection purchaser. There is therefore an initial compromise between the basic risk and the degree of transparency for the investor.

This is not the only compromise: from cost to index, one creates a basic risk, but there is also an additional degree of complexity for the investor. In fact, the construction of the indexes used is not simple (see for example the construction of *LifeMetrics* documented by JP Morgan) and the use of these indexes subsequently made for the structuring of index products is also complex. There is clearly a risk of investors' cautiousness as many are not specialists in mortality and/or longevity and may lose interest in these products because they do not master the underlying randomness of their investment.

9.4 Impacts on the Solvency 2 Benchmark

Generally, the introduction of a risk transfer mechanism induces a decrease in the capital requirement for the underwriting risk and an increase in the capital requirement for the counterparty risk, depending on the rating of the counterparty.

One wishes to compare the capital requirement (SCR) in the following two situations:

- The insurer has not subscribed to a reinsurance treaty and is therefore faced with a loss charge Λ;
- It sets up a treaty as above, so it retains R, the amount to which is added, if any, of the amount ceded in the event of default of the reinsurer.

SCR can be evaluated as part of the standard formula described in regulation[2] 2015/35 or by means of "internal model" approach (which is the case in the illustration presented in Sect. 9.2.2).

In the absence of a retrocession, only the underwriting risk and, more specifically, the incidence and longevity risks (Art. 138 of the European Commission Delegated Regulation) are assessed.

As soon as the risk is ceded, consideration is also taken of the counterparty risk, calculated on the reinsurance asset equal to the amount of the ceded *Best Estimate* adjusted with the default risk. For the calculation of the default adjustment for counterparties applicable to the *Best Estimate*, a simplification of article 61 of the Delegated Regulation is used:

$$Adj_{CD} = -\mathbf{max}\left((1 - RR) \times BE_{Rec} \times Dur \times \frac{PD}{1 - PD}; 0\right)$$

with $BE_{Rec} = E(C)$ (see Sect. 9.2.2 notes). The recovery rate is taken at $RR = 0.5$. The default probabilities according to the reinsurer rating are provided in article 199. In this case, this is a type 1 counterparty risk (*see* Art. 138 of the Delegated Regulation), the calculation procedures of which are specified in articles 200 and 201 and simply lead, with one counterparty, to

---

[2]https://esurfi-assurance.banque-france.fr/uploads/tx_bdftechnicalinfos/Regl.2015-32-N2-Solva2_FR.pdf.

$$SCR = c \times \sqrt{PD(1 - PD)} \times (1 - RR) \times BE_{Rec}$$

where the coefficient c is a function of the coefficient of variation of the loss distribution in the event of a default[3]: $c = 3$ when $cv \leq 7\%$, $c = 5$ when $7\% < cv \leq 20\%$ and $c \times \sqrt{PD(1 - PD)} = 1$ otherwise.

So, we have two effects related to the implementation of the reinsurance scheme:

- A decrease in the net asset value of the amount Adj_{CD};
- Substitution of the counterparty SCR to a fraction of the underwriting SCR.

For the counterparty risk calculation, the level of reinsurance deposits with the ceding company affects the loss level in the event of a default for which an SCR must be evaluated. However, a compromise should be carried out between the gain in counterparty SCR and the increase in the reinsurance premium related to this collateralization.

9.5 Conclusion

Continuance risk transfer solutions are varied and can thus meet different objectives (smoothing results, protection against extreme risk, limitation of the need for capital, etc.).

Mechanisms underlying these solutions are, however, relatively homogeneous and are based on the observation that a Long Term Care insurance portfolio consists of the sum of independent risks, conditional on continuance assumptions.

The implementation of models to quantify the effects on equity and capital need is possible in a simplified way and allows the best characteristics of the treaty to be defined according to the objective pursued.

Appendix: Net Premium Distribution with Reinsurance

The random variable $R = \Lambda - [(\Lambda - F) \wedge P]^+$ is considered where F and P are constants and Λ is a random variable with distribution F_Λ; we show here that the distribution function of R is given by:

$$F_R(x) = \begin{cases} F_\Lambda(x) & x < F \\ F_\Lambda(x + P) & x \geq F \end{cases}.$$

We note that $\varphi(x) = x - [(x - F) \wedge P]^+$, (with $F = 1.5$ and $P = 1$) (Fig. 9.5):

---

[3] This is the volatility of the current value of the cash flows transferred to the counterparty.

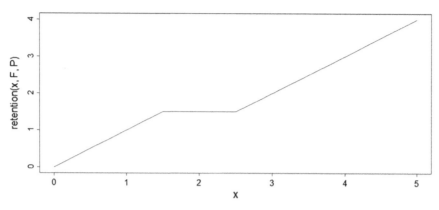

Fig. 9.5 Retention function

From the graph above, it is established that:

$$\begin{cases} \varphi(x) = x - P & \Leftrightarrow & x > F + P \\ \varphi(x) = F & \Leftrightarrow & F < x \le F + P \\ \varphi(x) = x & \Leftrightarrow & x \le F \end{cases}$$

and the expression of the cumulative distribution function follows directly as $R = \varphi(\Lambda)$.

References

1. Brouhns, N., Denuit, M., Vermunt, J.K.: A poisson log-bilinear regression approach to the construction of projected lifetables. Insur. Math. Econ. **31**, 373–393 (2002)
2. Haas, S.: Méthodologie d'évaluation économique des traités proportionnels en réassurance vie—application au swap de mortalité. Mémoire d'actuaire, ISFA (2006)
3. Planchet, F., Thérond, P.E.: Modélisation statistique des phénomènes de durée – applications actuarielles, Paris, Economica
4. Planchet, F., Tomas, J.: Uncertainty on survival probabilities and solvency capital requirement: application to LTC insurance. Scand. Actuar. J. (2014). https://doi.org/10.1080/03461238.2014.925496

Part IV
Prospective Vision of the Risk

Bob Yee

Introduction

Private long term care insurance in the United States has over 30 years of experience. France has a slightly shorter history, but the coverage of its public insurance program is broad. Long term care insurance is in its infancy in both the public and the private sectors for many other populated countries such as China and India. Experience in the United States and France are valuable lessons to countries contemplating such insurance. All countries should benefit from the evolution of actuarial analysis as discussed so far in this book. This final Chapter provides a glimpse of the universal future challenges for all countries.

Long term care insurance, as any other insurance line in France, is subject to risk management directives under the ORSA framework. This framework brings to the forefront the riskiness of long term care insurance. The French approach for modeling long term care product risks is couched in sound theoretical foundation. Stimulation techniques are used to quantify risk measures. In contrast, practicing actuaries in the United States have not yet fully embraced stochastic models. Their approaches are mainly grounded in detailed data analysis (perhaps data is more readily available than in France) to produce reasonable point estimates. Then sensitivity tests are performed to estimate the risk margins. The subchapter on Predictive Analytics illustrates a number of recent approaches to estimate incidence rates. Lastly, this Chapter explores some of the social and technological trends that will impact long term care insurance experience as we know it today.

Chapter 10
Solvency II Own Risk and Solvency Assessment for Long Term Care Insurance

Marc Juillard and Géraldine Juillard

10.1 Introduction

The purpose of this chapter is to develop a panorama of methods applicable in establishing an Own Risk and Solvency Assessment (ORSA) model for Long Term Care insurance products. As such, we address the following areas:

- Main generic principles related to ORSA
- Qualitative description of major risks of a Long Term Care insurance contract
- Main modeling constraints and built-in biases
- Proposals of models for insurance and financial risks.

For more details concerning the introduction of a quantitative approach to an ORSA process, the reader can refer to Planchet et al. 2014 "Prospective solvency—Quantitative methods for ORSA".

10.2 Main Principles

10.2.1 General

If the normative character of Pillar 1 of the Solvency 2 prudential system provides an exhaustive framework globally comparable between different insurers, in some situations it may lead to a valuation of risk poorly adapted to some products. This point is particularly evident in the case of long duration products, which present highly heterogeneous or evolving risks. Long Term Care insurance provides an enlightening example on this point: the challenge is not to develop an internal model to obtain a

M. Juillard (✉) · G. Juillard
19 rue ginoux, 75015 Paris, France
e-mail: juillard.marc75@gmail.com

© Springer Nature Switzerland AG 2019
E. Dupourqué et al. (eds.), *Actuarial Aspects of Long Term Care*,
Springer Actuarial, https://doi.org/10.1007/978-3-030-05660-5_10

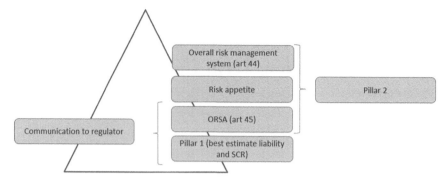

Fig. 10.1 Pillar 1–Pillar 2 links

reasonable capital requirement but to incorporate into reserve calculations justifiable management rules in terms of rate adjustments and/or management of a stability or revalorization fund.

This limit is partly corrected by Pillar 2 of Solvency 2 that encourages insurers to put in place an integrated risk management system. Pillar 1 and Pillar 2 links can be synthesized as follows (Fig. 10.1).

Based on an *Enterprise Risk Management* (ERM) approach, the integrated risk management system aims to detect, measure, monitor and communicate all major current and future risks of the insurer. Article 45, the real heart of Solvency 2,[1] provides that, within the framework of its risk management system, each insurer carries out an internal assessment of its risks and its solvency. This evaluation includes the following elements:

- Overall Solvency Needs,[2] given the specific risk profile, the approved limits of risk tolerance, and the insurance company strategy;
- Permanent compliance to capital requirements as well as reserving requirements;
- Analysis of the risk profile adequacy within the framework of the standard formulas.

These three steps are naturally linked and must be carried out jointly. Through the analysis of its risk profile adequacy, the insurer defines statistical models that

---

[1]For more details, the interested reader may refer to règlement délégué (ue) 2015/35.

[2]Starting from the regulatory framework, the following definition can be established: whereas Pillar 1 calculates the 1 year probability of ruin on the basis of portfolio *Run off* (closed) and through a set of risks defined by the European Commission, the overall solvency requirement corresponds to the level of capital necessary to ensure multi-year solvency of the insurer on the basis of a vision including a strategic development plan, by retaining assumptions adapted to specific characteristics of the company and taking into account the Administrative, Management or Supervisory Body (AMSB) risk appetite. It differs from the Solvency Capital Requirement (SCR) by its calculation horizon (that of the strategic plan), the risks taken into account (completeness of the major risks without being limited to the framework of the standard formula) and the safety margin retained.

will then feed the quantitative model for the calculation of Overall Solvency Needs.[3] This model is split into four major steps which will be described below.

10.2.2   Major Steps

10.2.2.1   Defining the Risk Appetite

The principle of risk appetite was defined in 2002 by COSO.[4] It is the maximum level of risk that an insurer agrees to take in order to achieve its strategic objectives. The risk appetite objective is to deliver a global and consolidated vision of the insurer risk exposure and to allow it to:

- Translate its global strategy within a global limit;
- Put this limit in relation to overall level of risks undertaken;
- Manage risk taking in accordance with this limit;
- Simply communicate this risk policy in order to share it.

Defining a company's appetite is a delicate exercise: beyond the problem of the level to retain, the real issues relate to current and future constraints:

- By increasing risk-taking, what is the capacity of the insurer to absorb the residual risk if it materializes? (This is referred to as ORSA within ORSA.)
- By decreasing risk-taking, what opportunities are missed if the risk does not materialize; what will be the impact on profitability and investment potential?

Ultimately, defining this appetite leads us to seek the optimization of risk-return. A parallel can easily be made with the Markowitz theory of an efficient portfolio in the sense that establishing an ORSA process leads to optimizing the portfolio's diversification potential, with the aim of lowering the risk without excessively degrading its profitability.

10.2.2.2   Calculating Overall Solvency Need

The overall solvency requirement represents the capital required after comparing the risk appetite of the insurer and the risk exposure of its strategic plan. In the case of "solvency" dimension, its calculation can be synthesized as follows:

- Starting from an initial risk profile (structure of liabilities and assets), the insurer will project the evolution of its balance sheet over the duration of its strategic plan. This evolution will be impacted by the strategic plan (evolution of asset allocation, changes in liabilities from new issues and new products), by management action (risk-hedging policy) and changes in market conditions. This change can be carried

---

[3]And analysis of continuous compliance.

[4]Committee of Sponsoring Organizations of the Treadway Commission.

out using a stochastic approach or by stress tests. The choice between these two approaches, both of which present advantages and disadvantages,[5] will be dictated by the complexity of liabilities and the nature of the risks to the organism. It should be noted that in the case of a stochastic approach, the use of proxies like those presented in the following chapters seems inevitable.

- Once this projection has been completed, the insurer can use a set of standard results[6] it will need to modify in order to calculate the dimensions that it wishes to follow. As an example, a 5-year strategic plan tested under five scenarios will lead to 25 hedging ratios. This step, particularly time-consuming, can lead to the development of methods to accelerate calculations. While many methodologies are retained by the market, the biometric risk burden of Long Term Care contracts will not permit to exclude risks of financial origins present in a typical portfolio and will lead to favor parametric approaches of the Least Square Monte Carlo or Curve Fitting type.
- Once this second step is carried out, the insurer has at its disposal the evolution of its coverage ratio under different scenarios and will then be able to define whether its risk appetite is verified. This step will define the level of free capital (positive or negative) that will be appropriate to allocate between the different risk bearers.[7] This will be done through its risk tolerance.

10.2.2.3   Defining Risk Tolerance

Complex and subjective, this stage of the process leads the insurer to allocate its overall solvency need between its different sources of risk. Based on a measure of capital allocation, this step[8] allows one to allocate the benefits of diversification between the different risk owners and contribute to its internal performance controls. While a variety of methods are possible, before all, this step remains a strategic decision:

- The proportional method;
- The marginal method;
- The Shapley method;
- The Euler method.

Summaries of these four methods are presented below [1].

The Proportional Method
This method amounts to allocating overall solvency requirement on the basis of the contribution of each risk to the overall economic capital, respectively annotated as $\rho\ (X)$ and $\rho\ (X_i)$:

---

[5]If the stress-testing approach seems simpler than the stochastic approach, it presents the risk of not testing the right scenarios.

[6]A few in the case of a stress-testing approach, an infinity in the case of a stochastic approach.

[7]Branches, subsidiaries, etc.

[8]Which must be consistent with strategic development goals.

$$\rho^{proportional}(X_i|X) = \frac{\rho(X_i)}{\sum_j \rho(X_j)}\rho(X).$$

If the proportional method has the advantage of being simple to implement, it presents the disadvantage of not properly managing diversification benefits (which in the case of a conglomerate will lead to inadequate measure synergies between the different entities). Let us consider the case of a multi-line insurer consisting of two entities A and B presenting respectively a free surplus need of 100 and 150 monetary units each, and such that the aggregation of activities leads to a free surplus need of 200 monetary units (i.e. a diversification benefit of 50). Application of the proportional method will lead to the allocation of 40% of the capital to entity A and 60% to entity B without properly analyzing the source of diversification benefit.

The Marginal Method
Offering a compromise between complexity and good properties, the marginal approach seems to win the market favor. Also called the incremental method, it makes it possible to allocate aggregate capital according to the marginal impact of each of the segments.[9] Denoting by

- $\rho(X)$, the global capital need;
- $\rho(X - Xi)$, the global capital need by removing unit i,

the application of the marginal method leads to the following allocation:

$$\rho^{marginal}(X_i|X) = \frac{\rho(X) - \rho(X - X_i)}{\sum_j \rho(X) - \rho(X - X_j)}\rho(X).$$

While it has the advantage of taking into account marginal impacts (and offers a better view of each entity contribution to global diversification), it suffers from the following criticisms:

- Based on the realization of a set of sensitivity tests aiming at measuring the marginal impact of each risk, it is more complex than the proportional method (especially if calculations are made on the basis of a partial internal model).
- By the standard formula approach, this method suffers from instability when changing the direction of the interest rate risk: diversification factors change abruptly and impact the benefit of diversification and therefore the marginal risk.
- It ignores cross-effects (only the impact of each unit is tested).

The Shapley Method
The Shapley method is a discrete method based on the theory of cooperative games describing a situation where players want to reduce their cost (solo vision) but pays attention to synergies between players that can reduce global cost, hence this problem is adapted to the framework of capital allocation. The Shapley method is an extension of the marginal method that aims to take into account the marginal impact of each actor at the aggregate level and at the sub-level. Given:

---

[9]It is therefore a sensitivity testing approach.

- $\rho\,(X)$: global capital need;
- D_i: the number of coalitions containing entity i operating within the company;
- S: a coalition;
- $\rho\,(XS\text{-}Xi)$: the capital need of coalition S if unit i is removed,

application of the Shapley method leads to the following allocation formula:

$$\rho^{Shapley}(X_i|X) = \sum_{S \in D_i} \frac{(S-1)! \times (n-S)!}{n!} \times (\rho(X_S) - \rho(X_S - X_i))$$

The Euler Method

The Euler method is an extension of the marginal method in which the impact of the units of each section is tested at the infinitesimal level (the marginal method being at the unit level):

$$\rho^{Euler}(X_i|X) = \lim_{h \to 0} \frac{\rho(X) - \rho(X - hX_i)}{h} \times \rho(X)$$

10.2.2.4 Defining Risk Limits

Truly an operational translation of risk tolerances, establishing risk limits defines specific and intelligible risk-bearing boundaries by risk takers (premium volume per product, target allocation, …).

10.2.3 Setting the Scenarios

ORSA leads to testing the impact of the Business Plan as part of a central scenario and stressed events. This requires us to:

- Model a central scenario, then stress scenarios around this central scenario;
- Project targets at the strategic plan horizon for each of the scenarios considered.

The central scenario definition does not call for a particular comment,[10] but the construction of the stressed scenarios offers several possibilities.

10.2.3.1 What Kind of Stress?

When setting up an ORSA process, two categories of stress may be considered:

Stress Tests

---

[10]The central scenario corresponds to the best estimate of the evolution of the economic environment. It also includes the set of all assumptions held in the context of the business plan.

Stress tests measure the sensitivity of an insurer to the occurrence of an unfavorable scenario. This sensitivity is expressed in terms of the observed impact on a dimension (such as solvency) and frequency of occurrence of the scenario (either in probability or in an event interval). The key factor upon setting the test is to ensure the adequacy between stress intensity and the associated event interval.

Reverse Stress Tests
Reverse stress test measures the financial strength of an insurer. In a more precise way, the application of a reverse stress test aims to determine the lowest stress intensity that will lead an insurer to bankruptcy. The key factor when it is set up is to identify the proper risk factors that lead to ruin.

Stress tests and reverse stress tests are typically used in two situations:

- In an ORSA framework: these are scenarios that will be applied in order to calculate the overall solvency requirement of the insurer. It should be noted that stress tests are more amenable to this exercise.
- In a framework for validating an internal model: they are intended to ensure that the aggregating procedure of the internal model is efficient. It is really reverse-engineering of SCR by defining stress levels:

  - that the insurer can bear; or
  - that would lead to a loss of surplus identical to SCR.

Only stress tests will be addressed since they are easier to set up and to validate (reverse stress tests require defining the dynamics of risks while stress tests are limited to quantification).

10.2.3.2   Construction Process

The definition of a stress must meet certain constraints:

- It should be used to manage all major risks as well as dependence structures (cross-scenarios will therefore be preferred).
- It must further present a coherent event interval with that retained for the risk appetite.

To these few "good practices", the following points should be added:

- The path of the stress scenario: a sharp increase in interest rates will not have the same impact as a constant but with a weaker progression (these scenarios still lead to the same structural terms at the end of the strategic planning.)
- Taking a realistic account of management action. In the case of a trend risk, modeling of management actions aimed at correcting the degradation of a combined ratio, for example, should be consistent with the time required to detect this change in trend.

Starting from these findings, three principles of modeling can be retained:

- Top-down scenarios. This approach is to test set scenarios communicated by international organizations (IMF, FSAP, EIOPA). Mainly financial, these scenarios have the advantage of providing a benchmark, but they present the disadvantage of not being in line with the insurer risk profile.
- Bottom-up, so-called expert, scenarios. These are plausible and coherent multi-year macro-economic scenarios. They are built based on the history of past crises and the anticipation of the changes in the insurer risk profile. If they have the interest of being pragmatic, they have the disadvantage of making it difficult to quantify risks in terms of limits (often expressed with VaR-type indicators). It should also be stressed that they presuppose that the "at risk" situation is known (risk factors, dependency structure, ...) but do not allow one to precisely determine the stress level associated to the scenarios under consideration. However, they do make it possible to manage risks which are difficult to model (disaster scenario, operational, ...).
- Bottom-up, so-called model, scenarios. These are scenarios built on the basis of the statistical properties of risk factors. They can be expressed "solo" or combined. More delicate to communicate internally, they cannot be sufficient by themselves: some "classical" expert scenarios should be tested in any case (the Japanese scenario, for instance). Moreover, scenario calibration requires access to a significant amount of data.

Given the risk profile of Long Term Care contracts, implementation of the ORSA process will be based mainly on *Bottom-up* scenarios. However, it would be illusory and potentially dangerous to limit oneself to quantitative approaches. First of all, the calculation of overall solvency need must be accompanied by the development of a set of major risk monitoring indicators which allow the initiation of emergency plans and associated management actions.[11] Finally, quantitative approaches should be retained only if they provide a realistic view of stress situations for the organism. In the absence of quality data or in the presence of poorly adapted models, it would be appropriate to turn to expert-based approaches.

10.3 Qualitative Description of Inherent Long Term Care Insurance Risks

The objective of this section is to develop an overview of major risks encountered in Long Term Care insurance products. Beyond their qualitative description, positioning within a Pillar 1 framework can be summed up schematically as follows (Fig. 10.2).

While emerging risks certainly impact Long Term Care products, they are not addressed in this chapter.

---

[11] Also to detect any unfavorable and unforeseen developments when calculating the overall solvency requirement (in order to trigger, if necessary, a "non-regular ORSA").

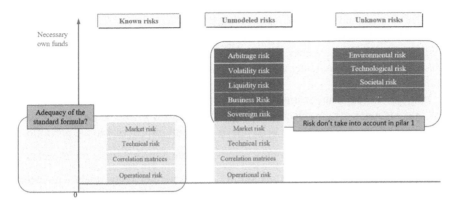

Fig. 10.2 Risk Categorization under Pillar 2

10.3.1 Major Risks

The stakes for the insurer is in the best estimate of the incidence and continuance. Long Term Care is a long-term risk for which a loss is the result of a combination of distributions (incidence and continuance). If age and gender are two discriminating factors, many other parameters must also to be taken into account. As an example, the cause of incidence (accident, senile dementia, Alzheimer's, or other diagnostics) has a significant impact on incidence rates and on continuance rates.[12] Long Term Care risk requires building experience distributions by diagnostic in order to reduce estimate bias when valuing the *Best Estimate*. This point is made delicate by two phenomena:

- Portfolio duration is often too short to have a volume of data ensuring estimator convergence (or at least a volatility risk within well-defined boundaries).
- If diagnostic information is known by the insurer, it is often stored in a Document Management System (DMS) file in PDF format that makes its use challenging.[13]

Active lives mortality is also to be followed carefully. With increasing incidence by age, an increase in active lives longevity will have an upward impact on the number of incidences. This means designing models to keep track of trend risks in the portfolio.

French insurance portfolio experience, even if enriched year after year, does not yet allow one to completely observe incidence rates and active and claimant mortality

---

[12] The health status of beneficiaries classified in the same state of dependence is very heterogeneous. Cause of incidence, accident or illness, has a direct impact on life expectancy in dependence. The nature of the disease is also critical because if some evolutionary pathologies are rather slow-running; others, on the contrary, can evolve rapidly. In Total Dependence (see Chap. 3), there are, for example, beneficiaries suffering from Alzheimer's whose life expectancy is much higher than beneficiaries with terminal cancer.

[13] It is then necessary to turn to handwriting recognition algorithms in order to process this initial information.

rates for all ages and all levels of dependence (see Chap. 3). For Total Dependence (see Chap. 3) for example, experience from the first generation of contracts[14] is now credible enough only for ages 60–90. The knowledge of Partial Dependence (see Chap. 3), which appeared later with the 2nd Generation of contracts,[15] is much more limited.

In the event the insurer turns to a reinsurer to capitalize on its knowledge of the risk, the adequacy of risk-selection policy of both parties should be checked, and to manage the basic risk due to the behavior of distributions external to the portfolio.

To conclude, Long Term Care contracts impose three biometric risks on insurers:

- A sampling risk in the case of small portfolios or in the case of too finely segmented experience distributions.
- A volatility risk in the case of distribution sets which are too weakly segmented.
- A trend risk to take into account long-term deformations of the longevity risk for active lives and claimants.

Beyond these biometric risks, two sources of risk can be added:

- Interest rate risk: Long Term Care being a long-term risk,[16] reserves are particularly sensitive to investment income assumptions.
- Inflation risk associated with maintenance and utilization costs.

10.3.2   Positioning with the Standard Formula

In the standard formula, the impact of these risks is taken into account through the application of the following shocks:

- Longevity risk: reduction in active and dependent mortalities by 20%.
- Morbidity risk: 35% increase in incidence in the first year and 25% in the following years.
- Severity risk: 10% increase in costs and 1% increase in annual inflation rate.

These shocks, which have been calibrated from retirement portfolios, are only weakly adapted to Long Term Care insurance contracts. In particular, assumptions for longevity shock, leading one to consider that active lives life expectancy will increase in the same proportion as that of dependents, are questionable and potentially dangerous (causes of trends in healthy and dependent life expectancies are strongly different). Pillar 1 provides here a set of guidelines that should be appropriate to challenge when setting up the ORSA process.

---

[14]The first Long Term Care insurance contracts were marketed in France in 1985. These were individual contracts with an annuity guarantee in case of Total Dependence. With an average issue age of 60 years, the maximum insured attained age is now just over 90 years.

[15]In the middle of the 1990s.

[16]For a 60-year-old the risk is not likely to materialize, if at all, before 15 or 20 years.

10.3.3 Risks not Covered by the Standard Formula

The ORSA model must cover the risks included in the standard formula, but should also cover significant risks not considered in Pillar 1, such as the image risk. Long Term Care products are potentially very sensitive to image risk in the sense that they have few constraints in terms of rate review and benefit revalorization. If this freedom given to the insurer significantly reduces its surplus requirement (and hence increases its profitability) its unlimited use can lead to a highly degraded image. This risk can be summed up, gradually, through two levels:

- Level 1: the policyholder comes to consider that the price paid for his or her Long Term Care coverage is disproportionate and decides to ask for a benefit reduction. Depending on the general conditions of the contract, two cases are possible:

  - The insurer communicates to the insured the amount of the reduced benefits at policy anniversary. By fixing the coverage level on that date, the insurer will no longer have the capacity to transfer part of its risk to the insured in the event of future underwriting losses.
    Or
  - The insurer communicates the amount of benefit reduction when a claim is approved, which allows the insurer to integrate the most up-to-date information about the evolution of Long Term Care risk and thus minimizes its risk of future underwriting losses (but without being able to reduce it completely because it will no longer be able to use a future premium rate revision as a mitigation measure). The French market leans more toward this practice.

- Level 2: the policyholder considers that the history of premium and benefit revalorization is to his or her disadvantage, which creates a degraded image of the insurer and leads to the cancellation all of his or her policies with this insurer.

The level 1 image risk may be limited by incorporating into the ORSA Model a *Management Action* which will, within a realistic timeframe, apply a rate revision over several years when it is important, so as not to exceed the maximum rate considered commercially acceptable. The level 2 image risk is more difficult to translate and should rather be integrated by qualitatively monitoring the satisfaction level of the policyholders.

10.4 Model Constraints for a Long Term Care Policy

Reserve *Best Estimate* calculations, through the central scenario, remain one of the major issues in the development of a management process for risks related to a Long Term Care policy. This valuation goes through the definition of the following elements:

- Modeling cash flow probabilities (i.e. modeling of the various health states).

- Modeling of claim.
- Construction of Best *Estimate* distributions feeding the model.

10.4.1 Transition Modeling

Given the nature of the Long Term Care risk, transition projections rely on multi-state models. The types and number of states to consider, as well as the transition rules, depend on the dependence level[17] from which a benefit is triggered. Selected states are associated with a transition probability to another state and a probability to remain in that state. In practice, the difficulty lies in estimating these probabilities.

In the absence of credible experience at all levels of dependence, it is difficult to calibrate all incidence and continuance probabilities. Models are often simplified by considering in particular that no recovery is possible (a dependent person will never again be active nor shift from a higher level of dependence to a less severe one), neglecting the impact of future external factors such as pollution or the progress of medicine,[18] and by grouping some levels of dependence.

Consider for example a Long Term Care insurance product providing for the payment of an annuity based on the level of dependence. This product guarantees 100% of the payment in case of dependence at 3 ADL out of 5 (total) and only 50% in case of dependence at 2 ADL out of 5 (partial). In this case, the annuity is modelled by considering two separate and independent benefits: the first providing for the payment of 50% of the annuity at any level of incidence, and the second the payment of a supplement of 50% in case of total dependence. For the first benefit, only a single state of dependence is retained, limiting modeling to the following three states:

- Not dependent at any level
- Dependent
- Died.

Transition and retention probabilities associated with these three states are:

- Probability that an active life enters into any level of dependence
- Probability that an active life dies without going through a state of dependence
- Probability of death of claimant
- Probability of staying active
- Probability for a claimant to remain in the same state.

For the second benefit (50% complement in case of total dependence), the three states considered are:

- Not totally dependent (active, or partial dependence)

---

[17]Total, partial or light.

[18]Therefore, the likelihood of a person who will be 75 years old in 20 years is assumed to be equal to that of a person who is now 75 years old.

- Totally dependent (total dependence)
- Death.

Transition and retention probabilities associated with this are analogous to those quoted above by replacing "not dependent" with "not totally dependent".

The fact of retaining this approach, however, leads to a skewed valuation of *Best Estimate* reserves through a heterogeneity bias: in the case of the second guarantee paid in the case of total dependence, the partially dependents and the active population are assimilated as "not totally dependent" with a risk of underestimation of mortality. If, at first glance, overestimating the volume of contributors would tend to diminish the value of the provision, it would lead to an increase in the population that could become claimants. This population remaining less well known than active population, the ORSA model will associate a higher cost.[19]

10.4.2 Modeling Probable Claim Amounts

Due to the uncertainty associated with the evolution of the Long Term Care risk, underwriting and financial results of Long Term Care contracts do not directly materialize in favor of "prudential funds". Some insurers set up two funds: a stability fund and a revalorization fund, funded by underwriting and financial results. However, maintenance of these funds (size and funding pace) depends on complex *Management Action* that should be calibrated through an ORSA type of *Run*. This can be explained as follows:

Elements Calling for a Large Stability Fund
Beyond its ability to cushion a worsening loss trend, the stability fund makes it possible to pilot the SCR and thus the RORAC (Return on Risk-Adjusted Capital) of the insurer. Being deemed to belong principally to the insurer, it can be used in full to cushion the SCR and can drastically diminish the need for the product's surplus. As such, the higher the level of the stability fund, the more effective an attenuating effect it will have.

Elements Calling for a Small Stability Fund
Since funding of the stability fund is carried out to the detriment of the immediate profitability of the product, it leads to weaken (if not erase) product profitability when it is launched (or during a large increase of new issues). This leads to limiting its size, especially if the ORSA process integrates a short-term profitability dimension in its metrics (we would then refer to a risk appetite process rather than an ORSA process). Juxtaposition of these two constraints then leads to careful testing of the stability fund setting (size and timing of funding) when launching a Long Term Care product. The objective of these tests will be to define the ideal proportion of the

---

[19]Through stress (Pillar 1 or Pillar 2) but also through conservative assumptions.

underwriting and financial results that should be allocated annually to the stability fund.[20]

Taking into account the revalorization[21] of premium and benefit flows in the calculation of the *Best Estimate* requires the revalorization interest rates to be used for all projection years. To do this, the availability of the revalorization fund should be modeled based on the balance of fund and reserve levels (active and claim).

To limit calculation time, active life reserve is combined with claim reserve in order to follow the different generations of claimants. This choice to model only a global provision generates bias but has the advantage of not having to manage a simulation within a simulation.

10.5 Modeling Biometric Risks

As stated above, insurance risks of Long Term Care are related to incidence and continuance[22] and their nature, and modeling, differ strongly:

Trend Risk

This risk is due to external variables interfering (most of the time leaving a residual effect) with the curve of the table over time. This translates into a source of uncertainty that disturbs (and therefore skews) the average value of the reference table. The introduction of antibiotics in 1950 can be cited as an example. This medical advance resulted in a sharp decrease in contemporary mortality, which was not predictable. As a result, external variables (such as global warming), or the evolution of a treatment for Alzheimer's, may be considered to have an impact on future mortality rates.

Volatility Risk

In contrast to trend risk, volatility risk does not put into question the average of the experience probability. This is the risk of fluctuation around the mean resulting from uncertainty associated with parameter estimates. This translates into an oscillation of actual outcomes around the mean defined by the experience table. This occurs during the two calculation steps of the overall solvency requirement: during the definition of *stress tests* (due to the increase in volatility) and during the calculation of the *Best Estimate* (which is often the application of a convex function on the experience probability curve).

---

[20] Within the limit of the stability fund ceiling, generally expressed as a percentage of active lives reserve and as a function of expected profitability of the product.

[21] Even though the revalorization of Long Term Care contracts is not regulated in France, the general terms and conditions of contracts generally provide for the presence of an annual revalorization mechanism, contract benefits and open claim benefits. This revalorization is generally determined by reference to an annual index performance (of the type of an annual social security maximum benefit) and within the limit of the capabilities of the revalorization fund. Premiums are re-valued in the same proportions.

[22] As the shift from one dependent state to another dependent state is difficult to quantify, its stress-based modeling will not be addressed in this chapter.

Sampling Risk or Pooling of Risk

Long Term Care insurance portfolios are often small and may present a risk of not being able to spread the risk. The pooling assumption being the basis of insurance premium rates, it is important to measure the risk associated with the small size of the portfolio when setting up the Long Term Care insurance ORSA model.

10.5.1 Sampling Risk

Depending on the size of the portfolio, it is by nature risk-pooling. As soon as the portfolio matures, sampling risk becomes more credible and less of a priority than systemic risks. Assuming independence among policyholders allows the application of the central limit theorem (a constant maximum benefit amount is a realistic assumption). Also, if Λ is the sum of discounted claim cash flows:

$$\frac{\Lambda - E(\Lambda)}{\sigma(\Lambda)} \underset{|I| \to \infty}{\to} N(0, 1)$$

the Alpha level quantiles can then be expressed by the following formula:

$$\text{VaR}_\alpha(\Lambda) \approx E(\Lambda) + \phi^{-1}(\alpha)\sigma(\Lambda).$$

Knowing the first two moments of Λ then allows one to measure without approximation the sampling risk impact. For instance, the Alpha level confidence interval for Λ is given by:

$$IC(\Lambda) = \left[E(\Lambda) - \phi^{-1}\left(1 - \frac{1-\alpha}{2}\right)\sigma(\Lambda), E(\Lambda) + \phi^{-1}\left(1 - \frac{1-\alpha}{2}\right)\sigma(\Lambda) \right].$$

As an example, these two moments are subsequently calculated by considering, to simplify the formulas, a portfolio of similar benefits and triggers, and a homogeneous risk profile. The present value of benefit payment cash flow at time t becomes:

$$F_t = \sum_{j \in J} r_j 1_{]t,\infty[}\left(T_{x(j)}\right),$$

where:

- J = all claimants in the portfolio.
- TX (j) = the time of death of the Tth claimant.
- r_j = the benefit amount of the Jth claimant (revalorizations are not taken into account).

The sum of discounted benefits is then obtained through the following formula:

$$\Lambda = \sum_{t=1}^{\infty} e^{-rt} \sum_{j \in J} r_j 1_{]t,\infty[}\left(T_{x(j)}\right)$$

$$\Lambda = \sum_{j \in J} r_j X_j$$

It remains to calculate the mean and variance of Λ. Under the hypothesis of claimant independence, this is equivalent to solving the following equations:

Mean

$$E(\Lambda) = \sum_{j \in J} r_j E\left(X_j\right)$$

where:

$$E(X) = \int_0^{\infty} e^{-rt} S_x(t) dt \approx \sum_{t \geq 1} e^{-rt} S_x(t).$$

Variance

$$V(\Lambda) = \sum_{j \in J} r_j^2 V\left(X_j\right).$$

In the absence of systemic risks (financial risks are ignored), the variance of X is calculated as follows:

$$E(X^2) = \frac{2}{r} \int_0^{\infty} \left(1 - e^{-rt}\right) e^{-rt} S_x(t) dt \approx \frac{2}{r} \sum_{t \geq 1} \left(1 - e^{-rt}\right) e^{-rt} S_x(t),$$

$$V(X) = \frac{2}{r} \sum_{t \geq 1} \left(1 - e^{-rt}\right) e^{-rt} S_x(t) - \left(\sum_{t \geq 1} e^{-rt} S_x(t)\right)^2.$$

Introducing systemic risks can be carried out by holding a conditional Gaussian approximation. This is equivalent to considering that these systemic factors can be synthesized in a random variable of state Z from which we are able to determine its impact on cash flow average and volatility. In other words, it is assumed that the above reasoning is conditional on Z. We then have:

$$\frac{\Lambda|Z - E(\Lambda|Z)}{\sigma(\Lambda|Z)} \xrightarrow[|I| \to \infty]{} N(0, 1)$$

which can also be written as follows:

$$\lim_{|I|\to\infty} E\left(e^{-u\Lambda}|Z\right) = \exp\left(-uE(\Lambda|Z) + \frac{\sigma^2(\Lambda|Z)}{2}\right).$$

The Laplace transform of Λ is then:

$$\lim_{|I|\to\infty} E\left(e^{-u\Lambda}\right) = E\left(\exp\left(-uE(\Lambda|Z) + \frac{\sigma^2(\Lambda|Z)}{2}\right)\right).$$

By specifying the distribution of Z and the functions $E(\Lambda|Z)$ and $\sigma^2(\Lambda|Z)$, we can then calculate the asymptotic distribution of Λ in this conditional model.

10.5.2   Incidence: Volatility Risk

Volatility risk occurs during the construction of the experience mortality table especially during the smoothing process (parametric or not) of observed rates. If the purpose of the latter is to remove fluctuations from the noise contained in the initial data, it can contain assumptions that are too strong when the adjustment sample is too small, a case frequently encountered in the construction of incidence or continuance distributions.

Taking into account these sampling fluctuations can be managed through a logit regression of observed data on the smoothed logits (See Planchet and Therond 2011):

$$\text{logit}\left(q_{x,t}\right) = \text{Ln}\left(\frac{q_{x,t}}{1 - q_{x,t}}\right) = a \times \text{Ln}\left(\frac{q^{Ref}_{x,t}}{1 - q^{Ref}_{x,t}}\right) + b + \epsilon_{xt} \ .$$

Starting from this equation, it is possible to simulate new mortality tables or to calculate a stressed version of the table. It should be noted, however, that the adjustment of regression parameters should be carried out with caution:

- A pair (a, b) different from the pair $(1, 0)$ will result in a bias risk that must first be corrected.
- The Gaussian character of regression estimators should be verified. If this point is not verified, it will be better to use a *Bootstrap* approach to generate new tables.

On the basis of the linear model presented above (and by assimilating the residues to Gaussian variables), the confidence interval is projected to 95% of the number of deaths after taking into account the risk of sampling. The projection is carried out not through randomness due to noise but through the uncertainty bearing on the regression coefficient:

$$\widetilde{\text{logit}\left(q_{x,t}\right)} = \left(E[a] + \sigma_a \cdot N(0, 1)\right) \cdot \text{logit}\left(q^{Ref}_{x,t}\right).$$

This interval is naturally narrower than the portfolio pooling interval in the sense that these are deaths that could have been predicted by a *Best Estimate* table (hence, it is a sampling risk but not a pooling risk).

10.5.3 Modeling the Trend Risk

Modeling the overall solvency requirement of a Long Term Care insurance product leads to projecting the survival of active and dependent policyholders, on a stressed basis. This leads to an understanding of the trend risk in these two populations. Even though these two risks are inherently different,[23] they are both impacted by a trend risk. After highlighting this phenomenon on a French population, a generic modeling framework is subsequently proposed.

10.5.3.1 Presenting the Risk

Recent studies show that force of mortality presents erratic variations at different ages around emerging trends; variations not explained by sampling fluctuations. This is reflected in the following, based on published data by *The Human Mortality Database*. It covers the period[24] 1946–2013 and presents the annual changes of lives and deaths by age for the French population. Based on this historical data, by taking the ratio $\frac{\mu_{age}(t)}{\mu_{age}(1945)}$ we observe that the empirical evolution of French mortality (both genders combined) presents a random behavior (Fig. 10.3).

A generic model is proposed below.[25]

10.5.3.2 Projection Model

It is proposed to project the evolution of population mortalisty on the basis of the Lee and Carter model [1], which offers the following estimation formula:

$$Ln(\mu_x(t)) = \alpha_x + \beta_x \times k_t$$

where:

- α_x mean value of the logarithm of the force of mortality on the study interval for age x;

---

[23] As an example, claimant lives are highly sensitive to medical advances for some pathologies such as Alzheimer's, which may not be as true for active lives.

[24] To avoid catastrophic changes to the model, deaths observed before and during the Second World War were not retained.

[25] In order not to skew the study by including obsolete phenomena, the observation interval was limited to 1971–2013.

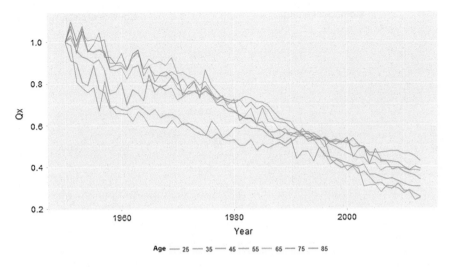

Fig. 10.3 Relative evolution of force of mortality by age

- k_t change in the logarithm of the force of mortality on the interval;
- β_{age} sensitivity, at age x, of the force of mortality to overall mortality decrease.

Model parameters are obtained by applying ordinary least squares:

$$\left(\widehat{\alpha_x}, \widehat{\beta_x}, \widehat{k_t}\right) = arg \min_{\alpha_x, \beta_x, k_t} \left[\sum_{x=x_m}^{x_M} \sum_{t=t_I}^{t_F} (Ln(\mu_x(t)) - \alpha_x + \beta_x \times k_t)^2 \right].$$

In order to make the model identifiable, two constraints are usually added[26]:

$$\sum_{x=x_m}^{x_M} \beta_x = 1,$$

$$\sum_{t=t_I}^{t_F} k_t = 0.$$

The equation to minimize being nonlinear, it is solved by application of the Newton–Raphson algorithm, which leads to the following results (Fig. 10.4).

10.5.3.3   Dynamic Time Warping

Parameters having been defined to reproduce past mortality, a dynamic time warping can then be applied to a Kappa series to give a prospective dimension to a Lee

---

[26]These constraints allow the interpretation of alpha coefficients.

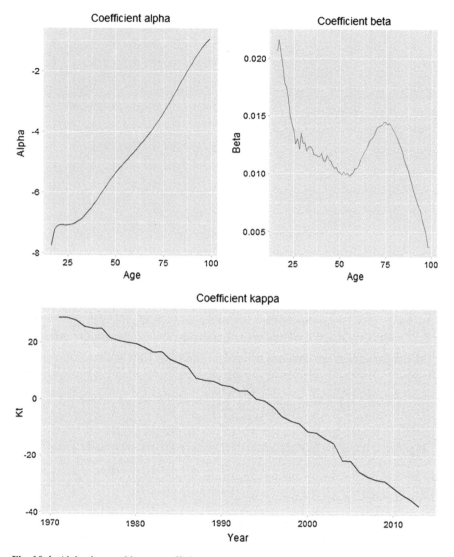

Fig. 10.4 Alpha, beta and kappa coefficients

and Carter model. This extrapolation can be done through linear regression or an Autoregressive Integrated Moving Average (ARIMA) process. However, beyond the randomness of the different coefficients, the analysis of mortality rate changes presented in Fig. 10.3 shows three zones (almost linear) corresponding to periods of important trends in medical advances.

- 1946–1960: impact of antibiotics, primarily on the young population.
- 1961–1980: stable mortality rates.

- 1980–2000: impact of advances on cardiovascular diseases, mainly on the older population.

Thus, given the high sensitivity of Long Term Care insurance portfolios to medical advances, it does not seem likely that a model based only on past observations is efficient. It may then be better to start from a constrained shift model,[27] medical experts setting a constraint on the evolution of residual life expectancy at pivotal ages. A model proposal (see [6]) is presented below:

- Starting from a logit model for which the following equality is available:

$$q_{xt} = \frac{\exp(\alpha_x + \beta_x \times k_t)}{1 + \exp(\alpha_x + \beta_x \times k_t)}.$$

- Life expectancy at age x is estimated in year T:

$$e_{xt} = \sum_{h \geq 1} \prod_{k=0}^{h-1} \frac{1}{1 + \exp(\alpha_x + \beta_x \times k_t)} = \Phi_x(k_t).$$

- At this point, it remains to specify the age x_0 retained as the "pivot age" to reflect expert opinion into the model, as well as the shape of future changes in the life expectancy at this age. Future shift of mortality is found by solving:

$$\min \left(\sum_{t=t_m}^{t_M} \left(e_{x_0,t} \Phi_{x_0}(k_t) \right)^2 \right).$$

10.5.4 Financial Risk Models

10.5.4.1 Important Considerations for Managing the Interest Rate Risk for Long-Term Products

In the case of Long Term Care products, particular attention should be paid to the scope of modeling yield rates. Given the duration of Long Term Care products, it is worthwhile to list a number of properties that the model should check:

- Management of the Ultimate Forward Rate (UFR) to ensure a certain coherence with neutral risk modeling (and monitoring yields on a long-term basis.)
- Ability to properly manage yield curve changes that are "dangerous" for the insurer (shift, slope, and convexity).
- Ability to model negative rates.
- Ability to integrate economic constraints in the model.
- Ability to link the ORSA model to the EIOPA model.

---

[27]Volatility around this shift can be inferred from the time series volatility.

Starting from these properties, it may be preferable to turn to models that fit a long-term scope and which respect macroeconomic balances while taking into account short-term changes in asset values. An example of a model which satisfies these criteria is presented below.

10.5.4.2   The Nelson–Siegel Model

In view of the properties listed above, we present the model derived by Nelson and Siegel [7]. It is a model with three initial-condition constants and a scale factor, such that the instantaneous *forward* rate is written as follows:[28]

$$f_t(\tau) = \mu_1 + \mu_2 \exp\left(-\frac{\tau}{\tau_1}\right) + \mu_3 \frac{\tau}{\tau_1} \exp\left(-\frac{\tau}{\tau_1}\right).$$

The zero-coupon rate $R_t(\tau)$ can then be written:

$$R_t(\tau) = \beta_1 + \beta_2 \varphi\left(\frac{\tau}{\tau_1}\right) + \beta_3 \psi\left(\frac{\tau}{\tau_1}\right)$$

with: $\varphi(x) = \frac{1-\exp(x)}{x}$ and $\psi(x) = \varphi(x) - \exp(x)$.

Starting from the dynamics of these two functions, we can make the following interpretation:

Factor	Interpretation
β_1	Long rate
β_2	Slope (-Spread)
$\beta_1 + \beta_2$	Short rate
β_3	Curve factor
λ_1	Inflection point

The Nelson–Siegel model has the following good properties:

- It correctly reproduces the first three French Prudential Control Authority (PCA) [2] requirements. It is therefore possible to adapt it to the insurer's investment policies (shift, slope, and level) and to compare the model with EIOPA developments during the calibration of the fifth Quantitative Impact Study (QIS5).
- The three β variables are independent, keeping the model simple.
- All risk factors are economically interpretable and make it possible to manage the UFR through the coefficient β_1.

---

[28] For more details, the interested reader may refer to Bonnin et al. [4].

Calibration of the yield rate model is done via the least squares method:

$$\left(\widehat{\beta_1}, \widehat{\beta_2}, \widehat{\beta_3}, \widehat{\beta_4}, \widehat{\lambda_1}\right) = \underset{(\beta_1, \beta_2, \beta_3, \beta_4, \lambda_1)}{\text{argmin}} \left(\sum_{\tau_{min}}^{\tau_{max}} \left(R_t(\tau)^{NS} - R_t(\tau)^{emp}\right)^2\right).$$

With the exception of the lambda parameter, the optimization problem is a linear problem in $(\beta_i)_{i=1,2,3,4}$. By setting the lambda parameter, the model becomes a problem of ordinary least squares and is easily solvable. The choice to set the coefficient λ_1 is also justified by the desire to resolve collinearity problems:

- If $\lambda_1 \to 0$ then φ and $\psi \to 0$ hence β_2 and $\beta_3 \to \infty$. The slope and curvature parameters β_2 etand β_3 become difficult to identify.
- If $\lambda_1 \to \infty$ then $\varphi \to 1$ and $\psi \to 0$. The slope and level parameters β_2 etand β_1 are almost collinear and are therefore difficult to identify.

This collinearity problem does not preclude calibrating a yield curve with a good match to the market rate curve. However, this distorts the value of the marginal impacts of each coefficient and their economic interpretation. The literature associated with the Nelson–Siegel model advises to set the curvature parameter or to restrict it to a reduced interval [0, 2.5]. Once this point is resolved, the model will be able to reproduce almost all forms of curves seen in the past (except for cases where the yield curve has several turns, as in 2008—a Nelson–Siegel–Svenson model would allow one to more accurately reproduce this curve).

Once the coefficient is calibrated to historical rates, it is possible to adjust prediction models. Being the reproduction of the first three vectors from the Prudential Control Authority (PCA) [2] for the structure by terms of interest rates, the three beta coefficients are independent. This point is validated by the analysis of correlations observed historically. Adjustment of the time series is somewhat tricky in the sense that the stationarity of the three time-series is often rejected by Augmented Dickey–Fuller, Philipps–Perron and KPSS tests. However, the presence of a unitary root would be equivalent to stating that the series of $(\beta_i)_{i=1,2,3}$ follow random walks and are therefore divergent series. However, due to their economic interpretation this point cannot be envisioned. Following the literature, it is often proposed to adjust a return-to-average process whose dynamics are governed by the Stochastic Differential Equation (SDE): $dX_t = \lambda(\mu - X_t)dt + \sigma dW_t$. By making the variable substitution $Y_t = \exp(\lambda t)X_t$, the previous SDE has the solution:

$$X_t = e^{-\lambda t} X_t + \mu\left(1 - e^{\lambda t}\right) + \int_0^t e^{\lambda(u-t)} dW_u,$$

$$X_{t+\delta} = e^{-\lambda\delta} X_t + \mu\left(1 - e^{\lambda\delta}\right) + \sigma\sqrt{\frac{1 - e^{-2\lambda\delta}}{2\lambda}} N(0, 1).$$

10.6 Conclusion

Based on the points presented above, it appears that setting up an ORSA process for Long Term Care insurance must be done with caution. In fact, this is a long-term risk presenting several sources of hazards (incidence, continuance, active lives termination) which turn out to be difficult to calibrate given the relative novelty of this type of insurance product. As such, and even when multiple theoretical approaches can be devised to measure the overall solvency requirement of a Long Term Care product, these approaches should be used with caution.[29] It would be better, in many cases, to turn to so-called expert approaches that, although less accurate than an approach based on portfolio experience, present less uncertainty in the event of insufficient data.

Beyond all these methods, one should remember that the central element of Solvency 2 (be it Pillar 1 or Pillar 2) is the calculation of the *Best Estimate* for contract reserves.[30] Verifying the *Best Estimate* nature of this valuation is therefore the first step (and probably the most important) to carry out during the implementation of the ORSA process. This process must go through an analysis of the biometric distributions used and their *Best Estimate* character. The interested reader can refer to Croix et al. [2]—"Mortality: A statistical approach to detect model misspecification" which discusses an applicable methodological framework.

End Notes

[1] For more details, the interested reader may refer to Derien and Le Floc'h 2011 and Decupere [2011].
[2] The French Prudential Supervision and Resolution Authority (Autorité de contrôle prudentiel et de résolution—ACPR, 2010) is an independent administrative authority which monitors the activities of banks and insurance companies in France. It operates under the auspices of the French central bank Banque de France.

References

1. Lee, R.D., Carter, L.: Modelling and forecasting mortality, with various extensions and applications. J. Am. Stat. Assoc. **87**, 659–671 (1992)
2. Croix, J.C., Planchet, F., Therond, P.E.: Mortality: a statistical approach to detect model misspecification (2015). hal. archives-ouvertes

---

[29] The level of caution depends on volume and quality of available information.

[30] This point is all the more important if we apply ORSA scenarios in the form of stress that disturbs the distribution underlying the *Best Estimate* (as the biometric shocks are calibrated under Pillar 1). In this case any discrepancy on the best estimate is amplified at the level of calculation of the overall solvency requirement.

3. Guibert, Q., Juillard, M., Nteukam-Teuguia, O., Planchet, F.: Solvabilité prospective en assurance—méthode quantitative pour ORSA. Economica (2014)
4. Bonnin, F., Planchet, F., Juillard, M.: Applications de techniques stochastiques pour l'analyse prospective de l'impact comptable du risque de taux. *Bulletin Français d'Actuariat* **11**(21) (2010)
5. Règlement délégué (UE) 2015/35 de la Commission du 10 octobre 2014 complétant la directive 2009/138/CE du Parlement européen et du Conseil sur l'accès aux activités de l'assurance et de la réassurance et leur exercice (solvabilité II)
6. Planchet, F., Thérond, P.E.: Modélisation statistique des phénomènes de durée—applications actuarielles. Economica (2006)
7. Nelson, C.R., Siegel, A.F.: Parsimonious modeling of yield curve. J. Bus. **60**(4), 473–489 (1987)

Chapter 11
An ERM Approach to Long Term Care Insurance Risks

Néfissa Sator

11.1 Risk Identification

If we consider the European Solvency 2 regime (Standard Formula) or the American RBC regime, we can identify some relevant risks that are either not included in those regulatory solvency regimes, therefore not quantified, such as the regulatory risk that should be considered in the ORSA quantification, generally adopting a scenario approach; or which are included in the regulatory solvency regimes but should be taken into greater consideration as they are significant and preponderant in the LTCI policies, such as the reputation risk, the risk related to the evolution of the medicine and the model risk.

11.1.1 Regulatory Risk

Numerous evolutions in the legal, regulatory and tax fields observed in recent years in the insurance industry have led to the emergence of a 'regulatory' risk. As aging and elderly people's care is a concern of most governments, this risk cannot be considered insignificant. Subjective and difficult to quantify, it is often absent from the regulatory solvency regimes' quantification.

Here are some examples of events that could impact the risk's assessment:

- Imposed market prices that would require some players to significantly increase their reserves and regulatory capital. This risk is already an issue on the American market where rate increases need to be submitted to the regulator (in each state where the company operates) for approval. Many players have exited the market, including large insurance companies, after regulators didn't approve the needed

N. Sator (✉)
46 East 21st Street, Apt 5, New York, NY 10010, USA
e-mail: nefissa.sator@gmail.com

© Springer Nature Switzerland AG 2019 271
E. Dupourqué et al. (eds.), *Actuarial Aspects of Long Term Care*,
Springer Actuarial, https://doi.org/10.1007/978-3-030-05660-5_11

rates' increases to manage the LTCI portfolio as the experience was emerging and showing deviation from the original pricing assumptions.

- A law that would require the transferability of LTCI policies between different insurers, thus reducing the insurer's ability to monitor its portfolio and amortize its acquisition costs.

Regulatory developments should be anticipated by identifying scenarios in order to apprehend their possible quantitative consequences.

Actions for risk mitigation: Participate in industry conferences and public and regulatory debates and organize a market watch to anticipate changes in the LTCI market and try to influence decisions using expertise, negotiation and lobbying.

An insurer that has a significant LTCI portfolio of long-term contracts provides it with an important position and weight, through credibility and listening, in the event of regulatory projects and consultations.

11.1.2 Reputation Risk

The reputation risk is generally part of the operational risk. As mentioned above, let's take the example of the rate's increase. It can have negative impacts on the insured population (e.g. complaints, litigations, threats from consumer associations) and on the distribution of the products from the distribution channels.

The evolution of internet rates comparators could increase this risk.

Actions for risk mitigation: Use the long-term duration of LTCI portfolios to apply moderate, smooth, but continuous periods of rate increases, as in health or automobile insurance, to avoid chocs and brutal adjustments. In addition, increase in parallel the level of guarantees/benefits to a lesser extent.

Moreover, transversal actions involving several business units can be set up to anticipate and measure the rate increases' impacts, such as: communicate to the distribution channels to inform them and explain the reasons for the rate's revisions, alert the back-office services (administration of policies) and prepare an explanation customer pitch, involve the legal department to pay particular attention to complaints (an insured person who threatens to seize consumer associations) and to policy terminations, and establish a frequent quantified tracking of complaints and terminations (KRI).

All the players—the insurance company, its distribution channel and the insured population—will have to get used to the frequent but moderate rate increases policy on this type of product, such as health insurance or automobile insurance. The culture of this risk will have to evolve.

11.1.3 Risk of Evolution in Medical Science

Let's consider here major developments in medical science (Alzheimer's treatment, for example) and the emergence of new pathologies (related, for example, to our lifestyles) resulting in a deferred but more severe disability or in an early disability or in an increase of longevity. This would change the loss ratio and the distribution of probabilities between the different states (active lives, disabled). These changes of trends related to the evolution of diseases and their treatments are slow moving phenomena, often requiring decades of observation. Given the significant duration of the Long Term Care products and the horizon of the projected experience, these risks must be considered.

Actions for risk mitigation: Carry out scientific research on this subject by communicating with subject matter experts (for example, geriatricians, researchers), take part in scientific conferences, follow the progress at an international level and work on innovations (for example, financing programs of research, conducting epidemiological surveys that provide a medium-term prospective vision of the LTCI risk) to anticipate changes and support the assumptions.

11.1.4 Model Risk

Models only partially reflect the reality, the model risk is always present. Because LTCI risk is recent for insurance organizations, and it mixes disability and longevity with strong interactions between different states, the risk of modeling error is even more important than in traditional life, protection, or health insurance products.

Actions for risk mitigation: Fully document models and computer programs developed, train users, set up internal and external reviews of model work (audit, certification), test the consistency of the model on a few lives/policies by creating easy-to-use files (on Excel© for example) and compare its results with those of the market.

In addition, the risk of error in the estimation of the parameters is important considering the little past experience available for the calibration of the parameters and the natural evolutions of the risk. Academic and market references for LTCI modeling and probabilities are too few and, when they exist, generally have data aggregation issues. Expert judgment is often used, which can be a source of a systematic risk.

Actions for risk mitigation: Conduct sensitivity studies on the various parameters to evaluate the volatility around the assumptions, automate the calculation of key indicators and monitor/analyze them regularly, communicate with multiple experts (doctors, reinsurers, academics, etc.) to discuss market trends and share best practices, break down and analyze the different sources of gains or losses and explain the variations in values between two calculation dates.

11.1.5   Base Risk and Non-mature Portfolios

Emerging experience for LTCI products is very long, it requires the portfolio to age. While the usual average underwriting age is 60 years (in France, for example), severe dependency usually occurs after 80 years. In France, in many cases, the pricing assumptions come from the reinsurers since most insurance companies don't have enough experience to build their own, and therefore from non-homogeneous portfolios (different generations of products, different underwriting practices and claims management that depend on each insurer). This constitutes a high base risk, accentuated by the long duration of the LTCI portfolios.

The claims distribution by pathology should also change considerably over time: the proportion of cancers in the population of dependents is greater in an immature portfolio, for example cancers appear more at younger ages than dementia. Pathologies modify the dependency continuance probabilities.

Actions for risk mitigation: Apply a prudent risk management monitoring at the beginning of the LTCI product distribution by putting in reserves a large part of the gains obtained in the first years until attaining a better knowledge and assessment of the risk. Develop experience studies by creating an automatic and regular monitoring of the claims with appropriate indicators. Sensitivity studies will help to quantify the impact of a change in claims experience and to assess safety margins.

11.1.6   The Issue with the Administration of LTCI Policies

Insurers must rely on expert geriatric doctors who are at the heart of the LTCI policies' risk management for underwriting and claims acceptance rules. This medical expertise requires very specific and uncommon skills that creates a recruitment challenge, as well as a risk of significant increase of medical expertise costs, which is generally beyond the shock that is retained in the solvency regimes (for the Solvency 2 standard formula, it is 10% over one year, for example). Also, insurance companies have to rely on expert geriatric doctors' skills, which is not something insurers are comfortable with, due to lack of familiarity.

In addition, the stability of the claims acceptance policy is a fundamental element of the experience studies and statistics. A change of an expert doctor must be supervised to avoid any unfavorable evolution of the experience because of a possible impact on the claims acceptance policy, which could become more flexible, for instance.

It is the same challenge and risk one encounters if the administration of the LTCI policies are outsourced to a third-party provider. This may represent an additional risk if the data reporting provided by the third party is not enough in terms of granularity, completeness and quality to be used for risk management purposes (experience studies for reserving, repricing, capital management etc.).

Actions for risk mitigation: Be very careful and selective with the recruitment of expert doctors and third-party administrators and draw up detailed arrangements that describe the expected quality of services and needed reporting using dedicated and appropriate KRI and KPI to implement and monitor over time. For example, set up automated indicators to monitor the underwriting and claims policies and carry out reviews and audits.

In addition, document the detailed rules that govern the underwriting and claims acceptance policies, and update this manual as needed.

Finally, use Internal Control to verify the correct implementation of the administration processes and improve their effectiveness.

11.2   Governance

The governance could be articulated through the following governing bodies, who participate in the risk management in an interactive manner, from the strategic point of view to the various operational levels.

11.2.1   Board of Directors

The Board of Directors defines the overall risk policy based on proposals from Senior Management and the Risk Committee members. This risk policy includes the minimum rate and target rate for the solvency ratio, considering the development and profitability objectives of the strategic business plan. The Board of Directors must be informed of any breaches/limit violations and the corresponding corrective measures taken.

11.2.2   Risk Committees

The Risk Committee's role is to validate the guidelines, assumptions, and models used for the risk assessment and risk management. It formalizes the risk management policy's proposal that is submitted to the Board of Directors for validation and allocation of risk budgets.

The Risk Committee is alerted by the operations if the risk limits are exceeded and validates the corrective actions with the business units involved. It also validates management actions such as the revaluation strategy, the rate increase, the reinsurance policy, as well as the design of new products etc.

The decisions made are based on observed statistics, sensitivity studies and stress testing results.

Regular Risk dashboards will make it possible to follow the different risk indicators, monitor the risk trends and the effectiveness of the corrective actions.

11.2.3   Steering Committees

The Steering Committee brings together the several businesses that functionally roll out and embed the risk policy into their operations (pricing, underwriting, reserving, asset management, administration, reinsurance, etc.). The recommendations and decisions made are based on statistics and control testing results, carried out through indicators and dashboards. The Steering Committee reports its recommendations and decisions to the Risk Committee.

Actions for risk mitigation: Obtain risk budgets/limits by business unit and break them down between different risks to deduce operational indicators that can be used directly by the businesses to drive the day-to-day operations and monitor the business. This involves segmentation and aggregation of risks issues where correlation between risks needs to be considered.

Also, develop automatic exception reports based on the operational indicators and risk limits mentioned above and set the alert levels below the limits in order to have an escalation and decision process before the breaches occur (for anticipation purposes).

11.3 Key Risk Indicators

11.3.1 Data Management

Data management is an essential part of the risk management framework. Anticipating it will make it possible to manage the risks in an automated and effective way.

First of all, it will be necessary to express the risk management related needs to build a suitable data model architecture that allows the storage of the historical data necessary for the experience studies, for developing the risk indicators, for the pricing, reserving and solvency calculations, and for preparing the various reporting statements (regulatory and internal). This data warehouse would need to be updated regularly.

Actions for risk mitigation: Data quality and audit trail criteria must be taken into account to justify the assumptions and management actions. Similarly, consistency checks should be formalized via IT and business reports.

Among the frequent data quality issues, we can cite the reliability of the death information. Indeed, with LTCI policies, the death of the insured population results in the termination of the policy without any paid compensation, therefore its record in the database is often partial. This raises the issue of construction and justification of the death probabilities, especially for the active lives population. Another difficulty could consist in convincing the Senior Management to allocate the important necessary budgets for the implementation of a quality data management system and the development of the indicators.

Actions for risk mitigation: Improve the quality of the death information by using other sources of data (such as looking at a more general insured database, population database or the 'retirement' database if the insurer is also providing retirement products that often target the same elderly clients).

11.3.1.1 Model Points

Model points are used in models' input and are generated by aggregating the available insured population's information in the database.

Actions for risk mitigation: Develop specific controls to verify the accuracy of the model points by generating automatic and dedicated data indicators and exceptions reports with statistics on the portfolio and systematically compare and explain the evolutions between two extractions.

11.3.1.2 Public Data

To enrich the portfolio information and statistics (experience studies), create a documents library and database with public data that will have to be cross-referenced with the insured database for benchmarking and accuracy checks. In addition, be aware of the drawbacks and weaknesses of the public studies by reading their limitations and discussing with other market experts.

In France, regarding public data, we can cite the example of the "HID" survey, the studies provided by DREES (Direction de la Recherche, des Etudes, de l'Evaluation et des Statistiques), INSEE (Institut National de la Statistique et des Etudes Economiques) and CNSA (Caisse Nationale de la Solidarité pour l'Autonomie), which contain information on incidence rates for dependency and life expectancy with and without disability. In the US, the inter-company study done on behalf of the Society of Actuaries provides market experience studies on the LTCI portfolios.

11.3.2 Key Risk Indicators 'KRI'

There should be a documented and automated process to regularly produce and store the KRI on a mature portfolio. The value changes of the KRI need to be analyzed, they can mirror a risk evolution or a modification in the way the policies are administered, or the claims are accepted, that would need to be reflected into the assumptions or corrected with the corresponding business units.

All the KRI are examined and analyzed by the Steering Committee members who escalate their important variations (depending on the risk budgets/limits) to the Risk Committee members for validation of the assumptions' adjustments, new pricing's basis or administration procedures' updates or other possible management actions.

When the KRI are produced automatically and in a standardized format, their interpretation by the different stakeholders is facilitated and the reactivity is augmented. The decision-making process is therefore enhanced in case deviations are observed. The governing bodies, relying on the monitoring of the KRI evolutions, can then make the appropriate decisions (management actions) with more confidence in an objective and quicker manner.

11.3.2.1 Example of KRI to Manage the Underwriting and Market Risks

Age Distribution
Split the insured (payees and reduced paid-up) and the claimants' populations by age and by gender to observe its distribution's evolution. The percentage of females tends to increase in the claimants' population when compared to the insured population.

Movement Analysis on the Portfolio
This indicator allows one to observe the number of underwritten policies, active lives, cancellations, non-forfeitures, terminations, deaths and claims, with the transitions within the different possible states (enhancement or aggravation). It is a basis of the construction of the underlying probabilities used for risks modelling.

Biometric and Behavioral Distributions
For the experience studies, use the pricing actuary or appointed actuary approach by validating the probabilities by using confidence intervals to apprehend the robustness of the observed statistics.

A segmentation by underwriting generations can be useful to study the evolution of the different observations with the maturity of the portfolio (volatility will reduce overtime), and in order to consider pricing basis or underwriting policies that might have change overtime. The KRI below (actual values) would need to be compared to their expected values coming from the assumptions:

Active lives mortality rate: by age and by gender,

Incidence rate: by age and by gender, can then be separated between the premium paying policies and non-forfeitures,

Claimants mortality rate: by age and by gender and claims continuance period (in number of months, for example). In addition, this indicator can be split by group of pathologies (cancers, dementia, strokes, for example) in order to refine the rates and consider the impact of the pathologies on the longevity. This will help to better apprehend the longevity risk and its evolution. For both the active lives and claimants' population, the observed mortality rates can be compared to the rates from the available public data and to appropriate regulatory mortality tables used in the insurance industry.

Transition rates: by age and by gender to model the remissions and aggravations,

Forfeitures rates and cancellation rates: by age and by policy duration. Can also be done by underwriting generation to monitor the impact of rate increases on those rates.

The implementation of the KRI and their monitoring allows one to justify the assumptions. The observation of values that sustainably exceed the limits of the confidence intervals should lead to the adjustments of the corresponding probabilities and assumptions. The impact of those adjustments on the solvency of the company needs to be evaluated. If the solvency ratio limits are approached, correction actions would have to be taken such as rate increase decisions.

It is recommended to perform frequent reviews of the experience studies' methodology and results by an independent expert party or by the reinsurers for assurance and benchmarking purposes.

Discounted Rate

A monthly reporting of the actual discount rate versus the one assumed during the pricing exercise and the quarterly reserving exercises can be developed. This monthly indicator should be completed by projections based on scenarios that include increases and decreases, to anticipate any gain or loss on the asset portfolio (profitability and solvency).

A prospective dimension, on a long-term horizon, aligned with the long duration of the LTCI policies, should be considered when setting the discount rate for the pricing and reserving, especially when the reserving assumptions cannot be adjusted overtime (locked assumptions). Stress tests should be developed to illustrate the impact of a range/vector of discount rate's increases and decreases on the profitability and solvency of the LTCI portfolio. Correlations with other risks, such as lapses, should be included in the stressed projections.

11.3.2.2 Example of KRI to Manage the Risks Related to the Administration of Policies

Underwriting Rate

The underwriting rate, defined as the ratio of the number of applications/requests over a period to the number of underwritten policies over the same period, allows one to monitor the stability of the underwriting policy, mainly of individual policies. This indicator can be split by gender, age band, and generation of products.

Claims Treatment Processing
The claims treatment processing indicator is to measure and monitor the portion of the accepted claims by month of claims submission. This indicator gives information on the claims administration process and is helpful when determining the IBNR reserves. It can be split by type of pathologies, as they influence the duration of the payments. In France, the claim is accepted after the definite consolidation of the disability state. This consolidation timeframe can be quite long for dementia, for instance (up to 3 years).

Administration Fees, Including Medical Expertise
The number of administration tasks and the corresponding average cost, needed for the general administration of the LTCI policies, as well as the medical expertise, should be monitored to detect any deviation of the administration fees. An increase of the medical expertise fees, which is usually higher than the administration fees for other insurance policies, projected over the long duration of the policies, can have an important impact on the profitability and solvency of the LTCI policies. Its present and discounted value should be added to the reserves in the financial statements. These reserves will follow the same path as the active lives reserves: they will build up progressively, over a long duration, with the aging of the population in the portfolio, since the administration fees will significantly increase with the increase of the claims' frequency.

11.3.2.3   Risk Culture, Adapted Communication

In order to promote a risk culture that is adapted to LTCI risks within the company and to communicate effectively with the Risk Committee members and Board of Directors, it is critical to select relevant, but also simple and available indicators, that give both the retrospective (observed) and prospective (via projections and sensitivities) visions of risks. LTC risks behave differently from risks related to traditional life insurance policies, so educating the decision-makers is key.

For a more efficient communication, graphic illustrations are preferable (see the following examples), and the number of indicators should be limited and stable over time to observe their evolution and help to draw conclusions.

Expected to Actual's Evolution Illustrated with Traffic Lights
For instance, the experience observed of the biometric and behavioral distributions of the insured population can be summarized with a green dot if it corresponds to the expected values (where variations remain within the confidence intervals), with an orange dot if it is starting to deviate from the expected values (variations start to be outside of the confidence intervals), and with a red dot if it is robustly outside of the confidence intervals (visible trend).

Active Life Mortality	●	Expected to Actual within CI level
Incidence	●	Expected to Actual within CI level, not on all ages
Disable Life Mortality	●	Old ages not observed

Present Value of Future Premiums, Claims and Prospective 'Loss Ratio'

The Management is used to apprehend traditional accounting values such as the technical reserves, calculated using mandatory methods or market practices. For instance, the Management is familiar with the Active Lives Reserves 'ALR'. However, the ALR is the result of two large amounts that are subtracted: the present value of the future claims (which represents the amount of risks) minus the present value of future premiums (which represents the amount of risk mitigations). Claims deviation will increase the amplitude of the first term of the equation (the risk), while rate increase and pricing management will increase the amplitude of the risk mitigation, i.e. the risk monitoring capacity.

Therefore, these two amounts constitute two key indicators that should be communicated and reported separately in order to make Management aware of the risks undertaken and the monitoring possibilities. For example, an ALR that is equal to 300 M USD (900–600) could seem relatively 'small' as a total, but 'hides' an amount of 900 M USD of risks undertaken, for which 5% of deviation will cost 45 M USD, i.e. 15% of the total ALR. This is due to the very long duration of the LTCI policies, where liabilities take decades to develop into claims, and when the premiums are leveled.

In addition, it is possible to build a loss ratio adapted to these LTCI policies that is not retrospective, but that includes a **prospective** dimension of the premiums and claims. Indeed, the traditional loss ratio calculation, based on accounting values that are sometimes adjusted by short-term reserves (INBR for example), could wrongly indicate that the portfolio is appropriately priced, especially in the first years of the LTCI products, before the portfolio matures, when very few claims are declared and paid.

Sensitive studies on the variation of this prospective loss ratio should be made and communicated to better apprehend the risk limits and the impact of corrective actions (increase of premiums, decrease of benefits). This prospective loss ratio should take into account the underwriting generations of products to adapt the monitoring to each type of product, underwriting policy and pricing policy.

Future Administrative Cost Reserving

The same logic as above applies to the administration fees. Since the claims development will be long on LTCI policies, the claims administration fees (which include the medical expertise fees, and therefore are costly) will be flat during the first years of the portfolio, misleading the analysis of the future costs. Management needs to be sensible to the future amount of the administration fees that will be observed at the maturity of the portfolio, and that needs to be anticipated in the reserving: the present value of future administrative costs minus the present value of the premium load.

An adapted prospective ratio, as described above, can be produced and its impact measured with sensitivities studies.

Claims Acceptance Policy

The claims acceptance rate, described earlier, is a simple key indicator that can be communicated to the Management governance bodies (committees) and be graphically illustrated (curves, with a time axis to observe its stability or evolution over time).

Re-evaluation Policy

In France, for example, benefits, and therefore premiums, can be annually reevaluated (augmented) to ensure that the value of the benefits over time remains in the market's range. The decisions regarding revaluation of the premiums and benefits are often taken with a short-term vision, based on the observed annual increase of the inflation. The objective is to make Managements and governance committee members aware of the important long-term financial impact of the re-evaluations by including a prospective dimension and quantitative assessments. Re-evaluation scenarios need to be implemented in the actuarial models. For each scenario, it will then be possible to calculate its prospective loss ratio and solvency impact.

Portfolio Stabilization/Maturity

The stabilization of LTCI portfolios is usually a concern. For example, the French market experience does not provide robust observations at all ages. Older ages have not yet been statistically observed. It is necessary to supplement the statistical approach with prospective scenarios/stress tests to get an idea of the possible trends and their impact on the liabilities, profitability, and solvency. This characteristic of LTCI portfolios can be illustrated by calculating the present value of future claims by age bands and showing the age segments that were statistically observed and the others that have not been observed yet (relying on theoretical assumptions and subject to more uncertainties):

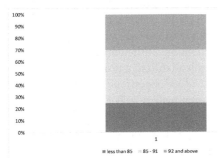

In the provided example, about one third of the future claims (i.e. risks) are estimated to occur after age 90, which is more or less the age band that has not been robustly observed yet in France. This amount can represent several hundreds of millions for a large LTCI player. This would constitute a large amount of uncertainty and unknowns that should be communicated to Management and governing bodies to make them aware.

11.4   Some Other Risk Mitigations Actions

11.4.1   *Risk Transfer via Reinsurance*

For LTCI portfolios, reinsurance offers are traditionally not towards the coverage of excess of loss to assure a stability over the amounts paid or the frequency of the claims (CAT treaties), but towards a partnership between the reinsurance and insurance companies to share the experience and expertise of LTCI risks. Therefore, in the French market, the main type of cession for LTCI policies is the quota share treaty.

The ceded shares are variable, depending on the risk appetite of both organisms. On the French market, they are above 50% for players with experience of LTCI policies, and between 80 and 90% when it comes to a new player that cannot use an internal experience and expertise to apprehend the LTCI risks.

If there are no market or regulatory tables/probabilities to be used as assumptions for pricing, it is likely that a new player has no other choice than relying on the reinsurer(s) expertise. It is recommended to diversify the number of reinsurers that intervene on the LTCI portfolio and make them participate in other reinsurance programs than LTCI solely, to anticipate diversification between profitable and less/non-profitable treaties and increase the "client" value and relationship.

11.4.1.1   Assumptions Concerns—Base Risk

Reinsurers will propose a set of assumptions to be used in the pricing and modelling of the LTCI risks. They could also provide expertise in the risk management practices and with the underwriting questionnaire and claims acceptance policy.

Those assumptions are usually built on the reinsurer's experience, which is based on several LTCI portfolios, managed and monitored differently, and on aggregated information coming from different countries with different health care systems. The access that reinsurers have to a large number of LTCI portfolios with different maturities provides them with an important and legitimate role, but the risk coming from aggregating the experience of different portfolios needs to be acknowledged and communicated.

The challenges related to the aggregation of experience/statistics coming from LTCI portfolios managed and monitored differently (underwriting policy, claims policy, maturity of portfolios etc.) should be acknowledged when considering the reinsurers' assumptions as best estimate assumptions, and requires that each insurer implements an appropriate data and risk management system to collect and analyze the experience related to its own products and administration policies, in order to enhance his knowledge and apprehension of its risks over time.

11.4.1.2   Importance of the Treaties Contractual Clauses

It is necessary to pay particular attention to the contractual clauses of the reinsurance treaties, since the treaties usually cover the entire duration of the policies. When considering a contractual relation that would last more than 30 years, some particular clauses become even more fundamental, especially those that can be triggered by either party in case of risk deviation.

A quota share treaty on TLCI policies generally ends with the extinction of the entire portfolio, covering all the reinsured policies until their expiration, even if the treaty has been terminated.

Insurers and reinsurers can have a different vision of the LTCI risks over the long period of the reinsurance treaty. The reinsurer could anticipate an unfavorable evolution of the risks, necessitating an increase of the premiums, and the insurer could disagree with this vision. In addition, the insurer could consider other factors than solely the risk deviation in its management decisions, since it has the direct relationship with the insured, therefore facing a strategic risk and a reputation risk. A disagreement between the insurer and reinsurer could have important consequences on their relationship, possibly leading to the prematurely termination of the treaty, in conditions where the in-force is not reinsured until its extinction. This would have an impact on the solvency of the insurer and its risk diversification efforts.

It is necessary to pay particular attention to the treaty clauses, especially those related to the rate increases agreement and termination.

11.4.1.3   Risks Transfer and Solvency Capital Needs

In addition to the expertise brought by the reinsurer, reinsurance offers solutions for risk diversification and solvency capital requirements reduction. Within the Solvency 2 regime, this risk transfer is recognized by the regulatory regime without limiting the percentage of the quota-share to be considered/recognized, as in the previous solvency regime. The default risk (counterparty) would need to be taken into account, which makes the choice of the reinsurer player important, with high solvency position and therefore high rating. The risk transfer operation is then more secure in case of important risk deviation, and the reinsurer will be more likely to sustain it.

11.4.2 Hedging with Other Risks on Similar Populations

The longevity risk is naturally hedged with the mortality risk. This hedging is not perfect as it is not applied to the exact same insured population. Nevertheless, solvency regimes usually acknowledge this natural hedging by allowing a negative correlation factor between those two risks. In the Solvency 2 regime, the correlation factor between longevity and mortality is −25%. It is the unique negative correlation factor between the risks.

> Hedging LTCI risks with mortality risks coming from death benefits on similar targeted insured populations (cross selling LTCI with burial insurance, for example) and building combo products that offer longevity and mortality risks, represent an additional risk mitigating action.

Some new LTCI products are already packaging dual offers, LTCI with whole life insurance.

11.4.3 Use of Accounting Reserves to Stabilize the Earnings' Volatility

In some countries (as in France), statutory accounting rules allow one to generate special reserves by putting a percentage of the earnings (technical results and financial results) into reserves. This accounting mechanism is particularly adapted to the risks that can fluctuate, such as LTCI products, to smooth the earnings over the duration of those products.

Usually this reserve does not benefit from a favorable taxation regime, like the other technical reserves (e.g. ALR, IBNR), and they are limited in their dotation by percentage of earnings, for example, and have to be released in a limited period of time (8 years in France).

This type of reserve does not match with an economic balance sheet view. Its best estimate can be considered null. However, it constitutes an equity for the company and a risk management tool in the medium term. Therefore, it is important to continue constituting this accounting reserve, especially when the portfolio is not yet mature. It will help to absorb a moderate deviation of the claims experience or a need for additional reserving and will mitigate and smooth the rate increase's needs, allowing the insurer to not over-react to an unfavorable experience that can be transitory, not confirmed over the long-term.

Bibliography

1. This article is largely inspired by the paper: "Approche Solvabilité 2 et ERM du risque dépendance—Guide de bonnes pratiques" written by Néfissa SATOR & Grégory SOTHER, Actuaires Experts ERM from the 2012 class. It has been translated to English, and slightly amended for updates for completeness purposes.

Chapter 12
On Long Term Care

Marie-Sophie Houis

12.1 The Long Term Care Paradox: The Reality of a Need, the Inability to Make It a Market

12.1.1 From Costs in the Hundred of Millions to Tens of Billions

In 2010, the Long Term Care Insurance (LTCI) market[1] was estimated at €530 million.[2] In 2016, the same market was valued at €751 million.[3] Premium income had grown by 41% in 6 years.

Less than €1 billion, to cover expenses related to Long Term Care for the elderly…

Should it be concluded that the benefit level and the average premium are too low and that they could grow with a better awareness of the reality of the risk and help with obtaining Long Term Care coverage?

Should it be concluded that insurance is not the right way to finance this extra cost of expenditure in the long term?

Today, coverage level is estimated at 15.4% of potential LTCI buyers.

In the following projection for 2060, we assumed that only the coverage rate would vary between 2016 and 2060 assuming that the average annual contribution remains the same over that period.

When coverage level is increased by 10 points to 25%, the LTCI market would almost double in 2060 (€1.250 billion).

M.-S. Houis (✉)
PMP Conseil, 32 Boulevard Haussmann, 75009 Paris, France
e-mail: mshouis@mx-conseil.com

© Springer Nature Switzerland AG 2019 289
E. Dupourqué et al. (eds.), *Actuarial Aspects of Long Term Care*,
Springer Actuarial, https://doi.org/10.1007/978-3-030-05660-5_12

Three LTCI evolution assumptions to 2060 as related to coverage level

	Population (million)			Coverage level (%)	Average annual premium (€)	Premium income (€million)
	20–75	Over 60	Covered			
2016	44.1	16.7	6.80	15.4	110.4	751.0
2060	45.4	23.6	6.99	A1: 15.4	110.4	771.7
2060	45.4	23.6	9.08	A2: 20	110.4	1,002.4
2060	45.4	23.6	11.35	A3: 25	110.4	1,253.0

The three central assumptions selected to establish this projection are as follows[1]:

- Compulsory insurance from age 50
- Unique €500 annual premium for all insureds
- Persons recognized as dependent (from IRG 1 to IRG 4)[4] are exempt from paying premium.

On the basis of these assumptions, the market went from €530 million to almost €11 billion.

Potential LTCI market under a mandatory insurance scenario

12.1.2 Wanting, Knowing, Doing: The Few Insurers Which Went There

12.1.2.1 A Market that Stagnates

Premium growth of the top 10 insurers is either low or negative with the exception of AXA France, the only company that achieved a double-digit growth (+16%).[5]

12.1.2.2   2012 FFA[3] Statistics: Dominance of Insurers[6] in Premium, the Dominance of Mutuals[6] in Insureds

In 2012 the French Insurance Federation (FFA) and the Senate had prepared a market survey. These statistics highlight two market profiles, that of mutuals (mutuelles[6]), with low LTC benefits integrated with other benefits such as health or retirement, for a large number of participants (due to the mandatory nature of contract) and that of insurers with individual policies providing higher annuity benefits in case of dependence.

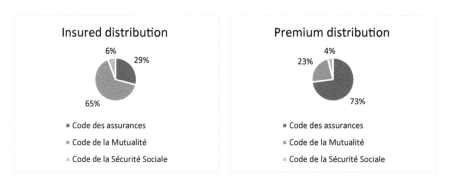

12.1.3   In Search of the Right Level of Mutualization

To cover additional expenses and to be cared for in case of dependence, several types of solutions exist:

- Insurance or savings
- Individual or group insurance
- Stand-alone or combination with other insurance, notably health or savings.

The market is generally structured by type of insurance family[6]:

- Insurers and bancassurance for individual insurance and savings;
- Mutuals for inclusions in complementary health contracts;
- Prévoyance[6] for standard or tailored group contracts.

Complete statistics are lacking to analyze the completeness of this market in all possible aspects.

Below are FFA statistics on the basis of 2015 figures which give a breakdown of the number of insureds covered based on type of coverage.

In 2015, insurance companies covered 3.4 million insured for a premium income of €611 million.

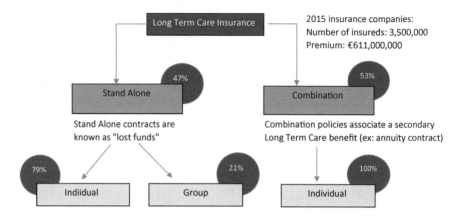

12.1.4 Toward Fair Contracts

A 2013 article[2] listed "the 7 flaws of Long Term Care Insurance":

- Medical questionnaires are considered "trappings";
- Rates may be increased "without limits";
- The definition of dependence, whether Partial[7] or Total,[7] lacks uniformity;
- Different medical viewpoints are often dismissed, usually relying solely on the insurer's medical examiner;
- Underwriting procedures are considered tedious, expensive, complicated and often incompatible with the condition of the applicant;
- Waiting Periods are considered "too long";
- Some clauses present are even "illegal" such as the exclusion of legal recourse in the event of disagreements between the company's medical examiner and a third-party opinion.

Hence, to be more transparent, fairer, in 2013 insurers via FFA[3], created the industry label LTCIBG[8] (Long Term Care Insurance Basic Guarantees).
This label is granted to LTCI policies respecting 9 conditions:

- A common vocabulary for clarity in the definition of benefits;
- A common definition of Total Dependence based on Activities of Daily Living (ADL);
- A lifetime benefit, regardless of the date of incidence of Total Dependence;
- A minimum $500 monthly annuity in case of Total Dependence;
- Terms of revalorization of the guarantees defined in the policy (benefit and premium);
- No underwriting before 50 years (except already dependent or suffering from a major long term illness);
- Preventive programs or assistance services for the insured person or relatives to be available as soon as the policy is issued;

- Noncancellation rights in the event of an interruption premium payment;
- Annual report.

Among contracts labelled LTCIBG, we find 'Entour'Age' of Axa, 'Libre Autonomie 2' of Allianz, 'Vers l'Autonomie' of Credit Agricole, 'Assurance Dépendance' of Generali, or still 'Assurance Autonomie' of La Banque Postale.

12.2   Twenty Contracts Analyzed

12.2.1   Analysis Grid

To analyze a selection of LTCI policies that represents all types of coverage, we built a structured grid around:

- General characteristics;
- Benefits;
- Service.

  - Pre claim
  - Post claim.

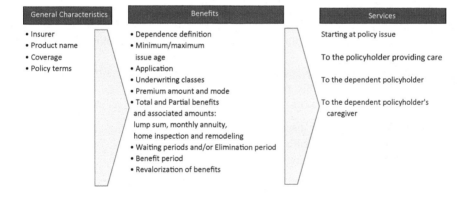

12.2.2   Contract Analysis

More than twenty contracts are analyzed to cover all the types of coverage proposed by the three types of insurance companies[6] from the three major insurance branches.

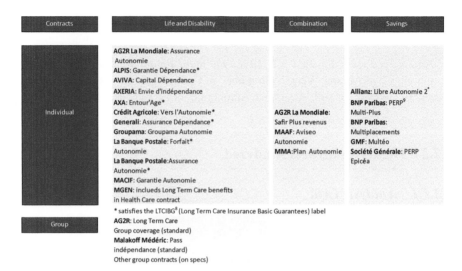

Contracts	Life and Disability	Combination	Savings
Individual	**AG2R La Mondiale:** Assurance Autonomie **ALPIS:** Garantie Dépendance* **AVIVA:** Capital Dépendance **AXERIA:** Envie d'indépendance **AXA:** Entour'Age* **Crédit Agricole:** Vers l'Autonomie* **Generali:** Assurance Dépendance* **Groupama:** Groupama Autonomie **La Banque Postale:** Forfait* Autonomie **La Banque Postale:**Assurance Autonomie* **MACIF:** Garantie Autonomie **MGEN:** inclueds Long Term Care benefits in Health Care contract	**AG2R La Mondiale:** Safir Plus revenus **MAAF:** Aviseo Autonomie **MMA:**Plan Autonomie	**Allianz:** Libre Autonomie 2* **BNP Paribas:** PERP[9] Multi-Plus **BNP Paribas:** Multiplacements **GMF:** Multéo **Société Générale:** PERP Epicéa
	* satisfies the LTCIBG[8] (Long Term Care Insurance Basic Guarantees) label		
Group	**AG2R:** Long Term Care Group coverage (standard) **Malakoff Médéric:** Pass indépendance (standard) Other group contracts (on specs)		

12.3 From Insurance to Savings, from Individual to Group

12.3.1 To Each One's Own Solution

Protecting oneself in the event the risk of dependence materializes is a matter important to everyone.

But the solution must be considered according to the characteristics of the insured: age, income, wealth, family, employment…

We analyzed the market by structuring it around five contract categories

- Stand alone Individual insurance;
- Benefits as part of a Health insurance contract;
- Benefits included in savings and retirement contracts;
- Combination contracts;
- Group contracts.

12.3.2 Individual Insurance… Which Arrives a Bit Too Late

Individual Long Term Care insurance is arguably the most widely known and widespread form of Long Term Care coverage.

Historically, it suffers:

- Older issue ages … too late to lead to good coverage with affordable premium?

- A pricing structure which leads to premium increases according to attained age…
  and too often leads to lapse before reaching the age when the coverage could have
  been useful;
- Benefits are still too often temporary… with loss of all premiums if premium
  payments stop before incidence;
- Annuities payments are too low to cover the extra costs of dependence.

12.3.2.1   75 Is Generally the Maximum Issue Age, with Detailed Medical History Required After 50

Minimum issue age varies according to the policy, between 18 and 45. The maximum
issue age varies between 67 and 75. 9 offers analyzed show a maximum issue age
beyond 69.

Issuing a Long Term Care policy is most often subject to medical questionnaires.
In the majority of cases, after 50, these questionnaires become mandatory. Only one
of the policies studied offers to issue without a medical history questionnaire before
65. Two levels of medical history are observed. The first level is a Health Status Statement[10] or a Simplified Health Questionnaire[11] with three to five questions. Based on
responses to the first-level questionnaire, the applicant may have to complete a second level of medical questionnaire called a Health Questionnaire,[12] more complete,
and attend an interview with the insured about four main types of questions: general
behavior, lifestyles, future medical exams, and pathologies over the past decade.

Finally, fewer than one third of the policies explicitly express in their marketing
documentation the criteria for calculating the premium for the Long Term Care
coverage. If, in general, the policyholder expects the premium to be determined
according to the benefit level chosen (Total Dependence and/or Partial Dependence,
level of annuity, lump sum), little marketing documentation explicitly indicates that
health status may influence the premium level.

12.3.2.2   The Central Issue of Claim Assessment

In general, Long Term Care policies refer to the same analysis grids. The grids ADL,
IRG, and some scales to assess cognitive dependence.

But the central problem remains the need to analyze this level of dependence
without coordination between claims related to public benefits and claims related to
private insurance benefits.

The ADL grid (Activities of Daily living) includes: toileting, dressing, feeding,
continence, moving, and transferring. ADL grids are not unanimous in their interpretation. Thus, some insurers consider it to consist of only 5 activities (conflating
moving and transferring) and others of 6.

Partial Dependence is recognized if the insured cannot achieve autonomously 2
activities out of 5 or 3 out of 6.

Total Dependence is recognized as long as there is an inability to achieve at least 3 acts out of 5 or 4 out of 6.

The IRG grid is a common referential basis in the evaluation of dependence. It is composed of 6 levels, Total Dependence (GIR 1 and 2), Partial Dependence (3 and 4), and light dependence (5 and 6).

These two grids are often supplemented by cognitive tests: the Folstein test (6 out of 11 policies) and the Blessed test.

12.3.2.3   In Case of Total Dependence, Most Policies Offer an Annuity, Some Offer Only a Lump Sum, Amounts Vary Widely from Product to Product and Levels of Coverage Issued

Once a claim is approved by the insurer, the insured is entitled to different types of benefits:

- Annuity;
- Lump sum;
- Services based on the evaluation of living conditions, and the reimbursement of home remodeling expenses;
- Services available even before dependence, as detailed in the next section.

Monthly annuity benefits are very variable depending on products and selected options:

- Minimums vary between €200 and €500;
- Maximums vary between €1,800 and €4,000.

Within the same policy, and based on the options chosen, the annuity amounts may vary widely, for example from €300 to €4,000.

In ten policies analyzed, only three offered a benefit in the form of a lump sum. The benefit amounts vary significantly depending on the products and options chosen; one policy offers a benefit of the order of €5,000 while another offers up to €100,000.

The vast majority of insurers offering an annuity include at least one "lump sum option", most often to deal with first expenses, in particular those for home remodeling. This lump sum is presented at issue in the form of an option at 6 levels, or in the form of a formula. When it is included by formula, this lump sum corresponds to the highest-end formula. The minimum amount varies between €750 and €5,000 and the maximum between €5,000 and €13,000.

12.3.2.4   Benefits in the Case of Partial Dependence, an Essential Element for Differentiation Between Policies

All the policies analyzed offer a benefit in case of Partial Dependence. Products are distinguished by how this benefit is integrated in the policy. Sometimes Partial Dependence is included in the basic formula, but the majority consists of its inclusion in a higher formula. More exceptionally, Partial Dependence can be issued under a specific option.

Monthly annuity benefits are lower than for Total Dependence and vary according to products and benefit levels between €100 and €2,400.

Generally speaking, the level of services does not vary with the level of dependence covered (at the level of services at issue or services while on claim), except when it comes to home remodeling assistance, which is rarely proposed in the case of Partial Dependence.

12.3.2.5   Accommodation Benefits

Housing is a real concern, both for public authorities and for insurers. Home Care costs are much lower, especially to non-single households, than stays in specialized establishments. However, to allow for home care, housing often needs to be adapted. For example, in Individual Savings policies, some products offer assistance in home remodeling. This is part evaluation (assessment of the adaptability of the home to accommodate the claimant, always included in the policy) and part lump sum for remodeling expenses (present in some policies).

Evaluations are always included irrespective of the method of payment of the benefit (annuity and/or lump sum). The benefit schedules are included at the start of a claim according to the policy issued (Partial and/or Total).

Such evaluations include, for example, an occupational therapist's visit to the claimant's home to determine solutions to improve daily life without causing a change in housing. They are paid by the insurer and the amounts are sometimes limited (ex: €300). Evaluations are in conjunction with, depending on the policy, practical assistance on how to contact service providers.

In the majority of cases, home remodeling is cited as an example of expenses under "first equipment".

In some contracts, there are lump sum options for domicile adaptation that offer to reimburse the remodeling work on presentation of invoices and within a limit (ex: €5,000).

12.3.2.6   Premium Mode, Joint Policies, Waiting Periods, Revalorization… but the Main Thing Is in the Coverage: Lifetime or Limited!

First of all, let us look at premiums, which are mostly on a monthly or quarterly mode. Premiums may be revised according to the policy developing experience and to changes in the taxability of the policy. In all policies, the premiums are waived when on claim.

One of the features of savings contracts is the advantage of a joint policy, more often that with the spouse. Three quarters of the policies provide a "couple advantage" in two forms:

- Either a reduction on each premium (generally 10% and not more than 20%);
- A reduction on a single premium (20%).

This premium advantage is also accompanied by total waiver of premium when one of the two policyholders is on claim.

Waiting Periods[13] are rather homogeneous on all policies studied. Thus, in the event of an accident, there is no waiting period, and this is usually the case from the basic formula. In case of illness, the Waiting Period is one year after the issue date. Some have no Waiting Period for somatic diseases. Lastly, in the case of neuro-degenerative diseases like Alzheimer's, the Waiting Period is usually three years (two policies studied offer a two-year Waiting Period).

Annuities, showcase benefits of individual savings policies are revalorized each year according to different criteria among which we can cite the INSEE[14] reference index, a revalorization fund, evolution of the AGIRC[15] value, Dependence index, Consumer Price Index…

But the main thing is probably not there. The key point to be analyzed when you buy a Long Term Care policy is the nature of the coverage and the how premiums are handled the event of non-dependence. There are several cases. First, when an active insured dies, there is an option present in a few policies allowing beneficiaries to receive a death benefit or a refund of the base premium paid if the death occurs before 85. Finally, in the case of a premium lapse, some policies offer a benefit reduction after 8 years, for the great majority, or after 5 years for the most liberal. The benefit reduction systematically ends the other options, the annuity is reduced and recalculated. Sometimes this reduction is granted only in the case of Total Dependence and without assistance or service benefits.

12.3.3 Long Term Care Included in Health Care Contract… High Number of Persons Covered but Small Benefits?

In 2010, MGEN[16] offered to its 2 million members Long Term Care coverage to be included in their life/health/disability contract.

The MGEN life/health/disability contract is structured with 4 levels (initial, equilibrium, reference, integral) and 5 types of benefits (health, life, disability, services, social assistance).[17] The originality of this addition is the inclusion of Long Term Care coverage within these very complete health/savings contracts. It consists of a monthly annuity of €120 in case of Total Dependence and home care services in case of Total or Partial Dependence.

Since June 2011, the mutual company offers in addition an optional Long Term Care coverage which in this case is subject to underwriting and an issue age (under 75)-based premium. It provides for two levels of protection comprising an annuity and a lump sum for equipment. It covers both Total and Partial dependence.

12.3.4   *Savings Policies … The Right Solution for Some*

12.3.4.1   Five Policies Analyzed

Allianz: Libre Autonomie 2*
BNP[18] **Paribas**: PERP[9] Multi-Plus
BNP[18] **Paribas**: Multiplacement 2
GMF[19]: Multéo
Société Générale: PERP[9] Epicéa
(*): Offers labelled LTCIBG[8]

12.3.4.2   Innovative Features

In general, it is possible to allocate freely one's savings to cover Long Term Care expenses. This is a solution for households with savings who want to avoid the risk of losing their premiums in case they do not need Long Term Care.

To go further, some savings policies sometimes offer an option to insure Long Term Care. It is then a matter of receiving, at incidence, all or part of the Cash Value paid as an annuity.

The Allianz offer, Libre Autonomie 2, is a life insurance contract. The Cash Value can turn into an annuity in case of Partial Dependence, the annuity is doubled in case of Total Dependence. If the policyholder dies without filing a Long Term Care claim, the Face Amount is paid to the beneficiaries.

Multéo Series 2, GMF's combination life insurance policy, provides that the policyholder has the option of choosing Long Term Care at the start of an annuity payout provided that he or she is over 50 and under 75, and after medical underwriting. The choice of this option is irrevocable. In case of dependence, an additional annuity for dependence is paid and assistance benefits are available. Some assistance services are available during the active phase.

PERP Epicéa provides for various types of annuities including "retirement with dependence insurance". This means that the amount of the annuity is tripled in the case of Total Dependence. A lump sum is paid in the case of Partial Dependence, and assistance services are provided after issue for the policyholder and his or her caregivers.

The PERP BNP Paribas, provides 5 annuity options:

- Life annuity;
- Reversible lifetime annuity;
- Guaranteed payment annuities;
- Increasing annuity;
- Long Term Care annuity for which the cash value is increased in case of dependence.

BNP Paribas's life insurance product Multiplacements allows the policyholder, after the fourth anniversary of the policy and subject to the agreement of the beneficiary, to receive the Cash Value in the form of an annuity if, at the time of processing, the claimant is younger than 80. This modality is also a "less marketed" way to transform a Cash Value into a Long Term Care annuity.

12.3.5 Combination Products … Hard to Explain?

12.3.5.1 Three Offers Analyzed

AG2R[20] **La Mondiale**: Safir Plus Revenue
MAAF[21]: Aviséo Autonomie
MMA[22]: Plan Autonomie

12.3.5.2 The Best of Insurance and Savings

Combination contracts are an attractive solution to benefit from the insurance protection mechanism as long as the savings part is not sufficient to cover eventual Long Term Care expenses and the benefits of savings, that is, the opportunity to keep the premiums paid if dependence does not occur.

At issue, the policyholder selects a desired benefit amount upon reaching 75. On the basis of this amount, a premium is calculated. Thus, at issue and even without Cash Value, coverage is guaranteed at the level of the maturity amount at 75. If a claim does not occur before 75, it is the maturity value which will make it possible to meet future Long Term Care expenses related to dependence in the form of annuity and a lump sum for initial equipment purchase.

Of course, the overall performance of this product will have to be analyzed by putting in perspective premiums paid, Cash Value available at 75, Long Term Care insurance before 75, and the amount of the annuity paid in case of dependence after reaching 75.

In case of dependence, the contracts analyzed pay a monthly annuity and some pay an initial lump sum for equipment (2 of 3).

The policies analyzed do not cover Partial Dependence.

In the case of death without dependence, the cash value is paid to the beneficiaries. In case of death during the dependence, depending on the policies either the beneficiaries do not receive any benefit, or they receive the difference between the cash value and the benefits paid.

For one policy, a choice is to be carried out in case of non-dependence before reaching 85. This policy offers either to invest part of the savings to double the Long Term Care benefit, or to be paid the cash value and retain the initial annuity level.

In case of premium lapse, Paid Up benefits are available.

In the case of partial surrender of an annuity contract with Long Term Care benefits, the lump sum and annuity amounts associated with the LTC benefit are recalculated.

The ADL and AGGIR grids also serve as references to assess incidence and severity of an insured's dependent status. The Blessed grid is used in addition to these first two claim assessment tools.

12.3.6   Group Coverage … A New Cost for Employers?

12.3.6.1   Analyzed Offers

Ag2r[20]: Group Long Term Care contract
Malakoff Médéric: Pass Indépendance
Other contracts made on specifications: AXA Employees, Thalés employees.

12.3.6.2   Standard Products

Group Long Term Care contracts, designed for employees, cover the risk of dependence through the working life, often providing benefits to help employees who must assist a family member, and also provide favorable conditions to continue Long Term Care coverage after retirement. They can also provide preventive services to reduce the risk of dependence.

In contrast to Individual insurance, these Group insurance approaches:

- Allow optional employer's co-financing;
- Encourage earlier premium payments;
- Avoid the quasi-systematic medical questionnaires of Individual policies.

For the employer, this insurance coverage is an additional benefit for employees, providing tax advantages, and a way to reduce presenteeism or absenteeism of employees who are caring for dependents.

The employer has a choice of several formulas (2 and 3 in the offers analyzed) with a unique premium. Pricing is defined according to average retirement age, the employment demographics, the turnover and hire rates, and the number of employees not actively at work.

Total Dependence is covered by monthly annuity payments (between €300 and €2,000 depending on the contract). Partial Dependence is not covered in the contracts analyzed.

We did not find home remodeling, or first equipment lump sum benefits in case of dependence in any contract.

When the employee becomes a caregiver, the contract offers to help search for short or long term accommodation solutions.

Optional supplemental Long -Term Care Individual policies, include benefits for Partial Dependence and home remodeling.

Finally, Group pension contracts offer specific mechanisms at retirement.

Thus, in a contract, when cashing out the basic retirement benefits of an employee, a lifetime Long Term Care coverage can apply if the employee is working with the company at the time of retirement.

Other contracts offer a more specific mechanism to carry over vested rights to Individual coverage, without a medical questionnaire, without a Waiting Period and with a premium reduction of up to 70% depending on the length of the employee's participation in the Group contract.

12.3.6.3   Contract on Specification

For the benefit of their employees, large companies have provided Long Term Care insurance beyond health and savings. This is particularly the case for AXA, insured by AXA France Vie, and supplemented by an assistance plan offered by AXA Assistance.

This is also the case of Thalès whose employees are covered by a contract underwritten by OCIRP[23] and administered by Humanis.

AXA employees group plan

- Mandatory;
- A benefit in case of Total Dependence (IRG 1, 2 and 3), a fixed annual annuity of €4,800 per year;
- The annuity is revalorized according to the evolution of the value of the ARRCO[24] pension index;
- Dependents of the employee can be covered provided that they are not dependent on the date of application;
- An employee retiring, or leaving the group may apply to convert to an optional policy with the same lifetime annuity;
- At the end of 8 full years, the employee is guaranteed a reduced paid up benefit;
- Claim evaluation grids used are IRG, ADL, and "Blessed" test;
- Assistance benefits are available and structured around three levels:

  – Assistance benefits granted before incidence:
     Long Term Care information counseling;
     Legal information by telephone;
     Remote (video) assistance by specialized service;
  – Assistance benefits granted after incidence:
     • Initial evaluation of dependence requirements;
     • Implementation of a "personalized parent/caregiver living plan";
     • Home remodeling evaluation;
     • Follow-ups;
     • Social support;

- Messaging Service;
- Help with the search for specialized facilities;
- Custody and transfer of domestic animals;
- Legal aid for signing contracts;
- Organization of services;
- Finding suppliers;
- Providing legal experts in case of lawsuits;
  - Benefits in case the employee needs temporary leave of absence:
    Psychological assistance by telephone;
    Presence of living support;
    Help with locating and transportation to temporary care facilities.

Thalès employees group plan

- The contract is structured around a mandatory group insurance and two options, one to supplement the group coverage and the other to cover the spouse;
- Before retirement, the employee may apply for continuation of the coverage under an individual policy;
- At retirement, the employee may maintain to continue coverage with an individual policy;
- Group coverage is not subject to underwriting;
- Individual policies are subject to underwriting;
- Claims are assessed using IRG; if the claimant is receiving the APA[24] benefit, the classification adopted by the public plan is kept;
- Premiums purchase lifetime Long Term Care benefit units added to an individual account;
- The unit value of the point depends on the attained age;
- Under the mandatory plan, the minimum benefit is about €300 per month;
- In the event of unemployment, the benefit level is maintained without premium during the period of unemployment compensation;
- The benefit in the case of Partial Dependence represents 25% of the benefit in case of Total Dependence.

12.4   Services, Meeting the Need?

Services constitute a major area of differentiation between savings contracts. On one hand they reduce the appearance of paying premium to a "lost fund", and on the other hand, for some services, an attempt to limit the frequency or the severity of a Long Term Care claim.

These services can be categorized into four types depending on time of activation, trigger event, and beneficiary:

- Services starting at policy issue (a)
- Services to the policyholder providing care (b)

- Services to the dependent policyholder (c)
- Services to the dependent policyholder's caregiver (d).

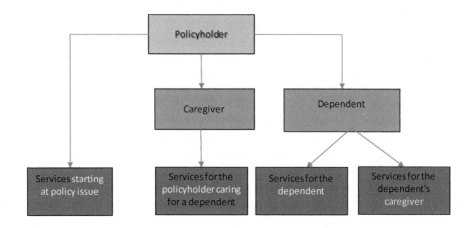

12.4.1 Inform and Prevent

Information and prevention services available as soon as a policy is issued are highly developed in Individual Long Term Care policies. Main benefits:

- Dedicated websites[25]
- Telephone information services
- Evaluation for:

  – Autonomy
  – Memory
  – Cognitive abilities
  – Hygiene care
  – Cardiovascular Risks

- After 70, a personalized prevention program if the lump sum amount is greater than €7,000.

On other savings, combination, or Group contracts, benefits are more limited and often the information is supplied without real interaction and/or follow-up, and is delivered via media (mail, internet, phone), such as:

- Information on health and aging well
- Wellness tips
- Review by phone to set up a personalized prevention program.

Two insurers offer Long Term Care dedicated websites.[26]

12.4.2   *Assistance for the Policyholder Caring for a Parent*

Overall, services to accompany the caregiver are identical regardless of the person's status, policyholder or relative of the policyholder. Some of the services cited are the same as those offered to the caregiver of the dependent policyholder. Two types of policies are observed:

- Contracts that offer a minimum base of services (note that some offers do not offer anything):

  - Information
  - Finding caregivers
  - Telephone support.

- More complete offers (whatever the kind of products):

  - Assessment of the parent's loss of autonomy
  - Residence evaluation
  - Caregiver 360° evaluation
  - Preventive assessment for the dependent person in the insured's care
  - Memory assessment for the dependent person in the insured's care
  - Dedicated internet site
  - Care training for the subscriber by a nurse
  - Caregiver psychological support.

12.4.3   *Assistance for the Claimant*

Services for the claimant are the core of service benefits. But here too, benefits are varied depending on the product and level of benefits. Overall, when it is covered, a contract provides assistance services and accessibility to the services at the time of Partial Dependence. Services are provided in France by assistance companies such as Europ Assistance, Filassistance, IMA, Mondial assistance.

In the case of group coverage, it consists of traditional assistance services, medical information and advice and guidance assistance. Combination contracts also offer traditional assistance services such as a housekeeping aid, teleassistance (via additional subscription fees), support for housekeeping, assistance in research of a nursing home, moving or travel assistance.

For Home Care, Prévoyance (life/health/disability) contracts provide a personalized diagnosis of the claimant's needs and advice on necessary accommodations. For housing accommodation, the claimant is assisted in finding and moving to an appropriate facility. The contract also makes available Remote Assistance based on different levels of subscription fees. This can be achieved by free 24/7 assistance for 3 months. Finally, other services are also there to supplement the cash benefit at the time the claim is approved, with lifestyle evaluation and advice in social resources.

Individual Savings policies, to differentiate themselves on this type of benefit, offer a multitude of services in addition to the services mentioned above. Among these, we mention access to psychological support, legal counsel, meal delivery, drugs, pet care, monitoring of the claimant's quality of life by regular contacts, availability of a family mediator and even a call on the 70th birthday. Services offered provide the claimant with real alternatives for daily challenges.

12.4.4 Assistance for Claimant's Caregiver

Services to caregivers do not generally differentiate who the caregiver is, whether the policyholder or not. A large part of caregiver services has already been identified in the previous section on policyholder services assisting a parent who has become dependent. Psychological support and medical advice/information are available in all contracts. A number of services complement the initial assistance. Thus, caregiver assistance (such as registered social worker, housekeeping, meal delivery, relocation) exist in both Individual and Group contracts. An evaluation and practical caregiver training are also available. In the same way, assistance services of dependent day care type, housekeeping assistance in case of dependent's hospitalization, or help with administrative procedures, orientation towards group therapy, Hotlines and dedicated websites[27] are mostly available in individual savings policies.

12.5 Is There Really a Long Term Care Market?

Contracts offering to cover the Long Term Care risk are many, with numerous services, and interesting features to find the best level of mutualization, notably in combination policies or Group contracts, or to encourage purchasing the coverage with promises of services offering benefits to the policyholder as a caregiver, and to develop preventive actions aimed at limiting the occurrence of dependence. Products are more liberal, with benefit levels more in line with the needs, offering more protection (lifetime benefits/indexation of benefits).

Yet the market does not develop, as public authorities through debates and announcements but without implementing real solutions contribute to a wait-and-see attitude, both among insurers and among prospective insureds.

At the same time, insurers continue to be afraid of this risk, which is capital intensive; a small number of players have credible databases, there is no pooling of data to construct tables. If Total Dependence is better controlled, Partial Dependence, which constitutes a critical commitment to the policyholder, is less understood.

Potential policyholders continue to ignore a benefit for such a far-off risk, in which they do not wish to project themselves, which, to be well protected, requires the incurrence of substantial premiums, with a level of benefit which probably will be not sufficient to be a real solution.

To get out of this impasse, we identify several paths:

- Encouragement by public authorities to implement individualized solutions to cover the additional costs associated with possible dependence;
- Pooling of data to better investigate the Long Term Care risk;
- Focus on prevention;
- Development of solutions allowing the maximum number of people to contribute as early as possible;
- Facilitate Long Term Care financing through savings and real estate;
- Development of quality coaching programs to facilitate home care.

On June 2018, the French government announced that it will study the "Long Term Care issue" in 2019 and will have new recommendations concerning the so-called 'Fifth Risk'.[28]

End Notes

1. "Que choisir?" (What to Choose?), Association UFC.
2. Association "60 million consommateurs" (60 million consumers, a kind of Consumer Report).
3. Fédération Française de l'Assurance (FFA); see Note 15, Chap. 3.
4. IRG: see Sect. 3.3.1 and Note 2 of Chap. 3.
5. See Sect. 3.5.1 of Chap. 3.
6. Insurers are categorized according to the three legal codes under which they operate:

   (1) Insurance (Code des assurances)
   (2) Mutual (Code de la mutualité)
   (3) Prévoyance (Code de la sécurité sociale).

   The great majority of insurers operate under, or follow, the insurance code. Prévoyance insurance is akin to Life and Disability coverage.
   Both Mutual and Prévoyance insurers operate under a non-profit status and are group insurers, without individual underwriting. Their statute does not preclude issuing Individual policies as long as the insurer follows the requirements of their respective code.
   Prévoyance products are mostly supplemental insurance for French Social Security benefits (see Note 28).
   Over the years the distinctions for insurance products have weakened as most insurers follow the insurance code, however the organizational and fiscal status have not changed.
7. Total Dependence: 3 ADL out of 5 or Level 1 or 2 for IRG.
   Partial Dependence: 2 ADL out of 5 or Level 3 or 4 for IRG.
8. LTCIBG, Long Term Care Insurance Basic Guarantees, translation of GAD, Garantie Assurance Dépendance.

9. PERP: Plan d'épargne retraite populaire
   Tax deductible (up to 10% of annual taxable income) defined contribution retirement plan defined per a 2004 regulation. Close to a US Individual Retirement Account, but funds cannot be accessed (except under exceptional circumstances) before retirement, the date at which a maximum 20% can be withdrawn (taxable).
10. DES or Déclaration d'État de Santé.
11. QSS or Questionnaire de Santé Simplifié.
12. QS or Questionnaire de santé.
13. Waiting Period: If a claim occurs during a Waiting Period, the policy is cancelled and premiums are refunded. Not to be confused with an Elimination Period.
14. INSEE see Note 6 of Chap. 3.
15. AGIRC Association générale des institutions de retraite des cadres.
    Retirement association which releases indices on which supplemental retirement benefits are based.
16. MGEN Mutuelle générale de l'Éducation nationale.
    Mutual insurance company associated with 3.5 million education professionals.
17. Social assistance: help with accessing and coordinating with social benefits, such as Social Security and lodging.
18. BNP: Banque Nationale de Paris.
19. GMF: Garantie Mutuelle des Fonctionnaires.
20. Ag2R: Association Générale de Retraite par Répartition.
21. MAAF: Mutuelle d'assurance des artisans de France.
22. MMA: Mutuelle d'Assurance des Armées.
23. OCIRP: Organisme Commun des Institutions de Rente et de Prévoyance.
    ARRCO: Association pour le Régime de Retraite Complémentaire des (Ouvriers, now Salariés).
    Will merge in 2019 with AGIRC (see Note 15).
24. APA: see Sect. 3.4.2 of Chap. 3.
25. See: i-dependance.fr or bienmangerpourmieuxvivre.fr.
26. I-dependance.fr or, by Crédit Agricole.
    Bienmangerpourmieuxvivre.fr, by Groupama Gan.
    Groupama Gan is a mutual insurance company from the 1998 fusion of Groupe Assurances Mutuelles Agricoles (1848 origin) and Groupe des Assurances Nationales (1913 origin).
27. Two examples of dedicated websites:
    Entreaidants.fr, by AXA Assistance.
    Vivreenaidant.fr, by La Banque Postale.
28. The four risks covered by the French Social Security, or sécurité sociale:

    1. Health Care
    2. Retirement
    3. Family care
    4. Workers' compensation.

Chapter 13
Predictive Analytics in Long Term Care

Howard Zail

13.1 What Is Machine Learning?

The craft of machine learning is making an ever-increasing impact across many industries and is changing the way businesses operate and companies compete. The insurance industry, which has always had a strong statistical underpinning to the development of its products and setting of rates, is certainly not immune to these developments. In this chapter, we develop a theoretical foundation for and practical approach to applying machine learning to develop morbidity assumptions for long term care products. In particular, we focus on analyzing incidence rates, namely the probability of a healthy insured going on claim. Similar and analogous methodologies can be applied to assessing claims termination rates and lapse rates.

The first question we must address is what is machine learning and how does it differ from traditional statistical analysis, or for that matter standard actuarial practice? Machine learning is a "set of methods that can automatically detect patterns in data, and then use the uncovered patterns to predict future data, or to perform other kinds of decision making under uncertainty."[1] It is an amalgamation of statistical techniques, computer science, algorithms, heuristics, data engineering and data management. Traditional statistics and actuarial science should be considered a subset of machine learning and not methods that stand in opposition to machine learning. Nevertheless, as we will see in this chapter, the focus, motivation and bias of machine learning practitioners is often quite different and might be considered strange to actuaries. Actuaries and statisticians typically require a strong theoretical basis for their analyses and they use methodologies that can be developed from first principles. Machine learning practitioners, on the other hand, are motivated by "getting the answer right" and are less concerned about why their models work. It is thus

---

[1] Murphy [1].

H. Zail (✉)
Elucidor, LLC, 305 East 40th St, Suite 21F, New York, NY 10016, USA
e-mail: hzail@elucidor.com

© Springer Nature Switzerland AG 2019
E. Dupourqué et al. (eds.), *Actuarial Aspects of Long Term Care*,
Springer Actuarial, https://doi.org/10.1007/978-3-030-05660-5_13

important that it can be shown that a machine learning model does in fact work and so robust validation techniques have been developed, and these will be described in this chapter.

Our focus is on four algorithms to develop incidence rates: a generalized linear model (GLM), the Lasso, Neural Networks and gradient boosting machines. We show how these models can be developed in practice and what tricks and techniques can be used to make these models more efficient. We compare and contrast the models and show the relative accuracy of the models.

13.2 Motivation

In 2015, the Society of Actuaries ("SOA") published a database of long term care incidence rates. The experience period covered 2000–2011 and the data was derived from 22 companies, comprising approximately 80% of the market.

The SOA then commissioned an actuarial consulting company, Towers Watson, to produce a set of experience tables using a machine learning approach (the "SOA report").[2] Towers Watson adopted a GLM approach and utilized its internal proprietary software to conduct the analysis. The report provided a list of predictors and interaction effects that were used, but beyond this, the technical details were fairly opaque. The software was not made freely available, and limited benchmarking was provided. Thus, it is difficult for users to test alternative machine learning algorithms against the Towers Watson model, or to update or enhance the base model.

This chapter provides the detailed technical means of using open source software to conduct machine learning on LTC experience data. We describes the core theory behind each method and provide the basic software to conduct the actual analysis. The core data snippets are explained, with the full programs available on https://github.com/howardnewyork/ltc (the "Github Site"). The methods described can be applied to the SOA data as well as smaller or proprietary datasets generated by individual companies. We focus exclusively on nursing home incidence rates. Developing home care or assisted living rates are analogous. Further, the approaches, as described, are general in nature, and can easily be extended to other life and health analyses and P&C studies.

13.3 The Data

From 22 companies' data, the SOA selected a subset of data from 12 companies that had data describing at least a certain number of "critical data elements". The resultant subset of data covered nearly 15 million years of total life exposure and 172,000 claims. The critical data elements are:

---

[2]"Long Term Care Intercompany Experience Study" [2].

Table: critical data elements of SOA 2000–2011 claims incidence study

Issue date	Underwriting type	Claim incurred date
Date of birth	Marital status	Claim type
Gender	Coverage	Paid amount
Underwriting class	Benefit period	

Together with a report, the SOA published the full data with policies being grouped together into summarizing "buckets". For example, the "IncurredAgeBucket" field is the age of the insured when the claim began, grouped into 5-year age bands. Full details of the database are provided in the SOA report.

13.3.1 Data Preparation

A key part of every machine learning project is preparing the data for analysis. This part of the project can make a material difference to the accuracy or appropriateness of the models and often can make a substantially larger impact than adopting a complex or advanced machine learning model over a simple model. This section will describe some of the key processes we adopted but recognize that our adjustments are incomplete and additional work can be done to materially improve the overall modeling process.

13.3.1.1 Grouping of Data

The SOA data is provided pre-grouped. Ideally, we would prefer to work with individual policy level data as grouping of data removes some of the signal from the data.

We have to make the choice whether any grouped "bucketed" variable remains as a categorical variable or a numeric variable. In our case, we convert all grouped variables back to numeric variables. For example, the issue age bucket "55–60" is converted into the numeric value 57.5. Setting a value for end categories is more difficult, for example, issue age bucket 90+, or the "unlimited" benefit period. In our case, we selected 93 for the highest age and 10 years as the equivalent predictor for the lifetime benefit period. It is not immediately apparent whether such an approach is optimal. It might, in some cases, be worth adding an additional flag such as a limited/lifetime benefit period, where "limited" is coded as zero and lifetime is coded as 1. In some cases, the buckets could be left as categorical variables. A combination of intuition and testing of alternative approaches is helpful in determining the most appropriate approach.[3]

---

[3]The full set of data adjustments to date is found in the `clean_incidence` function on the Github Site.

Certain categorical variables cannot be converted to an equivalent numerical value. In such cases, we adopt a "one-hot encoding" approach, by converting the categorical variable into a set of flags. For example, the predictor "Region" has five options: Midwest, Northeast, South, West, and unknown. One of these categories would be deemed the base case, e.g. Midwest, and the predictor would be then coded as:

Table: One-hot encoding of Region predictor

Category	Flag			
	RegionNortheast	RegionSouth	RegionWest	Regionunknown
Midwest	0	0	0	0
Northeast	1	0	0	0
South	0	1	0	0
West	0	0	1	0
Unknown	0	0	0	1

Using this approach, the single categorical variable with five options is transformed into four numeric variables that are coded as zero or one.

Similarly, Gender would be turned into a single flag "GenderFemale" which would be 0 for males and 1 for females.

13.3.1.2 Missing Data

Dealing with missing data is often a non-trivial exercise. One must first determine whether a particular field is "missing at random" or not. For example, if a company consistently leaves out a field, and that company's experience differed from the rest, then the missing field could be an indicator of differing experience. If a field is missing because the underlying product is differently structured than would typically be the case, then again that missing field could be indicative of differing experience. On the other hand, if "missingness" of a field does not provide additional information about the record, then the field is "missing at random".

In general, we assume that data is missing at random, and adopt a simplified approach: for numeric predictors, we set missing values as the mean of the non-missing values for that field. More advanced techniques are available including developing mini-regression models for each selected variable, where you create a linear regression model to predict the missing field from all the non-missing fields.

For categorical variables, any missing field is assigned its own category. Again, more sophisticated approaches could be adopted such as using a regression model (in this case a logistic regression model) to predict the category from the other fields or replacing a missing field with a weighted random sample of the other categories. The usefulness of any of these approaches can be checked under a cross-validation strategy described later in the chapter.

13.3.1.3 Data Problems

The data, as presented, contain numerous systematic issues that will likely persist through the machine learning process and would need to be dealt with separately. For example, the insurance companies may have reported the data inconsistently, and the underlying company products can be materially different in design and underwriting standards. The data are historic in nature and the machine learning algorithms will predict historic, not future experience. In fact, the SOA published a comprehensive report of caveats when using the data[4] which should be considered when applying the approaches described in this chapter. These caveats are not described herein, but should be considered when utilizing the techniques described in this chapter.

13.4 The Framework

13.4.1 The Time to Event Model

Each record in the database represents an annual period for a group of lives which all have the same covariates. The database contains two exposed-to-risk fields: "Total-Exposure" represents the number of life years including fractions thereof of exposure during the relevant period, not reduced by the period of time on claim; and "Active-Exposure" which reduces exposure for time on claim. Each record then contains the number of nursing home ("NH"), assisted-living facility ("ALF") and home-healthcare ("HHC") claims.[5]

We assume that during each annual period and for each record i, the time until a claim follows a piecewise exponential model with hazard function:

$$h(x_i, \theta) = e^{f(x_i, \theta)},$$

for some function $f(x_i, \theta)$, where x_i is the vector of predictors for record i, and θ is the vector of parameters of the model.[6] We will focus on time until an NH claim, and treat ALF and HHC claims as censoring events. It can be shown that the likelihood function for such distribution is proportional to the likelihood function for a particular Poisson distribution,[7] namely for each record i,

$$y_i = Claims\_NH_i \sim Poisson(\lambda_i),$$

---

[4]"Caveats for Use of Long Term Care Experience Basic Tables", 2015, Society of Actuaries.

[5]These claims count fields are called Claims_NH, Claims_ALF, and Claims_HHC, respectively, in the database.

[6]The selection of ActiveExposure as the measure, instead of TotalExposure, is described in Aalen [3], Chap. 5.

[7]See (i) Holford [4] and (ii) Rodriguez [5].

where

$$\lambda_i = ae_i \cdot h(x_i, \theta);$$
$$ae_i = ActiveExposure_i.$$

We can alternatively write λ_i as

$$\lambda_i = e^{\log(ae_i) + f(x_i, \theta)} \tag{13.1}$$

All the algorithms that follow will be structured around how best to determine the function $f(x_i, \theta)$.

13.4.2 Fitting the Model

In fitting all the algorithms, we will split the data into three subsets: a training set (80% of the records), a validation set (10%) and a test set (10%). The training set will be used to fit or train our model. The validation set will be used to test different varieties of our model. That is, each model will contain various tuning features and "knobs", choice of predictors, choice of predictor interactions and so on. The validation set allows us to determine the best tuned model. The test set remains untouched until the end of the process, whereafter it is used to judge the final accuracy of the model and provide an overall score for the model.

If there are N records in the training database, then the data consist of:

$$y_i = claims\ count\ for\ each\ record,\ i = 1\ to\ N,$$
$$x_i = a\ vector\ of\ p\ predictors\ for\ each\ record,\ i = 1\ to\ N.$$

Under the Poisson model:

$$
\begin{aligned}
Prob(y_i|x_i, \theta) &= \frac{\lambda_i^{y_i} \cdot e^{-\lambda_i}}{y_i!} \\
&= \frac{e^{y_i(\log(ae_i) + f(x_i, \theta))} \cdot e^{-e^{\log(ae_i) + f(x_i, \theta)}}}{y_i!}.
\end{aligned}
$$

Since λ_i is a function of θ, the goal is to choose a theta that maximizes this probability for all the data. That is, we want to maximize the likelihood:

$$\max_{\theta} \{Likelihood(y_1, \ldots, y_m|x_i, \theta)\} = \max_{\theta} \left\{ \prod_i \frac{\lambda_i^{y_i} \cdot e^{-\lambda_i}}{y_i!} \right\}.$$

This is known as the method of maximum likelihood. In machine learning, it is customary (and for numerical stability purposes), instead of maximizing the likelihood, to minimize the negative log likelihood. Thus, the negative Poisson log likelihood ("NPLL") becomes the scoring mechanism for all our algorithms. The lower the NPLL (with some critical caveats described in the next section), the better the training set fit.

The NPLL can be written as follows:

$$NPLL = -\log(likelihood(y_1, \ldots, y_m | x_i, \theta))$$
$$= -\sum_i \log\left(\frac{\lambda_i^{y_i} \cdot e^{-\lambda_i}}{y_i!}\right)$$
$$= \sum_i \lambda_i - y_i \cdot \log(\lambda_i) - \log(y_i!).$$

Since $\log(y_i!)$ does not change as θ changes, we can ignore the last term for purposes of minimizing NPLL. We also calculate the mean, rather than the sum of the log likelihood values, a change that does not affect the minimization problem. Finally, since under the Poisson distribution $E[y_i] = \lambda_i$, we can rewrite the NPLL as:

$$NPLL = \frac{1}{m}\left(\sum_i \hat{y}_i - y_i \cdot \log(\hat{y}_i)\right)$$

where $\hat{y}_i = predicted\ value\ of\ y_i = E[y_i] = \lambda_i.$ (13.2)

We use this form of NPLL to score all our models. In machine learning, this is called the "loss function".[8]

13.4.3 The Bias/Variance Trade-off

The models that we tend to use in machine learning are highly flexible and contain redundant parameters. This leads to the potential problem of overfitting the model to the training dataset. As we train models and make them more complex, we reduce the NPLL but the error rates tested on out-of-sample or validation datasets for such complex models will eventually begin increasing. What typically happens is summarized in the chart that follows. As the model becomes more complex, the error on both the training and validation sets will decline. However, at some point, the error on the validation set will begin to increase. The error on the training set is known as "Bias" and the error on the validation set is known as "Variance". This occurs

---

[8]This same loss function can be derived by using solely a piecewise constant hazard function assumption and does not require the y_i to have a Poisson distribution. See Aalen [3] Chap. 5.

because we are overfitting the model to the training data, and thus the Bias declines, but the Variance increases after a point.

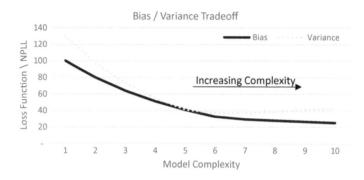

Ideally, we would like to decrease the Bias as much as possible, but not at the expense of an increasing Variance. Thus, we will repeatedly choose different models to train on the training set until we reach a minimum point of Variance on the validation set. It should be noted that this repetition results in "leakage" of information from the validation set into the training process. Even though we are not directly training on the validation set, we are using the validation set to inform our choice of the best model. Thus, the loss or error on the validation set is likely to be lower than the loss on a completely out-of-sample dataset. This is the reason for having the test set as a final untouched set to estimate the accuracy of the model.

13.5 The Algorithms

We now discuss the four main algorithms to analyze the database. We use standard off-the-shelf software and models. The analysis is conducted completely in R and the full software and data can be found on the Github Site.

The steps that we follow are:

1. Set the tuning parameters of the model.
2. Train the selected model on the training dataset to minimize NPLL.
3. Predict values on the validation set and calculate the NPLL on the validation set.

    – Recognize that the predicted values of the model for each record are just the expected values of the number of claims, namely $E[y_i] = \hat{y}_i = \lambda_i$.

4. Update the tuning parameters and repeat steps 2–3.
5. Continue this process until the validation NPLL can no longer be reduced.
6. Calculate the NPLL on the test set to produce a final score for the algorithm and model.

In the sections that follow, we use the following notation:

$$X = matrix\ of\ predictors\ of\ size\ N\ by\ p,$$

$$where\ N = number\ of\ records,\ p = number\ of\ predictors,$$

$$X = \begin{bmatrix} x_1' \\ \vdots \\ x_N' \end{bmatrix},\ where\ each\ x_i\ is\ a\ vector\ of\ p\ predictors,$$

$$Y = \begin{bmatrix} y_1 \\ \vdots \\ y_N \end{bmatrix} = vector\ of\ number\ of\ observed\ Claims,\ of\ length\ N,$$

$$AE = \begin{bmatrix} ae_1 \\ \vdots \\ ae_N \end{bmatrix} = vector\ of\ ActiveExposure\ for\ each\ record,\ of\ length\ N.$$

In the R program, the dataframe `incidence` represents the database of records, and is then split into three subsets: `incidence_train`, `incidence_val`, and `incidence_test`, being the training, validation and test sets, respectively. The matrix of predictors X will be derived from this database. The fields "NH_Claims" and "ActiveExposure" in the `incidence` dataframe correspond to Y and AE, respectively. We will show the detailed analysis only with regard to nursing home claims, but exactly the same process can be followed with ALF and HHC claims.

13.5.1 Generalized Linear Models

13.5.1.1 Model Specification

Recall from Eq. (13.1) that $\lambda_i = e^{\log(ae_i) + f(x_i, \theta)}$. In a generalized linear model, we set $f(x_i, \theta)$ to be a linear function, namely:

$$f(x_i, \theta) = x_i' \cdot \theta,\ where$$

$$\theta = \big[\theta_1, \ldots, \theta_p\big]' = vector\ of\ coefficients\ of\ length\ p.$$

In R, a lot of the hard work of selecting predictors and manipulating these predictors is automatically managed with R's `formula` syntax. For example:
Sample Formula

```
formula = Count_NH ~ (IssueYear + IncurredAgeBucket +
                       I(IncurredAgeBucket^2))
```

tells R that the response variable, or the variable to be predicted is nursing home claims (Count_NH), the variable on the left-hand side of the formula. The predictors are on the right-hand side, namely IssueYear, IncurredAgeBucket and the square of IncurredAgeBucket. "I" is a special operator in R that creates a new predictor, being the output of the formula in the parentheses. R also automatically creates an intercept predictor. This intercept predictor is a vector of ones. Categorical variables (coded as factors in R) are automatically converted to one-hot encoded flags.

An "interaction effect" is created when two or more predictors are multiplied against one another. R creates interaction effects among predictors in three ways using the special operators ":", "*" and "^". For example:

1. Count_NH ~ IssueYear:IncurredAgeBucket creates an interaction effect predictor being the multiple of IssueYear and IncurredAgeBucket and is equivalent to
    a. NH_Claims ~ I(IssueYear*IncurredAgeBucket)
2. Count_NH ~ IssueYear*IncurredAgeBucket selects *both* the individual predictors and the interaction effect and is equivalent to:
    a. NH_Claims   ~   IssueYear   +   IncurredAgeBucket   +
                          I(IssueYear*IncurredAgeBucket)
3. Count_NH ~ (IssueYear+IncurredAgeBucket)^2 extracts all predictors as well as all combinations of two way interactions of predictors within the brackets. It does not create quadratic terms (e.g. I(IncurredAgeBucket^2)) of any of the features.

Thus, it is important to see that both "*" and "^2" are treated differently in the formula notation depending on whether they are found inside or outside an "I" function. For more details, view the R help files for formula.

As an example, the matrix of predictors produced by the "Sample Formula" described previously will produce the following predictor matrix:

	(Intercept)	IssueYear	IncurredAgeBucket	I(IncurredAgeBucket^2)
1	1	2001	70	4900
2	1	1998	60	3600
3	1	2001	75	5625
4	1	2001	75	5625
5	1	1998	60	3600
6	1	2001	60	3600

To execute the GLM in R, we simply define the formula[9] and run the glm function using the following (called the "Full" formula):

---

[9]Each of the field names are described in the SOA Report [2].

```
# Define "Full" formula with Interaction Effects
formula = Count_NH ~ offset(log(ActiveExposure)) +
  (Gender + IssueYear + IncurredAgeBucket  + PolicyYear +
  Marital_Status + Prem_Class +
  Underwriting_Type + Cov_Type_Bucket + TQ_Status +
  NH_Orig_Daily_Ben_Bucket + NH_Ben_Period_Bucket +
  NH_EP_Bucket  + Region)^2   + Infl_Rider + I(Duration^2)

# Run GLM Model
mod_glm = glm(formula = formula, data = incidence$train, family = poisson)
```

The `offset` function in `formula` ensures that `log(ActiveExposure)` is a predictor in the model but no coefficient (or θ value) is multiplied against it. This differing treatment of `log(ActiveExposure)` can be seen to produce a consistent formula with Eq. (13.1). In other words, the `offset` ensures that $\log(ActiveExposure_i)$ is added to $x_i' \cdot \theta$ to produce an estimate of λ_i, namely $\lambda_i = e^{\log(ActiveExposure_i) + x_i' \cdot \theta}$.

The variables selected with the "Full" `formula` are all the critical variables selected by the SOA with all two-way interactions for these variables, plus a few additional variables.

The glm formula runs the generalized linear model and solves for the θ parameters, by minimizing the NPLL. The predictors are selected from the `incidence$train` data frame in accordance with the `formula`.

Once the model has been trained, we will predict the number of claims for each record in the validation set. This is achieved with the `predict` function. We then calculate the PNLL from the predictions.

```
# Define the PNLL Formula
poisson_neg_log_lik = function(y_pred, y_true, eps=0){
  mean(y_pred - y_true * log(y_pred+eps))
}

# Predict Values
preds = predict(object = mod_glm, newdata = incidence$val, type =
'response')

# Calculate NPLL
pnll = poisson_neg_log_lik(y_pred = preds, y_true = incidence$val$Count_NH)
```

The validation NPLL is calculated as 0.0384.

We next calculate the NPLL on the variables and interactions selected by the SOA for its model. We also analyze a model without any interactions.

```
# Full formula but without Any Interactions
formula_no_interactions = Count_NH ~ offset(log(ActiveExposure)) +
  (Gender + IssueYear + IncurredAgeBucket  + PolicyYear + Marital_Status +
   Prem_Class + Underwriting_Type + Cov_Type_Bucket + TQ_Status +
   NH_Orig_Daily_Ben_Bucket + NH_Ben_Period_Bucket +
   NH_EP_Bucket  + Region)   + Infl_Rider + I(Duration^2)

# SOA Formula
formula_soa = Count_NH ~ offset(log(ActiveExposure)) +
  NH_Ben_Period_Bucket + NH_Orig_Daily_Ben_Bucket:Region +
  PolicyYear:Prem_Class+ PolicyYear:Underwriting_Type +
  PolicyYear:TQ_Status + IncurredAgeBucket:Cov_Type_Bucket +
  IncurredAgeBucket:NH_EP_Bucket + IncurredAgeBucket:Gender +
  IncurredAgeBucket:Marital_Status
```

The results are compared against the original "Full" formula:

	NPLL	
	Validation set	Test set
Full formula with interactions	0.0384	0.037
SOA formula	0.0393	0.0385
Full formula no interactions	0.0389	0.0381

The following chart compares the actual to predicted incidence rates by duration, using the "Full" formula:

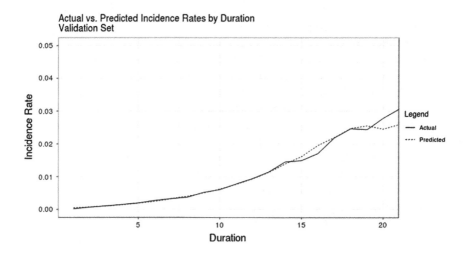

13.5.1.2   Variable Selection

We next analyze different data selections to find an adequate bias/variance tradeoff. This can be done manually. Alternatively, R has a systematic approach for conducting variable selection. This entails utilizing the step function which can conduct forward or backward stepwise regression. Forward stepwise regression entails adding one variable at a time to the `formula` until the single best addition is found. Once this is found, the next variable is added until no addition produces a better model. Backward stepwise regression is similar, except that we start with a large, overspecified model, and then systematically remove one variable at a time until the model cannot be improved.

Instead of using cross validation to test the strength of a model, the step function utilizes the Aikake Information Criterion or AIC.[10] The AIC is determined as:

$$AIC = 2.NPLL + 2 \cdot k, \ where$$
$$k = number \ of \ estimated \ parameters \ in \ the \ model.$$

The AIC thus contains two elements: The NPLL which measures goodness of fit, and the "2.k" component which measures complexity of the model. Thus, the AIC is penalized for each new predictor that is added to the model. The result is that AIC is a measure that balances model fit with model complexity. AIC is measured directly against the training dataset and not the validation dataset and is an alternative to the cross-validation approach.

In our case, we could start with the selected formula and then apply forward or backward stepwise regression, e.g.:

```
mod_step = step(mod_glm, direction = "forward")
```

Unfortunately, for this size of database the algorithm is prohibitively slow and can take days to run. A single selected formula can take over an hour to run, and thus the methodology is not practical to run on a CPU. As an alternative, a GLM can be run as an equivalent simplified neural network on a GPU (see Sect. 13.5.3), allowing a single model to be solved in minutes, making stepwise regression feasible for larger databases. Using this approach, we started with the SOA formula, used forward stepwise regression to find the best additional predictors (excluding interactions), and then from the selected subset of predictors, we again used forward stepwise regression to select the best model including interactions.

The resultant formula identified using this approach is:

```
Count_NH ~ NH_Ben_Period_Bucket + Underwriting_Type + I(IncurredAgeBucket^2) +
    Region + IssueYear + Marital_Status + IssueAgeBucket + Prem_Class +
    NH_EP_Bucket + Cov_Type_Bucket + IncurredAgeBucket + Region:NH_Orig_Daily_Ben_Bucket +
    IncurredAgeBucket:Gender + Prem_Class:PolicyYear + Underwriting_Type:PolicyYear +
    PolicyYear:TQ_Status + Cov_Type_Bucket:IncurredAgeBucket +
    NH_EP_Bucket:IncurredAgeBucket + Marital_Status:IncurredAgeBucket +
    Underwriting_Type:Cov_Type_Bucket + IssueYear:NH_EP_Bucket +
    Underwriting_Type:IssueAgeBucket + NH_EP_Bucket:Cov_Type_Bucket +
    NH_Ben_Period_Bucket:Underwriting_Type + Underwriting_Type:NH_EP_Bucket +
    IssueYear:IncurredAgeBucket + NH_Ben_Period_Bucket:Cov_Type_Bucket +
    IssueYear:IssueAgeBucket + NH_Ben_Period_Bucket:IncurredAgeBucket +
    Underwriting_Type:Prem_Class + IssueAgeBucket:Cov_Type_Bucket +
    offset(log(ActiveExposure)) - 1
```

---

[10] Aikake [6].

The validation NPLL is 0.0383 and the test NPLL is 0.0377.

The analysis suggests that forward stepwise regression can be used to good effect for selecting a good model.

13.5.2 Lasso

There is clearly a need for an efficient means of variable selection, especially for large models such as the LTC database. One popular option is the Lasso model. The Lasso is designed to automatically penalize large parameters in the model and "shrink" unnecessary parameters to zero. This is done by using an ingenious but simple mechanism: instead of trying to minimize the NPLL, we minimize a penalized version[11] of the loss function, namely:

$$\min_{w.r.t\,\theta} \left\{ NPLL + LAMBDA \cdot \sum_i |\theta_i| \right\}, \ where\ LAMBDA\ is\ a\ positive\ constant.$$

The second term is known as the "L1 Norm" and is the sum of the absolute values of the parameters of the model. As more predictors are added to the model, the L1 norm will increase in value, resulting in a penalty for more complex models. It turns out that the geometry of this minimization problem results in weak predictors having their coefficients shrunk to zero.[12] A weak predictor is a predictor that does not significantly benefit the model. Thus, in a single optimization routine, the Lasso is able to select the important variables and identify those variables that can be removed from the model. Unnecessary variables are identified as those with zero-value θ coefficients.

A complexity of the model is that the value of LAMBDA is not known and is an input to the model. In order to determine LAMBDA, we need to need to test a range of possible values and select the best LAMBDA through a process of cross-validation.

We use the glmnet R package to implement the Lasso algorithm. Glmnet does not use the R formula syntax so a design matrix of predictors must be created beforehand using the model.matrix function. glmnet automatically adds in an intercept term, so the intercept term should be removed from the design matrix. In some algorithms it is necessary to scale the data to have a zero mean and unit standard deviation. This is required for the Lasso so that the penalty term is not distorted by the relative scale of each of the predictors. However, this scaling process is done automatically by the glmnet routines and so can be ignored in our code. Given the large database, it is helpful to run the routines in parallel, which can be done using the doMC package. We use a formula that has fewer interactions than the "Full"

---

[11] We use the term "LAMBDA" to designate the constant in order to be consistent with the R package glmnet which will be used to execute the Lasso. This LAMBDA is not the same as and should not be confused with the Poisson mean parameter λ.

[12] Tibshirani [7].

formula used in GLM in order to speed up the calculations, and in particular, the cross-validation routine which can be time consuming.

```
library(glmnet)
require(doMC)
registerDoMC(cores=6)  # Set cores to the number of CPU cores available

# Base formula
formula = Count_NH ~ offset(log(ActiveExposure)) +
  (Gender + IncurredAgeBucket  + PolicyYear + Marital_Status + Prem_Class +
   Underwriting_Type + Cov_Type_Bucket  + NH_Orig_Daily_Ben_Bucket +
   NH_Ben_Period_Bucket + NH_EP_Bucket )^2 + TQ_Status + Region +
   IssueYear  + Infl_Rider + I(Duration^2)


# Create a matrix of predictors, excluding the intercept
design = model.matrix(formula, incidence_val)[,-1]

# Run the routine
cv_mod_glm= cv.glmnet(x = design, y = incidence_val$Count_NH,
                      offset = log(incidence_val$ActiveExposure),
                      nfolds = 5, parallel = TRUE,
                      family = "poisson")
plot(cv_mod_glm)
(cv_mod_glm$lambda.min)  # lambda with the lowest cross-validation error
(cv_mod_glm$lambda.1se)  # largest value of lambda such that error is within
                           1 standard error of the minimum.
coef(cv_mod_glm, s = "lambda.1se")  # plots the coefficients using selected
                                      LAMBDA values
```

The following plots the "Deviance"[13]:

Deviance Calculated by Cross Validation for Different values of LAMBDA

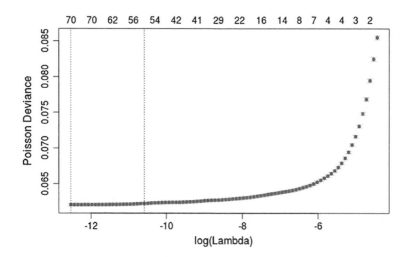

---

[13]"Deviance" is a measure of the accuracy of the model (lower is better), see the glmnet package vignette for more details.

The top x-axis shows the number of non-zero coefficients, the bottom x-axis shows the value of log(LAMBDA). The graph indicates that the minimum deviance is found at 70 non-zero parameters, but that a model with 55 non-zero parameters is almost as good. The LAMBDA value that minimizes the loss function is stored in `cv_mod_glm$lambda.min` and the largest value of LAMBDA such that the error is within one standard error of the minimum is stored in `cv_mod_glm$lambda.1se`

We now proceed similarly to the GLM approach of predicting the values on the validation set and checking the NPLL and plotting sample rates. We use the `lambda.1se` as the selected LAMBDA value.

```
# Create a matrix of predictors on the training set, excl. intercept
design_test = model.matrix(formula, incidence_test)[,-1]

#Calculate the predictions at the lambda.1se level
preds = predict(object = cv_mod_glm, s = "lambda.1se", newx = design_test,
                newoffset = log(incidence_test$ActiveExposure),
                type = 'response')
```

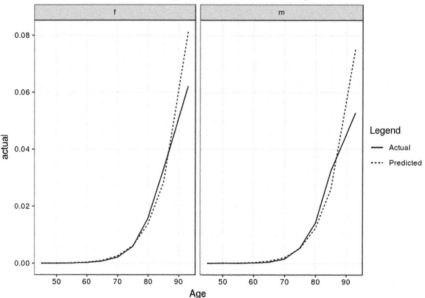

Lasso Model: Actual vs. Expected Incidence Rates by Gender

The calculated NPLL on the validation set is 0.0386 and on the test set is 0.0378.

13.5.3 Neural Networks

Neural network algorithms have achieved tremendous success across a wide variety of tasks, particularly with tasks like image processing, natural language processing, and voice recognition. They are also being used to achieve at or close to best-in-class error rates with large scale structured data problems such as the LTC database. The advantage of neural networks is that they automatically conduct a form of feature engineering and feature selection. On the other hand, neural networks are very flexible, and have many tuning parameters. It can be time consuming and difficult to determine the right neural network architecture for the problem at hand. Further, unlike GLM, it is difficult to interpret the parameters in a neural network model.

Although neural networks are often described as a mathematical approximation to how the brain functions, a more useful description is that they are a generalization of linear modeling. A single node in a neural network is simply a matrix multiplication of $x'\theta$ where x is a vector of inputs and θ is a vector of parameters for that node. A set of nodes at the same level is called a "layer". In fact, we will start by recreating the GLM model as a single node, single layer neural network. This is graphically depicted as:

Graphical Representation of a Network

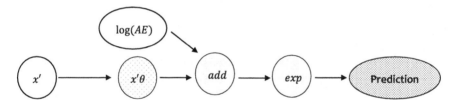

This is a graphical representation of each of the calculations in Eq. 13.1 with $f(x_i, \theta) = x_i' \cdot \theta$:

1. The inputs are x' and $\log(AE)$, AE = Active Exposure (White Circles).
2. x' is multiplied by θ. The Light Dotted circle represents a one-layer, one node neural network.
3. The result from (2) is added to log (AE) (Shaded Circle).
4. The exponent of (3) is taken (Shaded Circle).
5. A prediction is made (Heavy Dotted Circle).

We use the R package Keras to code the above network. Keras itself calls a software package called Tensorflow, which is Google's deep learning framework. Keras can be run on both a CPU and GPU. You will need a GPU to run models on the full LTC database,[14] given its size.

---

[14]Unfortunately, at the time of writing, setting up Keras on a GPU is not for the faint of heart. Only Nvidia GPUs are currently supported. Help can be found on the RStudio, Tensorflow and Nvidia websites on how to install the various components. An alternative is to use a cloud server through Google, Amazon where the server is pre-installed with R and Keras.

In Keras, the network is first specified in R but no calculations are actually done. The network is then compiled and the model is only finally run with the `fit` function. Keras uses "piping" syntax which is shown by the pipe symbol "%>%". This pipe takes the output of one function and feeds it to the next function. The pipe is equivalent to the arrow in the Graphical Representation. Each call to a `layer` function defines a node or a set of nodes in the neural network.

```
library(keras)

# The input layers are specified (white circles in chart)
input_offset <- layer_input(shape = c(1), name = "input_offset")
input_predictors <- layer_input(shape = c(N_predictors),
                                name = "input_predictors")

# Single node, single layer (light dotted circle in chart)
rate_mean = input_predictors %>%
  layer_dense(units =1, activation = 'linear')

# Prediction is made (shaded and heavy dotted circles in chart)
predictions = layer_add(c(input_offset, rate_mean)) %>%
  layer_exp()

# The model is defined linking the inputs to the outputs
model <- keras_model(inputs = c(input_offset, input_predictors),
                     outputs = predictions)

# Summary of the model
summary(model)

# Model is compiled
model %>% compile(
  loss = 'poisson', # Designates NPLL as the loss function
  optimizer = optimizer_adam(), # The type of optimizer
  metrics = c('mse', k_poisson_neg_log_lik) # Additional values to calc.
)

#Model is run
history <- model %>% fit(
  x = list(input_offset = offset, input_predictors = design),
  y= y,
  epochs = 10, batch_size = 2048*2,
  validation_split = 0.2,
  verbose = 1
)

#Predictions on the validation set are made
preds = model %>% predict(
  x=list(input_offset = offset_test,
         input_predictors=design_test),
  verbose=1
)
```

The various parts of the neural network are coded as a set of layers. A layer can be viewed as a set of nodes, but in this case each layer contains only one node.

- layer_input: designates an input node (White Circles in Graphical Representation)
- layer dense: designates linear multiplication layer, namely $x' \cdot \theta$, where x' is the input vector (Light Dotted Circle)
- layer_add: adds two layers together (Shaded Circle)

- layer_exp: is a custom designed layer which calculates the exponential value of each cell. See the Github Site for details. (Shaded Circle)
- batch_size represents how many records are fed into the neural network at one time
- epochs represent how many times the neural network is trained on the data.

With a neural network, a batch of data is fed into the network at one time. This differs from GLM, where the entire dataset is fed into the optimization routine at each step. The neural network training process involves making a slight improvement to the parameters given each batch of data. The parameter epochs represents how many times the model runs through the entire set of data. With neural networks, the optimization process must be stopped early to avoid overfitting the data. The number of desirable epochs to be used in training is determined using cross validation, as is shown in the following graph which measures the NPLL and the Mean Square Error. These decrease as the number of epochs increase.

Change in NPLL Loss Function and Error Rates with Increasing Epochs

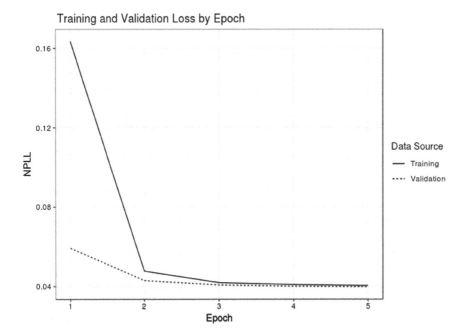

When we run the above, we produce an NPLL of 3.85, which is similar to what was obtained with the one line of GLM code.

Creating a more advanced neural network entails replacing the single Light Dotted node in the prior example with a more advanced structure. The rest of the network will remain exactly the same. For example, instead of one node, we could have thirty nodes, representing 30 mini regressions, the results of which are each fed into a second layer of 15 nodes, which are then fed into a single node. This is graphically depicted as:

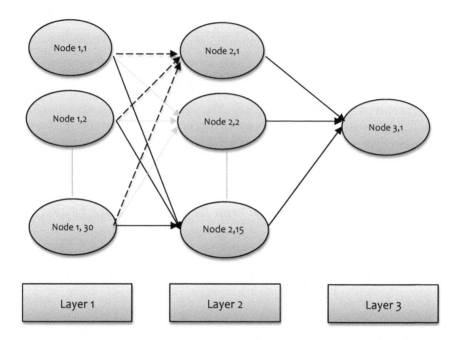

In this architecture, the network has three layers, 30 nodes in the first layer, 15 in the second and one in the third. The last layer will always have one node representing the single prediction per input record.

As can be seen the multiple layers in effect create interaction effects from the input variables.

A trick that has had enormous success in improving neural networks is the process of "dropout". Since neural networks have substantially more parameters than standard machine learning models like GLM, they have the tendency to very quickly overfit the training data. With "dropout", the network "drops out" or deletes a randomly selected percentage of the outputs of a layer. This means that the network builds in redundancies and has the effect of reducing the overfitting of the training data. In our code, 2% of all outputs are dropped from the first two layers. The dropout layers are only used in training the model and not in prediction, where the full power of the network is used. This simple process has had a huge effect in helping to build very large and deep neural networks.

The advanced neural network is coded as follows with the rest of the network remaining unchanged:

```
rate_mean = input_predictors %>%
  layer_dense(units =30, activation = 'relu') %>%
  layer_dropout(0.02) %>% #.01
  layer_dense(units =15, activation = 'relu') %>% #10
  layer_dropout(0.02) %>% #.01
  layer_dense(units =1, activation = 'linear')
```

When building this multiple layer model, we want the model to analyze non-linearities in the data. To do this, we apply an "activation" function to the output of each node. In our case, the function we will use is the "rectified linear unit" or RELU, which is simply the function $f(x) = \max(x, 0)$. The last layer has a "linear" activation function applied, which is $f(x) = x$, namely the output is unchanged. Thus, the output of each node is simply:

- RELU activation: $\max(0; x_i' \cdot \theta)$; or
- Linear activation: $x_i' \cdot \theta$.

With the above Neural Network, we were able to produce a validation NPLL of 0.380 and a test NPLL of 0.374.

13.5.4 Extreme Gradient Boosting

In this section, we will implement the Extreme Gradient Boosting algorithm[15] as implemented in the R package XGBoost. XGBoost provides an efficient form of gradient boosting machines, an idea originated by Friedman [8]. XGBoost is typically used for classification tasks, but can also used for regression, including Poisson regression tasks. We now describe the key elements of the algorithm.

13.5.4.1 What Is a Tree?

A tree is a very simple mechanism of iteratively dividing the data and then scoring the results. With our data, a tree could be:

---

[15]Chen [9].

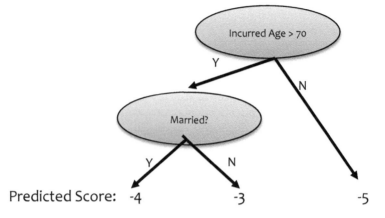

The predicted score is then fed into $f(x)$ in Eq. (13.1). Clearly, there are lots of ways to divide up the data into small trees and score the data record. An "ensemble" of trees does exactly that and calculates the score, $f(x)$, as the sum of scores from a number, say K trees, each with score $f_k(x)$ so:

$$f(x) = \sum_k f_k(x)$$

Each node is called a leaf and each split of the data is called a branch. The number of branches in a tree is called the depth of the tree. The XGBoost package builds the ensembles one tree at a time, by first selecting the best tree and then adding the next best and so on. The strength of the tree is determined by the Poisson log likelihood adjusted by a penalty or regularization term. This penalty is a similar in concept to the penalty used in the Lasso approach, although a different form of penalty used. The penalty is included to mitigate against overfitting the data. Although each tree is not very good at making a prediction on its own, the successive addition of trees, a process known as boosting, turns out to be a highly effective means of creating accurate predictions. Thus, a series of "weak" learners can be combined to create a "strong" learner. XGBoost provides a highly efficient means of building these trees, and is now recognized as one of the most accurate machine learning tools for large tabular data.

XGBoost learns by adding one tree at a time to the model. If too few trees are added, the model will have high bias and underfit the training data. Too many trees will result in high variance, with the training data being overfitted. The number of trees is controlled by the parameter nrounds, which is best determined through a process of cross-validation.

13.5.4.2   Tuning Parameters

XGBoost has lots of tuning parameters and knobs that are critical to its successful implementation on a task. For many of these parameters, we will need to use cross validation to select an optimal set. However, it is best to start with a reasonable set as the number of choices can become prohibitive to test because of the size of the data.

Some of the key tuning parameters are[16]:

1. **eta**: At each boosting step, the weights of the new features are found. Eta shrinks these values to make each additional tree even weaker. Eta ranges from 0 to 1. The default is 0.3, but we recommend starting with 0.015–0.05.
2. **max_depth**: Specifies the maximum depth of the tree. The default is 6, but we recommend starting points of 4 or 5.
3. **colsample_bytree**: When creating a tree, a subset of the columns (predictors) of data can be selected, effectively randomly excluding some features for a tree. The default is 1, namely no removal, but we recommend starting with 0.8.
4. **subsample**: This is similar to colsample_bytree, except instead of randomly selecting the columns of the data matrix, subsample randomly selects rows or records of the data. The default is 1 but we recommend starting with 0.8.

There are various methods of testing alternative sets of tuning parameters. The most straightforward is to use a grid search which systematically tests all permutations of a selected set of tuning parameters.

13.5.4.3 Coding the Model

Preparation of data for XGBoost has a number of idiosyncrasies:

- Data must be stored in a dense matrix using the `xgb.DMatrix` function.
- All predictors must be numeric, not categorical. One-hot encoding can be used to prepare the data, but typically categorical predictors are coded as a number, and XGBoost applies its tree routines directly to such number. It is often helpful to order the category numbers in a reasonable fashion, and there are various tricks for doing this.
- Missing data is easily handled by coding the missing value as either zero (or a very small or large number) and allowing XGBoost to automatically recognize the missing value as a separate predictor. Other missing value techniques described earlier can still be used.
- Interaction effects are not typically fed into an XGBoost model as separate predictors as we are relying on the routines to determine interaction effects.

---

[16] A lengthy discussion of tuning parameters can be found at http://xgboost.readthedocs.io/en/latest/ //parameter.html.

- Feature engineering, nevertheless, is an important part of the XGBoost model and including custom features can make a large difference to the overall results.
- As with GLM, an offset for the log(ActiveExposure) must be included, which can be done by specifying the "base_margin" parameter (see below).

```
# Set up data
offset_train =log (incidence$train$ActiveExposure)
y_train = incidence$train$Count_NH
offset_val =log (incidence$val$ActiveExposure)
y_val = incidence$val$Count_NH

dtrain <- xgb.DMatrix(data=as.matrix(map_df(incidence$train[,1:23],
as.numeric)),label=y_train)
setinfo(dtrain,"base_margin",offset_train) # Identifies offset
dval <- xgb.DMatrix(data=as.matrix(map_df(incidence$val[,1:23],
as.numeric)), label = y_val)
setinfo(dval,"base_margin",offset_val)# Identifies offset


# Define tuning paramaters for xgboost
params <- list(objective="count:poisson",
                eval_metric = "poisson-nloglik",
                booster = "gbtree",
                eta = .03,
                gamma = 1,
                max_depth = 4,
                min_child_weight = 1,
                subsample = .8,
                colsample_bytree = .8
)


#Establish the nrounds parameter
nrounds = 2000*1.1

## Run the Model
mod <- xgboost(data = dtrain,params = params, nrounds = nrounds)


# Perform validation set predictions
preds <- predict(mod, dval)
```

The validation NPLL is 0.03789 and the test NPLL is 0.372 using the above choices.

A comparison of rates grouped by Issue Year Bucket is shown below:

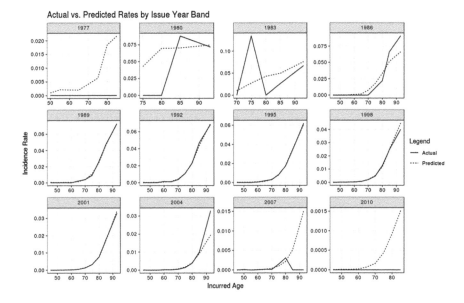

As can be seen, for the earliest and latest issue years, there is limited data resulting in volatile results. For the middle issue years, the fit is very good.

13.5.5   Development of Actuarial Incidence Rate Tables

An insurance company will typically require an incidence rate table or set of tables to be used for pricing its policies. It is thus necessary to convert the rates developed from the above processes to a format that more readily fits an insurance company workflow. This can be done using the following procedure:

1. Select a set of cells that represent the average demographics for the group you are seeking to price. Each cell will become a record in a new database or data frame (the "Pricing Cells"). Set the "ActiveExposure" field, namely ae_i, to one.
2. Use the created data frame to predict a new set of predicted incidence counts, $\hat{y}_i$.
3. Since $\hat{y}_i = \hat{\lambda}_i = ae_i \cdot \hat{h}(x_i, \theta) = \hat{h}(x_i, \theta)$, we have now estimated the hazard rates.
4. The incidence rates can now be found by converting the hazard rate to a probability, using the exponential distribution cumulative density function to calculate the probability of a claim occurring within one year, namely: $incidence_i = 1 - e^{-\hat{h}(x_i,\theta)}$.

Care should be taken about selecting demographic features that are not well represented in the experience data. For instance, "IssueYear" has limited data for the later years. It might be reasonable to choose an earlier IssueYear to predict the rates or a weighted set of IssueYear values, and then manually project the rates to future issue years using an improvement factor. If the latest IssueYear in the experience data is

2010, then most of the algorithms discussed would not provide credible rates, if we were to plug in a value of, say, 2018 into the Pricing Cells. Similarly, "edges" of the experience data might produce non-credible rates, for example, incidence rates for Issue Year 2010 appear to be too low by a factor of 10! Finally, you might want to smooth the developed tables, particularly for cells where there is limited exposure in the training data. In fact, it is often very helpful to plot cross sectional slices of the developed rates to assess where the data is non-smooth or areas where the predicted rates are likely not credible.

13.5.6 Comparison of Models

Key advantages and disadvantages of the various algorithms are now assessed:

GLM

- **Advantages**: The solved coefficients in the model provide a degree of interpretability which help the analyst understand the relative and directional effects of various predictors on the outcome of the model. This is particularly the case for "smaller" models with few predictors. However, in models with many predictors, including interaction effects, such as we have used here, it becomes quite difficult to interpret the factors.
- **Disadvantages**: The process of predictor selection can be tedious, and important effects and interactions can be missed.

Lasso

- **Advantages**: The model provides an improved degree of interpretability over GLM as the important predictors are automatically selected. We also see that improved accuracy over GLM can be achieved.

Neural Networks:

- **Advantages**: The algorithm is very good at analyzing interactive effects and produces very good overall accuracy.
- **Disadvantages**: The model is more of a black box than GLM and Lasso, and so other means are required to interpret the results.

Gradient Boosting Machines:

- **Advantages and Disadvantages**: These are very similar to neural networks, although the algorithm itself is very different. Namely the algorithm is highly accurate, but is very much a black box. The implementation through the XGBoost model produces excellent accuracy overall.

13.6 Summary

We have analyzed four powerful machine learning algorithms for determining nursing home incidence rates. The algorithms worked well on the very large database accumulated by the SOA. We believe all the algorithms would also be useful for analyzing smaller datasets such as those produced by an individual company. The methodology used in this chapter can be applied to calculating ALF and HHC incidence rates, persistency rates and claims termination rates without much change in process.

We have shown that standard, off-the-shelf open source software can be used to run the various algorithms. These algorithms can be executed with a small amount of code.

Both neural networks and Gradient Boosting Machines are not typically used for count data learning problems, but we have shown that these algorithms can be easily adapted for managing Poisson-regression type problems.

This chapter is not intended as an exhaustive treatment of all the algorithms, but provides a solid introduction to each of them. The GLM and Lasso algorithms are the standard workhorses of machine learning when analyzing structured or tabular data. XGBoost (and similar implementations) have shown to provide state of the art results across a wide variety of tabular data tasks. Well-optimized neural networks are also now producing state of the art results in structured data problems. The methods described in this chapter provide a base for producing highly effective models for a wide variety of insurance-related risk analysis projects.

13.6.1 Further Reading

Much of the art of setting the parameters and tricks for cleaning data and feature selection are found in non-traditional sources such as blog posts, websites and forums. Below is a list of interesting places to browse and books to read:

1. Select Blog Posts and Websites:

    a. XGBoost details: http://xgboost.readthedocs.io/en/latest/model.html.
    b. XGBoost parameter choice: https://www.analyticsvidhya.com/blog/2016/03/complete-guide-parameter-tuning-xgboost-with-codes-python/.
    c. Rstudio for setting up Keras: https://keras.rstudio.com/.

2. Kaggle Forums: Kaggle runs machine learning competitions across a wide range of tasks. The winners post reports on methodology used and these posts, as well as posts from other competitors, can provide a wealth of information.
3. Cross Validated forum contains a wealth of information on statistical related matters.
4. Arxiv.org contains the latest artificial intelligence papers.
5. Various Books:

a. GLM, Lasso and Gradient Boosting Machines: Hastie, T., Tibshirani, R., Friedman, J., "The Elements of Statistical Learning".
b. Neural Networks in R: Chollet, F., Allaire, J.J. "Deep Learning with R".

References

1. Murphy, K.P.: Machine Learning: A Probabilistic Perspective (Adaptive Computation and Machine Learning series). Massachusetts Institute of Technology (2012)
2. Long Term Care Intercompany Experience Study—Aggregate Database 2000–2011 Report, January 2015, Society of Actuaries
3. Aalen, O.O., Borgan, O., Gjessing, H.: Survival and Event History Analysis: A Process Point of View. Springer (2008)
4. Holford, T.: The analysis of rates and of survivorship using log-linear models. Biometrix **36** (1980)
5. Rodriguez, G.: Generalized Linear Models. Princeton University, http://data.princeton.edu/wws509/notes/c7s4.html
6. Akaike, H.: A new look at the statistical model identification. IEEE Trans. Autom. Control. (1974)
7. Tibshirani, R.: Regression shrinkage and selection via the lasso. J. R. Stat. Society. Ser. B (Methodological) **58**(1), 267–288 (1996)
8. Friedman, J.H.: Greedy function approximation: a gradient boosting machine. Ann. Stat. **29**(5), 1189–1232 (Oct., 2001)
9. XGBoost: A Scalable Tree Boosting System. Tianqi Chen, Carlos Guestrin. https://arxiv.org/pdf/1603.02754.pdf (2016)
10. McCullagh, P., Nelder, J.A.: Generalized Linear Models. Chapter 13, Chapman & Hall (1989)

Printed by Printforce, the Netherlands